Service Science

Contents

Foreword

The science of service is emerging. Undoubtedly, a journey of this complexity, striving to scientifically understand a phenomenon as fundamental and richly diverse as service phenomenon, must be explored along multiple pathways over multiple decades. Therefore, it is always a great pleasure for me to recognize and encourage those embarked on this journey. Truly, we are all students of service, learning from each other as we go.

In this volume, entitled *Service Science: The Foundations of Service Engineering and Management* authored by Robin Qiu, I would like to draw the careful reader's attention to three main aspects of this work.

The Pioneers: In Chapter 3, a brief overview of the evolution of service research is presented. The complexity and diversity of service phenomenon is reflected in part by the number of academic disciplines whose scholars have written on this topic. Scholars from schools of management, engineering, natural sciences, social sciences, as well as arts and humanities (service design), not to mention practitioners and policy makers in government, have all played a role in the exploration. Figure 3.5 entitled "A sustainable socio-technical process-driven service system" provides an excellent visualization of five types of capital (natural, human, social, financial, and infrastructural) and the processes that transform these resources over time. It is worth noting that each of the major scholarly schools has a primary focus on one of the five major forms of capital.

Putting People First: In Chapter 5, I especially enjoyed the section on putting people first. The book presents novel approaches to the mathematical formalization of service, without losing sight of this important fact—service is about putting people and their experiences first. Pay special attention to Figure 5.8 entitled "Service value diagram corresponding to GE's change effectiveness model"—for though it is one of the simpler diagrams in the book, it highlights that increasing value derives from increasing quality and increasing acceptance, when mutually agreed to and cocreated by providers and customers. Furthermore, with the global rise of smart phones and social media tools, there has never been such an exciting time in human history to gather and analyze big data aspects of service encounters. We are in the age of increasingly powerful tools for value cocreation. This work also makes the important point that value cocreation is also about cotransformation of providers and customers.

Education as a Service: Chapters 7 and 8 provide an excellent example of applying the theoretical developments in this book to the challenge of improving education as a service. Both chapters highlight the value of structural equation modeling techniques as well. Chapter 9 further distills the theoretical developments into a practical and iterative method for daily improvements to service business offerings. Figure 9.3 entitled "Engineering and managing competitive services: scientific perspective" conveys a tremendous amount of methodology quite concisely. Readers familiar with statistical control theory will find this chapter an especially nice summary of the developments in the book.

While much work remains to broadly establish a holistic and lifecycle approach to service systems, this book boldly suggests pathways and approaches to help researchers mathematically formalize service systems and networks in the age of big data, without losing sight of the importance of putting people first. In the coming years, I look forward to reading more along this pathway as the ideas presented are further tested and refined.

JAMES C. SPOHRER

Director, IBM University Programs World-Wide (IBM UP)
IBM Almaden Research Center

Foreword

Services. What do we think of? Taking cash from an ATM machine? Talking on our cell phones? Surfing the web? Watching TV? Picking up our mail? Yes, all these daily activities and much more. In fact, we would be hard-pressed to identify significant parts of our lives that are not service-related. About 150 years ago, over 50% of the US workforce toiled in agriculture. Today, it is about 2%, and we grow a lot more food. Agriculture is not a service, but its workforce has plummeted while the sector has become more productive. In the US post-WWII boom, in the late 1940s, the fraction of the US workforce in manufacturing peaked at about 35%. Today, it is a mere 9%. The percentage of gross domestic product (GDP) associated with manufacturing parallels these numbers, from being about 30% post-WWII to being about 12% today. What has filled the void? Answer: The US service sector. It has swelled to about 80% of the GDP!

What precisely is the service sector? Economists define it by subtraction. The service sector is everything in the economy that is NOT agriculture (including forestry and fishing) OR industry (manufacturing and also mining and construction). That subtraction leaves us with the majority of the world in which we live! In addition to the mundane day-to-day services chores, we have the health care system (about 18% of the economy), education (8–10% of the economy) and much more—government, transportation, entertainment, utilities, etc. The excellence or nonexcellence of services can literally mean the difference between life and death!

We are fortunate that Robin Qiu has written this book at this time. He reports that we in the United States have had a national obsession with manufacturing and our international competitiveness in that domain. Yet, it is services that comprise the largest part of the economy, by far. The service sector creates a net international trade surplus for the United States. Scores of books have been written about manufacturing, which is now 12% of the GDP. Far fewer books have been written about services, which constitutes 80% of the GDP. Robin has been a leader in pushing us, not to ignore manufacturing, but to move it upward to its rightful place focusing on the services sector. He is the principle founder of the new INFORMS journal, *Service Science*. This book represents another major contribution to service sector analysis.

At my home institution, the Massachusetts Institute of Technology, the graduate Masters program "Leaders for Manufacturing," founded in 1988, has recently been renamed "Leaders for Global Operations," reflecting the fact that many of today's industrial leaders are in services such as retailing and supply chain management and not manufacturing.

Robin says that a service is provided as part of a complex sociotechnical system. This broad nontechnocratic view is perfect from my point of view. Services cannot be meaningfully studied solely through sharply focused discipline-based glasses. To be effective, service sector analyses cannot be Tayloristic "time and motion" studies. We require an interdisciplinary approach, where aspects of the social sciences often dominate traditional narrow engineering-measurable quantities.

My favorite early example of this is the story of queueing at elevators in the 1950s in New York City. With the post-WWII economic growth, more high rise buildings were constructed in Manhattan—as office buildings, hotels, and apartments. People started to complain about delays for elevators in these buildings, especially at morning and late afternoon rush hours. A narrow engineering-focused queueing analysis might have concluded that some of these buildings should be destroyed and designed over, with more elevator shafts, as the current designs could not support peak load traffic. (I say this only slightly tongue in cheek!) But a colleague of Professor Russell Ackoff of the University of Pennsylvania was dispatched to study the situation. He indeed verified the numerous customer complaints about elevator delays. Then, in a moment of true creative thinking, he redefined the problem. He thought to himself, "What if the problem is not the *magnitude of the delays* waiting for the next elevator? What if the problem is the *complaints* about those delays?" He postulated that the elevator customers needed a distraction while they were waiting. So, in a spurt of lateral thinking, he purchased and had installed floor-to-ceiling mirrors adjacent to all the elevators in a test building. Guess what? The complaints about elevator delays plummeted to near zero, while the statistics of delay remained unchanged! Problem solved, but not with traditional queueing theory. Here a touch of psychology was needed. And so was born the psychology of queueing, the same year (1955) that Walt Disney opened his first amusement park—in Anaheim, CA. Over the years, the Disney Company has shown itself to be a true master of designing and managing complex sociotechnical service systems—including its queues. The arts and entertainment services industry comprises about 4% of the US GDP.

Reading Robin's book chapters, with its many useful framings of services provision, I started reflecting on my own personal services experiences and preferences. He says that trust and reliability are important aspects of services. Here is trust: I have used the same travel agent for 34 years! And, yes, I am happy to pay more than what is charged by an anonymous discount Internet-based travel service because I know it will be done right, changes will be easy, and that she 'has my back' if anything goes wrong during travel. Car repair: I have an 8-year-old Subaru WRX STI, a rally racing champion. No one touches it except my dealer (and me)! Eight years, one place for maintenance and on rare occasions—repair. They have my back on that car. And I would not trust a random person employed by some discount national auto repair chain to look after this car—which is an extension of me! Medical services: You guessed it, over 35 years with the same organization. Maybe I am too fixed in my ways. But trust in-services are of paramount importance.

Trust goes in the reverse direction as well: One bad service encounter can lead to a lifetime pledge never to patronize that organization again. The median age of students in a typical graduate class that I teach at MIT is 25. I ask them, "How many of you have had such a bad service encounter in your lives that you have pledged to yourself never ever to go back there again?" These are 25-year-olds, less than 10 years from living with their parents. And, invariably, over half the class raises their hands! How many providers of services are aware of this fact? That continual excellent quality service is required for customer retention. Customer loyalty may go only as far as the next service encounter. Robin Qiu drills the lesson home in this important book.

Services are nuanced, not readily quantified into various measurement bins. Robin describes this in many ways. From my life in Lexington, MA, a historical suburb of Boston with a population of 28,000: We have two Starbucks, one Peet's Coffee, and seven Dunkin Doughnuts! Plus various convenience stores and quick-stop shops located at gasoline stations—all serving coffee to go. I guess Lexingtonians are highly caffeinated! From my home I prefer to drive to the third closest Dunkin Doughnuts. Why? The coffee and food products and prices are identical to each other. Answer: Only in this shop do I get greeted each time with sincere friendly smiles, as if they really want to see me and are happy to have me as a customer. Plus, the place is a neighborhood hangout with many retired folks just sitting around, enjoying each other's company, and passing the time of day—a type of nice 'bar scene' in a coffee shop. The ambience is just right. My minute or two of extra driving is worth it! Again, how many "time-and-motion" type studies would ferret out these concerns? I do not think I'm unique in valuing such nuanced aspects of services as important. Robin Qiu hits the nail on the head. Many others miss it completely.

After reading Robin's book, we would know that there is only one topic he discusses for which I have a minor disagreement: Internet-based services. To allude to an 'alien' terminology, he equates these to a type of "Service Encounter of the First Kind," that is, rather distant and impersonal. (In the 1977 movie, Close Encounters of the Third Kind, an encounter of the first kind was an alien encounter beyond 500 ft—implying little closeness, complexity, or subtlety.) I agree with him for many Internet services, such as those associated with airlines, hotels, and rental cars. But, there are Internet-based services such as Etsy (https://www.etsy.com) that resemble personal face-to-face interaction. You might call these "Close Service Encounters of the Third Kind," that is, up close and personal, nuanced, and complex. In fact, I have found web sites such as Etsy better than shopping mall face-to-face interactions because I am dealing with the proprietor of a small artisanal business and his/her future success depends 100% on customer satisfaction. The email 'back-and-forth' between proprietor and customer often resembles a conversation of an old country general store of the 1800s! Writing reviews online for all to see can show each customer's satisfaction or dissatisfaction. It is difficult for the average customer to have that type of impact with impersonal national chain stores, with either face-to-face or Internet-based service encounters. My hunch is that Robin will agree with me and say that I may have misread the book with relation to all Internet-based services! And I am sure he would be right!

It is an honor that Robin has asked me to write this foreword. Enjoy the book. See all the many faceted aspects of services that Robin describes and structures. Also, reflect back on your own personal experiences with services, and you will see that Robin hits the mark virtually every time. In addition, if you are in a planning or managerial role in a service firm, you and your company can gain significant competitive advantage listening to what is said in this book. A service is a complex sociotechnical system, and those who recognize it as such are bound to prosper.

Engineering Systems Division
Massachusetts Institute of Technology

RICHARD C. LARSON

Preface

This book essentially introduces a new perspective of service study. By taking a holistic view of the service lifecycle, we discuss approaches to explore the real-time dynamics of service systems and networks. We advocate that a service must be defined as a value cocreation and transformation process. As such, we can holistically analyze the performance of service systems that enable and execute complex and heterogeneous services. By leveraging the advances in computing and network technologies, social science, management science, and other relevant fields, we present the concept and principles of putting people first in service and demonstrate that service networks in light of service encounters can be comprehensively explored in a closed-loop and real-time manner. The presented framework can be potentially applied in facilitating service organizations to understand and capture market trends, design and engineer service products and delivery networks, execute service operations, and control and manage the service lifecycles for competitive advantage.

Service research is not new. In fact, service research in the field of marketing has an over 30-year-history. In addition to research and development in service marketing, academics and practitioners have actively developed a variety of theories, methods, and tools and then applied them to address service delivery efficiency and effectiveness issues in service operations and management across the service industry for decades. Recently, significant attention in the service research is related to a variety of exploratory studies of service systems, focusing on how to leverage the advances of management science, systems and network theory, and computing and network technologies to help service organizations improve the overall performance of their service systems from both engineering and operational perspectives.

Note that the worldwide economy was dominated by manufacturing during the last couple of centuries. As a result, both academics and practitioners paid much attention to the design, development, production, and innovation of physical products. The economic shift from manufacturing to service made us rethink people's social, physiological, and psychological roles in the economic activities. However, inertial thinking is part of sociopsychological norms to the majority of human beings, resulting in many service organizations offering and delivering their services using manufacturing mindsets. Consequently, the advances in social science, management science, computing technologies, and others are not well integrated and thus leveraged in support of effective service engineering and management as needed in the service industry.

Change is the only constant in today's business world. The effectiveness (E) of a service as a solution to meet the changing needs of customers is equal to the product of the quality (Q) of the technical attributes of the solution and the acceptance (A) of that solution by the customers, that is, $E = Q \times A$. However, the customers' acceptance changes rapidly, varying with people, time, places, cultures, and service contexts. Because people's acceptance is largely subjective, manufacturing mindsets with a focus on physical attributes indeed become ineffective when applied in the field of service engineering and management. Hence, to address the discussed change acceleration phenomena with scientific rigor, we must rely on people-centric and service mindsets. In fact, the introduction of putting employee and customers first in the 1990s radically made a turn in the way how service organizations should develop, operate, and manage businesses and measure their successes. Indeed, people-centered service mindsets have afterwards been emerging and receiving more and more attention in the service industry.

Bearing this discussion in mind, we consider a service as a transformation process rather than an offered product. Truly, both provider-side and customer-side people are always involved in an interactive manner in service. Hence, we view a service as a value cocreation process. For a service, goods are frequently the conduits of service provision; the physical attributes and technical characteristics that specify the goods are indispensable to the service. The quality (Q) of the technical attributes in the service, indeed, mainly defines the quality of the goods. To a service customer, Q is frequently perceived in service provision as the quality of designated service functionalities that are defined in a service specification. As indicated in the equation of $E = Q \times A$, the value of E also depends on the value of A, which is largely related to sociopsychological perceptions of the customer throughout the service lifecycle. Although this is well understood conceptually, however, the service industry lacks methods and tools to explore and measure Q and A in service in a holistic, real-time, and quantitative manner.

Services are highly heterogeneous. For a given service, a specific customer and a service provider essentially cocreate the service values that meet the respective needs of the customer and the service provider. Thus, each service is unique. The variability of service and the need for measuring sociopsychological perceptions had made extremely challenging the full exploration of the service lifecycle, spanning market discovery, engineering, delivery, and sustaining, in an integrated and quantitative manner.

Indeed, the prior lack of means to monitor and capture people's dynamics throughout the service lifecycle has prohibited us from gaining insights into the service engineering and management in a service organization. Promisingly, we believe that the combination of the following advances in technologies has made possible the design and development of the needed means and methods that can help service organizations overcome the challenges:

- Digitalization
- Networks and telecommunications
- Collaborative methods and tools
- The fast advances in social network media
- Big data technologies and analytics methods and tools

In other words, real-time data on the dynamics of service cocreation processes from both service providers and customers could be comprehensively captured and analyzed if needed. (Surely, people's privacies must never be compromised, which are beyond the coverage of this book.) When the enabling technologies are appropriately implemented, we can create and execute smarter working and consuming practices so that we can make service cocreation processes not only beneficial but also enjoyable.

Simply put, enormous opportunities truly lie ahead of us. We quite often ask ourselves: "Do we have right methods and tools that ensure service systems to perform in such a way that the respective values for both service providers and customers can be optimally cocreated, at present and in the long run?" However, the question remains unanswered, partially or totally. By leveraging both systems methods and networks analytics, in this book we present one solution to address this unanswered question.

Holistically, a service organization is a service system, essentially consisting of service providers, customers, products, and processes. Different from a goods-producing system, a service system must be people-centered. Therefore, a service system surely is socio-technical. On the basis of the earlier discussion, we understand that it is the transformation process that ties all other system constituents together and cocreates the respective values for both service providers and customers. Whether the values can be fully met relies on the efficient, effective, and smart business operations that are engineered, executed, and managed across the service system.

Service is people-centric, truly cultural and bilateral. The type and nature of a service dictate how a service is performed, which accordingly determines how a series of service encounters could occur throughout its service lifecycle. The type, order, frequency, timing, time, efficiency, and effectiveness of the series of service encounters throughout the service lifecycles determine the quality of services perceived by customers who purchase and consume the services. Note that the people-centered, interactive, and behavioral activities in a service system essentially engender a service cocreation network. Indeed, as the velocity of globalization accelerates, the changes and influences are more ambient, quick, and substantial, impacting us as providers or customers in dynamic and complex ways that have not seen before. The understanding of service networks is essential for service providers to be able to design, offer, and manage services for competitive advantage in the long run.

Because of the sociotechnical nature of a service system, we use a systems approach to evaluate the performance of the service system, aimed at capturing both utilitarian functions and sociopsychological needs that characterize service systems. However, the true people's behavioral and attitudinal dynamics of a sociotechnical system requires conducting real-time, corresponding social network analytics. As a result, the insights of the service interactions in the formed service networks can be truly explored and understood, which assist stakeholders to make respective while cooperative informed decisions at the point of need to improve their service cocreation processes across the service lifecycles in an optimal manner.

To get a comprehensive understanding of this new perspective of service research, readers should read chapters sequentially. Brief introductions to all chapters in this book are provided in the following:

- *Chapter 1*. Introducing service by briefly reviewing the evolution of service, we emphasize that the holistic view of service is a must in today and the future's world economy.
- *Chapter 2*. Discussing the concept of cocreation in the service industry. A definition of service for this book is provided, which radically lays the foundation for the remaining chapters in this book.
- *Chapter 3*. Exploring cocreation transformation processes. We articulate that the increasing complexity of service research and development requires the science of service in a new perspective.
- *Chapter 4*. Looking into service science fundamentals. By analyzing the dynamics of service, we define laws of service in general. A holistic and sociotechnical view of service becomes essential for us to develop service science.
- *Chapter 5*. Revealing the digitalization of service systems and networks. We argue that putting people first should be a mindset. The mindset is what service organizations must bear when they design and develop their service systems. Through leveraging process-aware computing systems and sensor-based networks, people's behavioral and sociopsychological data and information can be well monitored and captured.
- *Chapter 6*. Showing computational thinking of service systems and networks. By taking advantage of the digitalization of service systems and networks, we demonstrate that the system dynamics of service cocreation processes can be fully modeled, analyzed, and controlled in a closed-loop, real-time, and quantitative manner.
- *Chapter 7*. Using education examples to show how service and service systems can be explored from a systems perspective. Specifically, we apply structural equation modeling to investigate mechanisms of improving educational service systems. By integrating cross-section and longitudinal analyses, we

demonstrate the tremendous potential of the applications of the proposed approach in the general field of service engineering and management.

- *Chapter 8*. Using an online education example to demonstrate the dynamics of service networks. The concept and principles of putting people first are illustrated in great detail in this chapter. When people-sensing mechanisms are well implemented in a service system, service networks that are essentially formed from service interactions within the service system can be fully investigated. We present effective data, network, and business analytics with a focus on looking into the insights of the service system in real time. Ultimately, once system performance modeling and service network analysis are well integrated in a closed-loop, real-time, and quantitative manner, we can truly help service organizations perform optimal service engineering and management throughout the service lifecycle.
- *Chapter 9*. Concluding the book with some final remarks. We articulate that innovative approaches to the development of Service Science are truly needed. However, we advocate that the service research and practice community must appreciate and continue to develop a variety of methodologies and tools that can be well derived and evolved from the known theories and principles in systems theory, operations research, marketing science, organizational behavior and theory, network theory, social computing, and analytics.

This book does not intend to cover the state of the art in the service research field. Instead, this book simply provides readers a new perspective of service research and practice. It could serve as a good reference book for scholars and practitioners in the contemporary service engineering and management field.

Disclaimer

No product or service mentioned in this book is endorsed by its maker or provider, nor are any claims of the capabilities of the product or service discussed or mentioned. Products and company names mentioned may be the trademarks or registered trademarks of their respective owners.

Professor of Information Science
Pennsylvania State University

ROBIN G. QIU, PHD

Acknowledgments

Since 2004, I have been involved in a variety of research projects related to the emerging field of service science. The presented concept and principles of service science in this book are surely derived from those funded studies. The funded projects include US NSF Grants (DMI-0620340 and DMI-0734149), Department of Education Grant (08JA630040, China), Nanjing University of Aeronautics and Astronautics Endowed Professor Scholarships (1009-905346 and 1009-908332), Chinese NSF Grants (70541007, 70772073, and 70902026), and IBM Faculty Award (2008–2009, China), and IBM Grants (SYRTHU5, D07009SUR, JLP201111006-1, NUAA-SUR-2012), Jiangsu Science and Technology Innovation Award (JSTIA269008, Jiangsu, China), and Penn State COIL RIG Grant (2012-14). Without their financial supports, I might never have had an opportunity of working in this emerging and promising field.

Portions of the content of Chapter 5 were previously published as a chapter entitled "Information Technology as a Service" in *Enterprise Service Computing: From Concept to Deployment*, which was published by Idea Group Publishing (Hershey, PA) in 2007, Copyright 2007, IGI Global, www.igi-global.com. Portions of the content of this book were previously published as chapters entitled "BPM" and "BPM/SOA, Business Analytics and Intelligence" in *Business-Oriented Enterprise Integration for Organizational Agility*, which was published by IGI Global (Hershey, PA) in 2013, Copyright 2013, IGI Global, www.igi-global.com. Reprinted by permission of the publisher.

Portions of the content of Chapter 6 were previously published in *Service Science*, 1(1), 42–55, entitled "Computational thinking of service systems: dynamics and adaptiveness modeling" and are used with permission from the Institute for Operations Research and the Management Sciences (INFORMS, https://www.informs.org). INFORMS is not responsible for the errors introduced in the translation of the original work. During the period of developing this project, the author wrote several editorial columns for *Service Science*. The author made cross-references between those editorial columns and this book.

The author would like to acknowledge the help of all involved in the collation and review process of this book, without their supports the project could not have been satisfactorily completed. Special thanks also go to all the staff at John Wiley & Sons, Inc., whose contributions throughout the whole process from inception of the initial idea to final publication have been invaluable. In particular, I am very grateful to Susanne Steitz-Filler, Senior Editor, who continuously prodded via e-mail for keeping the project on schedule. I want to thank anonymous reviewers whose insights and criticisms helped substantially improve the quality of this book.

ROBIN G. QIU, PHD

1

Evolving and Holistic View of Service

1.1 WHAT IS SERVICE?

The word "service" has many connotations, varying with domains and settings. We must understand and deal with its extant variability in order to decipher and capture its inherent nature in business (Morris and Johnston, 1987). This is particularly important for this book because we have to stay in focus to discuss one solution, namely our unique and innovative approach to address the challenges that we have faced in the service sector over the years or new challenges that we will confront for the years to come. Put in a straightforward manner, presenting the "BEST" solution to address all the challenges confronted by academics and practitioners in the service sector is surely not our intention as there will never be such a one-size-fits-all solution. Given that the business world becomes more integrated, complex, and interdependent than ever before, a systemic view of service is the mindset that we will hold throughout this book. In other words, by relying on systems thinking and holistic viewpoints (Flood and Carson, 1993), we will explore and accordingly decipher the inherent nature of service in the unceasingly changing business world.

Service is frequently defined as an act of beneficial activity. A service that is considered as an act of beneficial activity actually has a long history. If we retrospect to the simplest material exchange that occurred in ancient times, such as a bushel of wheat exchanged for a barrel of oil, we know that a very basic trading service was performed. No matter what units and containers were used and how the trade was done in ancient times, the exchange or trade, a performed service, was essentially an act of helpful and beneficial activity that met the respective needs of the involved exchangers.

Service Science: The Foundations of Service Engineering and Management, First Edition. Robin G. Qiu.

A food service in a restaurant is another good example of an act of beneficial activity. Similar to the above-mentioned simple trading service, we can also easily retrospect to ancient times in the early social and economic development stage thousands of years ago. A food service in ancient times certainly had no conceptual difference from a modern food service. Although the catering setting and foods in a restaurant at that time were limited and simple, a performed service was substantively involved with a list of necessary service elements, provider, consumer, resource, process, and value. The service provider was the owner who owned the restaurant and offered dishes as service products. A service consumer was a client who ordered and ate his or her selected foods. A typical service process started from the time the client entered the restaurant and ended when the client paid for the service and left the restaurant. The process was involved with a transformation with the support of operation resources. The client's order is the process's input. The value for the client and the owner is the process's output. The value could simply be the profit the owner made and the satisfaction the client had. The client's hunger stopped; he/she was happy to some degree. Surely, the service was mutually beneficial. Resources, largely natural and labor-based, were leveraged in an extremely simple manner throughout the simple catering process (Figure 1.1). Without question, the corresponding business operations at that time were radically experience-based.

Figure 1.1 illustrates the conventional classification of resources. By focusing on resource supply and demand in the social and economic activities, we understand how resources are leveraged in the transformation of goods and services to meet human needs and desires. As a result, we traditionally recognize three categories of resources: natural, human, and manufactured or infrastructural resources. Natural resources essentially are the source of raw materials. Human resources consist of human efforts provided in the transformation of products or services. Manufactured or infrastructural resources consists of man-made goods or means of production

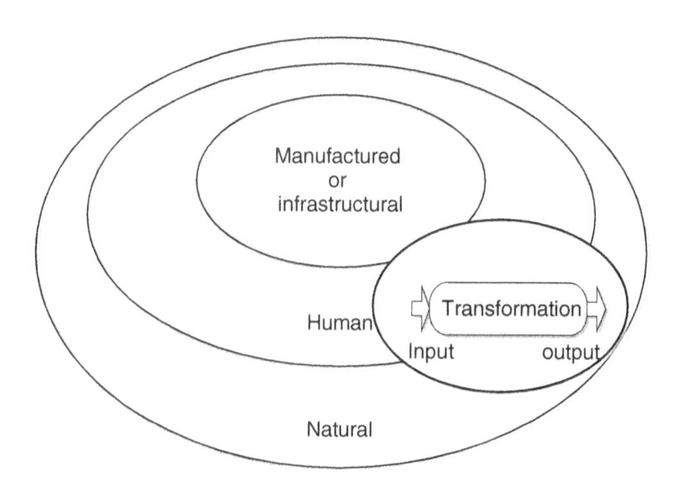

FIGURE 1.1 *A conventional resource model view of a food service.*

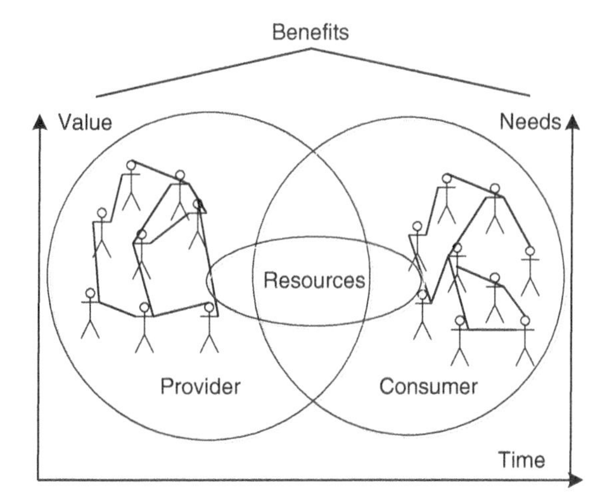

FIGURE 1.2 *A service involving certain fundamental elements.*

(machinery, buildings, and other infrastructure) used in the transformation of other goods and services (Samuelson and Nordhaus, 2009; Sullivan et al., 2011).

Regardless of what type of service is provided and consumed, five essential and core elements characterize a service in a conventional act of helpful and beneficial activity (Figure 1.2). More specifically, the five elements involved in services are resource, provider, consumer, benefit, and time, which can be described in an intuitive way as follows:

- *Resource*. Resources can be in a physical, soft, or hybrid form. For example, foods as a physical, transformable, and consumable resource or service product in a restaurant play a fundamental role in a given food service. Knowledge or experience in a focused subject area transferred in a training service seems to be a soft resource or service product. When a haircut service is performed, both barber's skills and haircut kits as a hybrid resource must be simultaneously applied or operated to make the service performed in a satisfactory manner. Essentially, with the help of resources, the act of performing a transformation task for a customer who asks for it in exchange for acceptable compensation is termed as service provision. Apparently, resources are the radical conduits of service provision to customers (Vargo and Lusch, 2004).
- *Provider*. A service is purposely performed by a service provider. A service provider as an entity can be an individual, group, organization, institution, system, or governmental agency.
- *Consumer.* A service consumer is usually a human being who consumes, acquires, or utilizes a service offered and performed by a service provider.
- *Benefits*. A performed service surely generates certain benefits. Typically, different benefits are pursued by the service provider and the service consumer as they have different value propositions in executing the service. The benefit for

the service provider could be value-based, such as a profit. The benefit for the service consumer might be need-based, such as desire and satisfaction.

- *Time*. Small or big, simple or complex, a service certainly takes time to get performed to realize the desired benefits. Interactive activities between the provider and the consumer could occur in an *ad hoc* or predefined, unattended, and/or well-controlled process.

Note that service provider-side employees and customer-side consumers should also be part of recourses if we strictly follow the resource model as is illustrated in Figure 1.1. To make the discussion vivid and people-centric, we have to emphasize the identity of active participants in the service model that will be developed and discussed throughout this book. Therefore, we will always make an exception from the general resource model by distinguishing the elements of providers and consumers or customers from the general human resource concept. The concept of human resource in Figure 1.1 will be needed only when the whole resource model is the focus in a related and focused discussion.

Indeed, no matter how small or big, simple or complex a service is, it surely takes time for its provider to perform and its customer to consume the service. Evidently, the consumer and the provider of the service shall interact with each other, directly or indirectly, consecutively or intermittently, physically or virtually, and briefly or intensively, during the process of performing the service. The interaction time accordingly can be short or long. All of these changing factors that largely characterize provider–consumer interactions vary with the types of services that are actually performed. Hence, there are a variety of perspectives on service in academia and practice.

1.2 DIFFERENT PERSPECTIVES ON SERVICE

Because of the existence of the above-mentioned variations in perception, a consumer's perception of one kind of service could differ considerably from another. Different forms of resources applied and operated in executing services and varieties of provider–consumer interactions create many different combinations of consumers' perceptions of services, which consequently complicate our service studies in academia and practice. Different consumers' perceptions of services then give rise to the existence of numerous definitions of service. As a result, different service industries have historically adopted different definitions of services to accommodate their respective needs. For example, service is also quite often defined as the supplying of utilities or commodities in the modern economic society. From an end user's point of view, consuming electricity as a service fits in this definition very well. If we consider the daily consumption of electricity as an example, we will see that there is very little interaction between its provider and customer. Typically, we as home owners or apartment tenants in the United States simply call a local office of an electricity service provider we choose, and then we inform the electricity service provider of the date we move in. When we move out, we simply do the same. The needed simple interaction serves only one purpose, which is basically to ensure that the monthly bill statement will accurately and correctly reflect the usages of electricity when we legitimately stay in the houses or apartments. Unless there could be a problem with power lines or a discrepancy in a monthly bill statement, we might not interact with the electricity service provider at all during our entire stay in the house or apartment. We as consumers feel that there is a very little interaction with a service provider; such a service is undoubtedly defined as the supplying of utilities from an end user's perspective.

The unceasingly increased online shopping in the twenty-first century presents another perfect example for a service that is defined as the supplying of commodities. It is well understood in the retailing industry that this supplying of commodities as a service includes commodities, related distributions, and retailing. However, to an end user, such an online shopping service is nothing but the supplying of the needed commodities. We as online shopping customers place orders from a website powered by an online retailer (i.e., a service provider). The orders will be delivered to us regardless of how the orders are fulfilled and how far away the orders have to be transported. Just like utilizing electricity at home, unless there would be a problem with the ordered commodities, we might not interact with the online retailer after the initial online order placement. Obviously, there is surely little physical interaction between a service provider and a service customer throughout the lifecycle of such a typical online shopping service.

Other forms of definitions of service include the providing of accommodation and activities required by the public or the supplying of public communication and transportation. A variety of services in the modern economic society fit in this category of service definition. A list of good examples will be trading, communication, transportation, tourism, hospitality, and health care services. Nevertheless, services provided by educational institutions, security and military, and governmental agencies can also be well classified in this category of definition.

As mentioned earlier, different consumers' perceptions of services have historically resulted in the existence of numerous definitions of service. At first glance, different forms of definitions of service seem to define different things. When we further examine these definitions, it is not difficult for us to find that regardless of how a service is defined in a given discipline in academia or in a specific service sector in practice, a service must include the five core elements shown in Figure 1.2. The differences felt or perceived by customers based on their perceptions of services come from user's experiences (Qiu, 2013) acquired from service encounters by the customers throughout the lifecycle of service executions.

1.3 THE LIFECYCLE OF SERVICE

In both academia and practice, we can find many versions of defined phases that compose the lifecycle of service. Depending on what we expect for a service or largely perceive during the process of consuming a given service, we might use different constituting stages to compose a service lifecycle. Quite often, we are subjectively or objectively impressed by certain phases or stages of the lifecycle of service. Then, we tend to ignore other phases that we prejudicially think they are less important. The information technology infrastructure library (ITIL®) version 3 (ITIL, 2011) is a nonproprietary and publicly available set of best practices for information technology (IT) service management. ITIL v3 defines five phases of service lifecycle, service strategy, design, transition, operation, and continual service improvement, and accordingly provides comprehensive guidelines throughout all the phases for aligning IT services with the needs of business. We also frequently derive the definitions from the ones widely applied in the manufacturing sector. As a result, just like service, a variety of terms or definitions of the lifecycle of service exist.

Regardless of how many versions we can find in the extant literature, all the described lifecycles of services should always be composed of four essential and classic phases, "learn," "develop," "perform," and "improve," from a service provider point of view. We use these four essential and sequential phases in a service lifecycle to define the fundamental service diamond relationship in a service organization, which are illustrated in Figure 1.3. These four phases are briefly discussed as follows:

- *Learn*. We have to know what the market need is before the concept of a new service product gets conceived. Regardless of the type of service, we have to learn the market to identify the needs of prospective customers through a variety of approaches. We understand that customer needs keep changing as time goes. Therefore, we have to learn and capture the changes and accordingly incorporate the changes into service provision throughout the lifecycle of services.
- *Develop*. We develop, transform, or leverage resources to serve customers and meet their needs. Frequently, the resources in service are mainly and paradoxically perceived as service products. Indeed, as we discussed earlier, the resources in the service industry can be in a physical, soft, or hybrid form. Leveraging all natural, human, and infrastructure resources are essential in service provision. The operand or operant roles of resources in rendering services

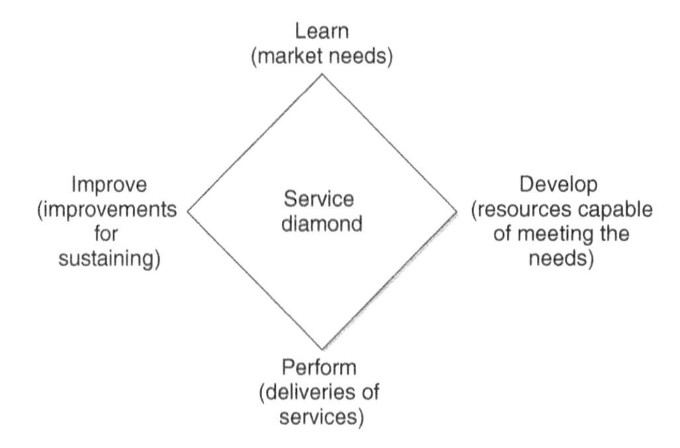

FIGURE 1.3 *The lifecycle of service: a classic service diamond.*

are significantly more sophisticated than the ones in the manufacturing industry. As a matter of fact, operant resources in service provision produce the effects that are largely perceived and truly appreciated by customers (Vargo and Lusch, 2004).

- *Perform*. This phase is largely highlighted by a process of service provision. Typically, the performing phase in a service lifecycle is known as the delivery of the service. For a service designated to a given customer, this is also the phase that the majority of service encounters occur from the customer's perspective.
- *Improve*. As we know that customer needs keep changing as time goes, we must continuously improve our services to stay cutting-edge in the business. Indeed, the improvements in all aspects of services are crucial to keep our services competitive (Qiu, 2013).

This classic service diamond relationship in a service organization clearly marks the four key milestones across the lifecycle of service. When the first version of services is conceived, developed, and offered by a service organization, clear and well-specified milestones that are explicitly based on the above-defined sequence might be created and followed from the operations and management perspectives. During each phase, the service organization usually has different business objectives set as the highest priority in management and operations. The diagram in Figure 1.4 shows a normal and classic view of managerial and operational priorities pursued in the service operations and management of a service firm, in which a milestone priority shifts along with the emergence of a new phase during service business operations. These typical four priorities that are logically identified throughout the lifecycle of service are briefly discussed as follows:

- *Innovation*. A service is not competitive unless it is creative and innovative. Service products are just part of a service. First of all, we should focus on the transformation of resources to innovate services that are aimed at positioning our services in the market for competitive advantage. Innovations must be thoroughly embodied in not only the service products, but the processes of delivering and improving the services.
- *Value Proposition*. The execution of a service by a service provider must create a value for the service provider. As a service takes time from beginning to end, we must have the value of a service clearly defined in order to have the value appropriately measured, monitored, and realized in the process of service provision.
- *Value Creation*. The targeted value is usually created in the process of service delivery. However, it is not uncommon in the service industry that we argue that the delivery of a service actually starts from its development phase.
- *Performance*. We know how a service meets the needs of its end user once the service is delivered. Quite often, we would like to have the deliveries to be monitored, so the real-time performance of our service businesses can be captured and then weakness can be identified for further improvements.

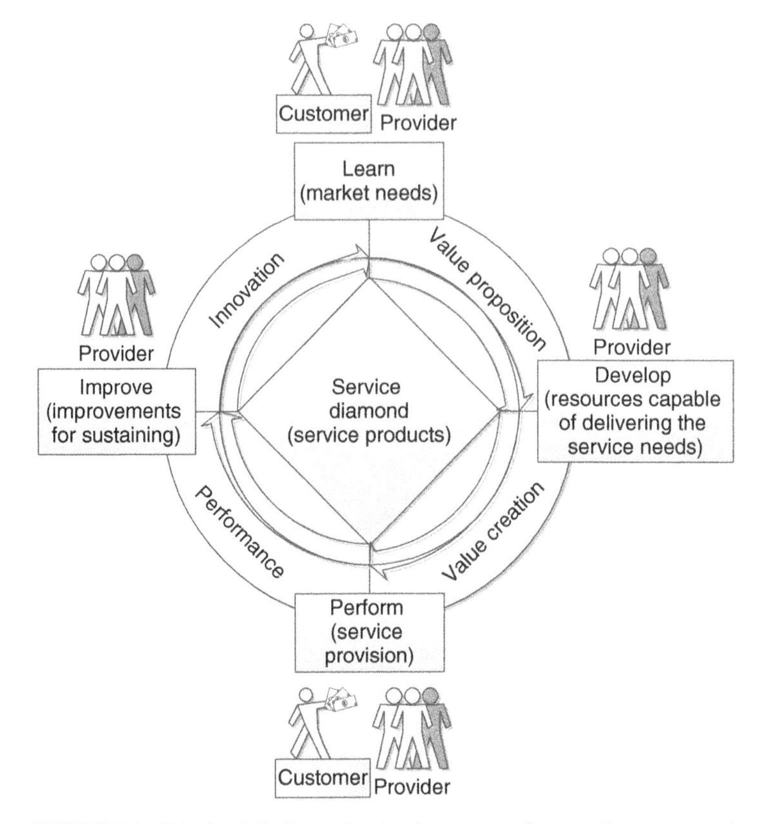

FIGURE 1.4 *Priority shifts in service business operations and management.*

Similar to the lifecycle of manufactured goods, the relationships defined in Figure 1.4 are most likely neither strictly linear nor purely sequential. In other words, the four priorities should not be separately considered during business operations in a competitive service organization. Frequently, a service organization will operate all the four phases in parallel as soon as the first batch of services gets completed. Please keep in mind, the first batch of services could simply be prototype or trial-based services. Competitive services are the results of both the coordinated and collaborated business actions taken by all the employees in the service provider, resulting in that satisfactory consumptions are realized and thus quality services are perceived by the customers.

1.4 SERVICE ENCOUNTERS THROUGHOUT THE LIFECYCLE OF SERVICE

Before the emergence of the Internet, a physical context type of interactions between a service provider and a service customer was radically necessary in the process of performing a service. We define a service encounter as an act where a customer interacts with the service the customer wants. Therefore, a service encounter essentially is a social and transactional interaction in which a service provider performs a service activity beneficial to its corresponding service customer (Czepiel et al., 1985; Czepiel, 1990; Bitner, 1992). Undoubtedly, each service encounter becomes a moment of truth. For a given service, we are the service provider and might perform "good" or "bad" services by rendering "good" or "bad" user/service experience. In other words, we have the ability to either satisfy or dissatisfy a customer when we are engaged in a service encounter. With a previously dissatisfied customer, surely we can rely on another service encounter to offer a service recovery that will be satisfying such a previously dissatisfied customer and potentially making him/her a future loyal customer (Surprenant and Solomon, 1987; Bitner, 1990; Tax and Brown, 1998).

Product, price, and place consist of the rudimentary marketing mix (Figure 1.5a) that is crucial when a product is set for sale. Marketers have reconstituted and/or expanded the mix by including different components to accommodate the differences derived from different goods and the changing customers' needs in the market, aimed at improving sales from time to time. Since the 1960s, product, price, place, and promotion (or simply called 4 Ps) have been widely and steadily used as the pillar components (Figure 1.5b) in the supply-side marketing management to define or describe the marketing mix that can be applied for identifying the niche of a physical product for sale (McCarthy, 1960).

In the business world, the effectiveness of marketing a product or service has traditionally and largely depended on how 4 Ps would be coordinated in the product or service marketing and sales process. The fundamental concepts of these components in marketing goods can be briefly summarized as follows:

- A quality physical product has long been the core in the goods marketing. The value of a piece of goods lies in its ability to satisfy the needs of a customer, which is mainly seen in the physical attributes and technical functions of the provided product.

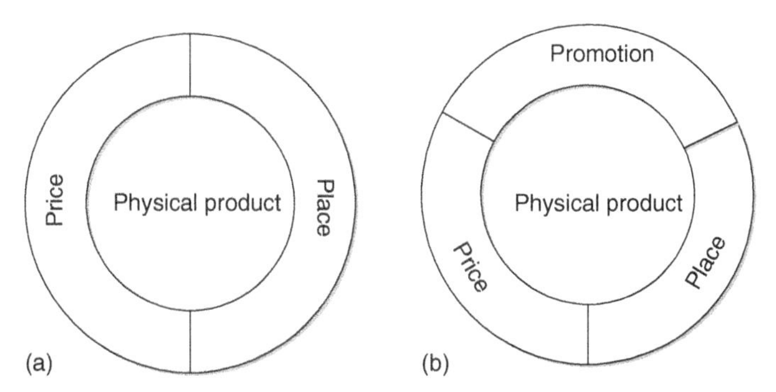

FIGURE 1.5 *The rudimentary and popular marketing mix. (a) Rudimentary 3 Ps and (b) Popular 4 Ps.*

- The price of a product has a lot of impact on its customer's satisfaction level. Quite often, right price is the first step to help push products into the marketplace to get quickly accepted by the customers.
- The price of a product for a designated marketplace should be appropriately set. Varying with socioeconomic statuses, customers in different places frequently have different affordability. They might also have quite different preferences to the physical attributes and technical functions of the provided product because of their cultural preferences and physiological characteristics.
- Promotion plays a critical role in attracting prospective customers in a given marketplace. It varies with marketplaces; it might also change with seasons. This is particularly true when a holiday is approaching. Manufacturers (or retailers) tend to take advantage of the increased number of shopping days if the products are primarily for consumers.

As the competition gets intensified over the years, organizations have shifted their foci to customers, resulting in a customer-focused marketing mix, which is termed as 4 Cs (commodity, cost, channel, and communication) (Tannenbaum and Lauterborn, 1993). The 4 Cs marketing mix model essentially replaces 4 Ps (i.e., product, price, place, and promotion), providing a customer-centric version alternative to the 4 Ps in the goods marketing. Commodity promotes the pleasure realized when a product is used by a customer. Cost considers not only the producing cost but also the use and social costs applied to the customer over time. Channel focuses on the convenience provided to the customer when the product is purchased. Communication highlights the interaction and education to help the customer use the product in an optimal and satisfactory manner.

The focus shift from supply to customer clearly shows that organizations know the increasing importance of inclusion of customers in business operations and management. This is especially critical in the service sector as service encounters bundled with additional distinguishing characteristics of service directly impact the corresponding service quality and satisfaction perceived by customers. Over the years, the academics and practitioners have expanded 4 Ps to 7 Ps in the service marketing and delivery model by including three more components, people, process, and physical evidence, to reflect the substantively changed market needs and the evolution of customer-centric service marketing and delivery (Booms and Bitner, 1981; Bitner, 1990).

- People are crucial in service provision. People are human actors centered at service encounters, including employees, customers, and other personnel who are directly or indirectly involved in the service encounters.
- Processes define and govern the procedures, mechanisms, and flow of activities in service encounters, extremely important for service providers to conduct effective marketing and deliver quality and satisfactory services.
- Physical evidence refers to the physical surroundings and tangible cues that could influence the customer's perception of services. As services quite often are intangible, customers intuitively rely on certain tangible cues that can assist them to assess the offered services.

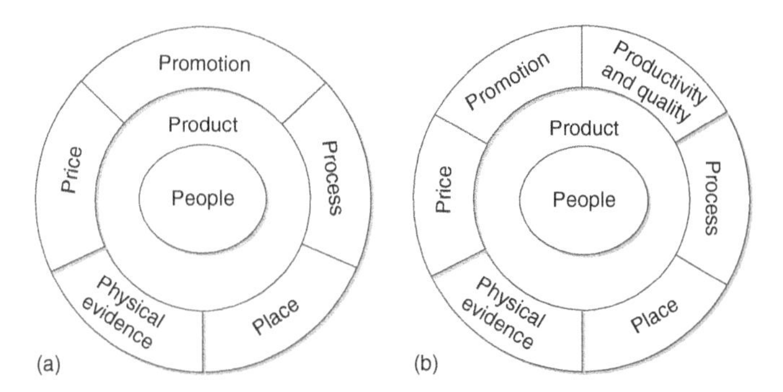

FIGURE 1.6 *The 7 and 8 Ps of services marketing and delivery model.*

Although we can learn quite a lot from the manufacturing industry, we have inevitably confronted unprecedented challenges in understanding people's roles in rendering services in the service industry. It becomes clear that a service organization must put *people* (customers and employees) rather than physical goods in the center of its organizational structure and operations to keep businesses competitive (Qiu et al., 2007) (Figure 1.6a). For example, service quality is highly regarded as a comparison of customers' expectations with performance perceived in service provision. Thus, service quality can be extremely subjective. As a result, service productivity and quality are extremely difficult to monitor and measure as they vary with circumstances. Lovelock and Wirtz (2007) include productivity and quality in the service marketing and delivery model, as shown in Figure 1.6b, to warrant that service productivity and quality are well considered throughout service lifecycles. To be competitive, service organizations must control and manage the total lifecycle of service in a cost-effective and efficient manner.

In exploring service encounters in the service industry, the literature has thus developed a series of concepts and models and applied different combinations of 8 Ps to meet the specific needs under different business circumstances, such as marketing, operations and management, and business strategic planning. As discussed earlier, when a service is performed, its consumer and provider interact with each other, directly or indirectly, consecutively or intermittently, physically or virtually, and briefly or intensively, during the process of performing the service. We illustrate a series of service encounters in Figure 1.7 to highlight a variety of possible social and transactional interactions throughout the lifecycle of service.

It is worthy to mention that this book promotes a new look of service encounters. Instead of focusing on the interacting activities between providers and customers during the process of service deliveries, we explore all the interactive activities between service providers and customers throughout the service lifecycle, from service conceiving to service termination. Consecutive service encounters form a service encounter chain (Svensson, 2004), which can be modeled using an event-based time series. Furthermore, highly correlated service encounter chains thus create a service encounter network. A comprehensive discussion on service encounter networks across the lifecycle of service is provided in Chapter 3.

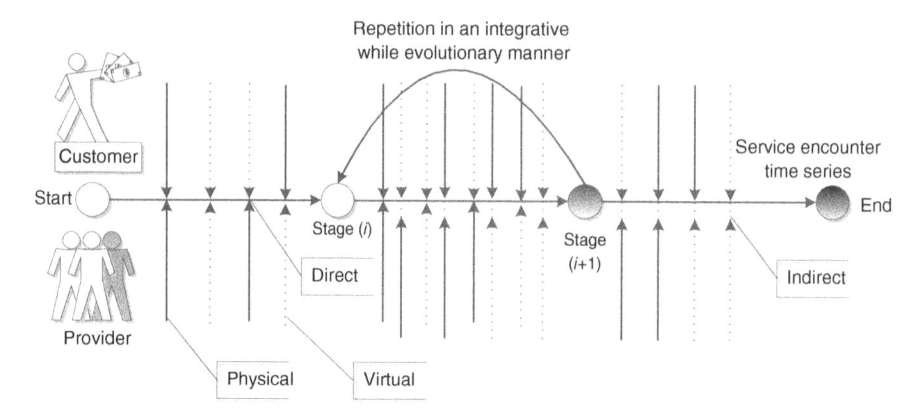

FIGURE 1.7 *A series of service encounters throughout the lifecycle of service.*

At first glance, the concepts that are illustrated in Figures 1.3 and 1.4 seem to have no difference from any other illustrations of lifecycles of businesses in any industry. Just like a manufacturing firm, the lifecycle phases in a service organization can be recursive, nested, repetitive, or in parallel during business operations. Indeed, a moment of truth for a service is an instance wherein a customer and a provider-side employee interact to execute the service. An instance might be considerably different from another as the number of involved Ps would change and the constituents of the involved Ps and their relationships could also change (Chase, 1978; Booms and Bitner, 1981; Czepiel et al., 1985; Czepiel, 1990; Bitner, 1990; Bitner, 1992). As indicated in Figure 1.7, various instances could constitute moments of truth in completing the total performance of a designated service. As time goes, to an end consumer, satisfactory services shall evolve with further improved user experiences, while to the service organization services shall evolve iteratively in rendering further enriched and pleasing moments of truth to loyal and new customers.

Physical interactions describe interactions in which a service consumer and a service provider perform service activities to realize the mutual benefits with certain physical evidences that are directly and real-time related to the service required by the consumer. Daily service examples that largely depend on physical interactions include in-person meetings in a physical facility, depositing checks in a bank branch office, eating food in a restaurant, attending a class at school, shopping for merchandises in a shopping mall, or seeing a doctor in a clinic office or a hospital. Without question, physical interactions are radical and key parts of service encounters.

By contrast, virtual interactions describe interactions in which a service provider performs actions to serve a service consumer without providing the physical evidences that are directly and real-time related to the service requested by the consumer. The service encounters are essentially telecommunication or cyber based, such as checking an order status by phone, tracking an order by accessing a website, e-banking, online shopping, online education, or playing computer games over the

Internet (Bitner et al., 2000). Virtual interactions have unceasingly increased their roles in service encounters. In particular, self-service systems have received tremendous attentions. On one hand, a service provider can considerably reduce the cost of service management and operations while maintaining a uniformity of services when toward uniformity makes more sense in the services. On the other hand, a service customer can take advantage of the convenience that is entailed by the self-service systems as this kind of service can be consumed anytime and anywhere. Virtual interactions essentially are those interacting activities that are mediated by technical devices (e.g., phones, webs, and social networks).

When interactions occur between a service consumer and a service provider without any help or assistance from a third party, they are essentially direct interactions. For example, a patient sees his/her family physician; or a customer has his/her lunch in a fast food restaurant. The services are directly performed between a service provider and a service customer. By contrast, when services that a service provider promises to offer to customers are actually delivered by a partner of the service provider, the incurring service encounters are described by indirect interactions as the customers indirectly interact with the service provider. A perfect example for an indirect interaction in a service encounter will be a service that a customer buys a set of lovely furniture from a local furniture dealer. Many furniture manufacturers contract many local dealers to sell their famous brands. A set of furniture will be delivered to a customer house only after it has been purchased by a customer. To the customer and the furniture supplier, the interaction occurs indirectly. Figure 1.8 graphically shows direct and indirect interactions in service encounters that most likely occur in service operations from an organizational point of view.

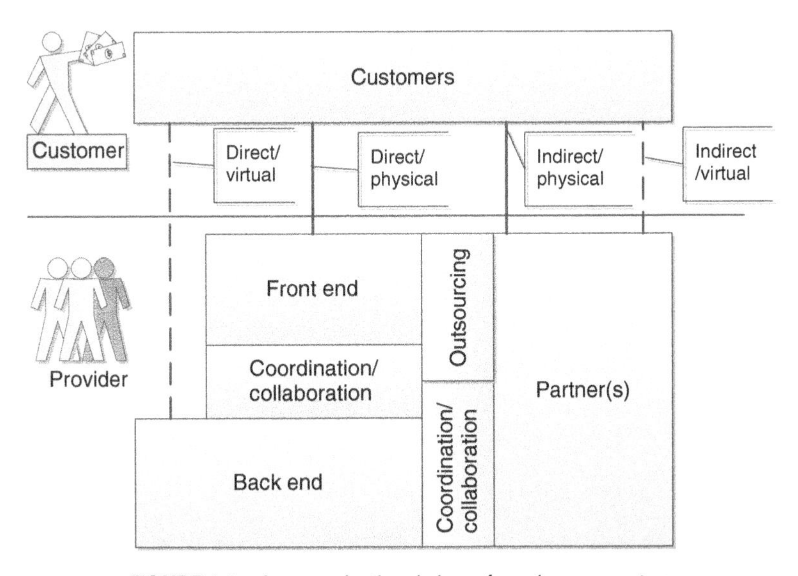

FIGURE 1.8 *An organizational view of service encounters.*

1.5 THE ECONOMIC GLOBALIZATION

Globalization is the phenomenon that highlights the process of integrating nations. Essentially, globalization refers to the exchange of world views, products, services, and cultures around the world (Deardorff and Stern, 2002). The world economy has indeed made extraordinary improvement since World War II. It has been largely credited to the fast advancement in science, engineering, and technology, such as material science, electronics, computers, networks, transportations, and telecommunication technologies over the past half century or so. In particular, the role and power of IT has been exceedingly increased, consequently transforming the ways the business works and people live around the world. Accordingly, people, production systems, computing resources, and information are effectively linked, resulting in the accelerated globalization that has precipitated today's indispensable interdependence of economic and cultural activities.

According to Deardorff and Stern (2002), "At the most basic level, globalization is growth of international trade. But it is also the expansion of much else, including foreign direct investment (FDI), multinational corporations (MNCs), integration of world capital markets, and resulting financial capital flows, extraterritorial reach of government policies, attention by (nongovernmental organizations) NGOs to issues that span the globe, and the constraints on government policies imposed by international institutions." On the basis of the data published by (the World Trade Organization) WTO, the fast growth of international trade has indeed occurred since the 1980s. The international trade growth keeps its fast pace in this new millennium. Figure 1.9 shows the worldwide (gross domestic product) GDP from 2000 to 2011 in US dollars.

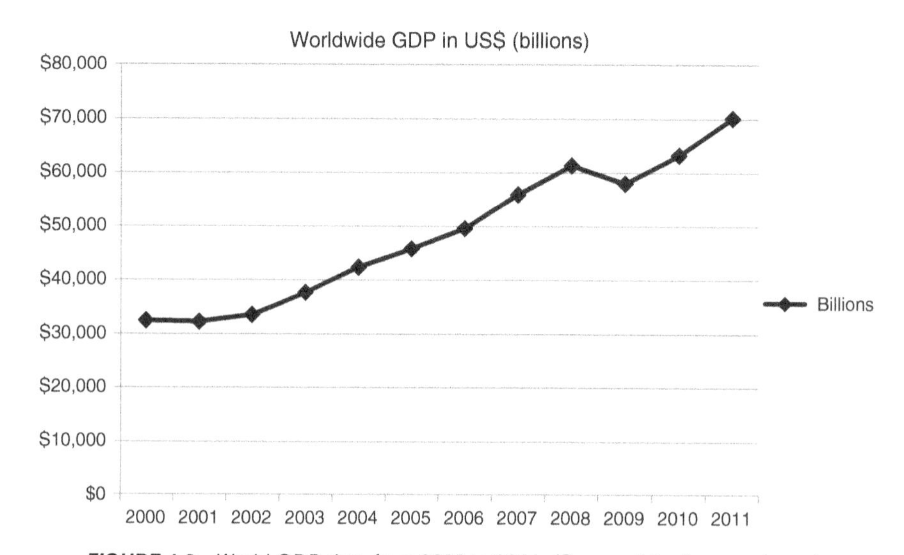

FIGURE 1.9 *World GDP data from 2000 to 2011. (Source: http://www.wto.org).*

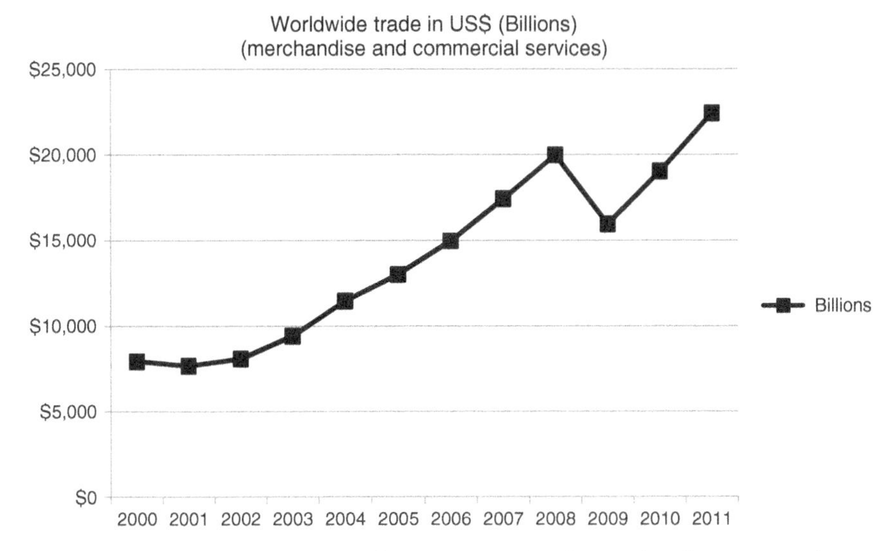

FIGURE 1.10 *World trade data from 2000 to 2011. (Source: http://www.wto.org).*

Although the worldwide GDP dropped in 2009 because of the worldwide financial crisis, the GDP growth in general is the trend. The worldwide economy in 2011 doubled the size of the economy in 2000, appropriately growing 116% in numbers. The international trade had been tripled over the same period, growing from 7687 billion US dollars to 22,424 billion US dollars (Figure 1.10). The percentage of the overall international trade in the worldwide GDP grew at a relatively moderate speed, appropriately from 25% to 32% that resulted in about 28% growth from 2010 to 2011 (Figure 1.11).

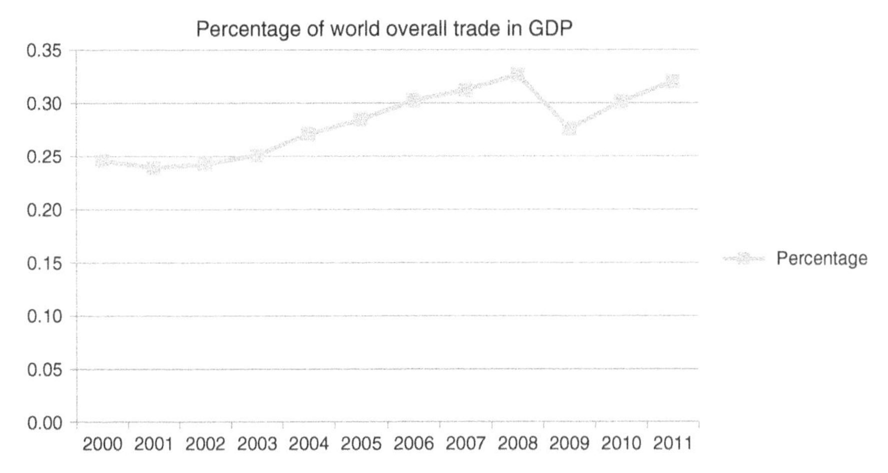

FIGURE 1.11 *The percentage of world overall trade in GDP from 2000 to 2011. (Source: http://www.wto.org).*

The direct effect of the growing international trade will be surely a more integrated global market. Regardless of where physical products are made, they are made readily available for customers around the world. Because of the accelerated globalization, it is well understood that a typical consumer with an average income in the developing economies would have the increasing opportunity and affordability of purchasing products and services that are traded internationally. Figures 1.12 and 1.13 provide the world merchandise trade in 2011 by region using export value and import value,

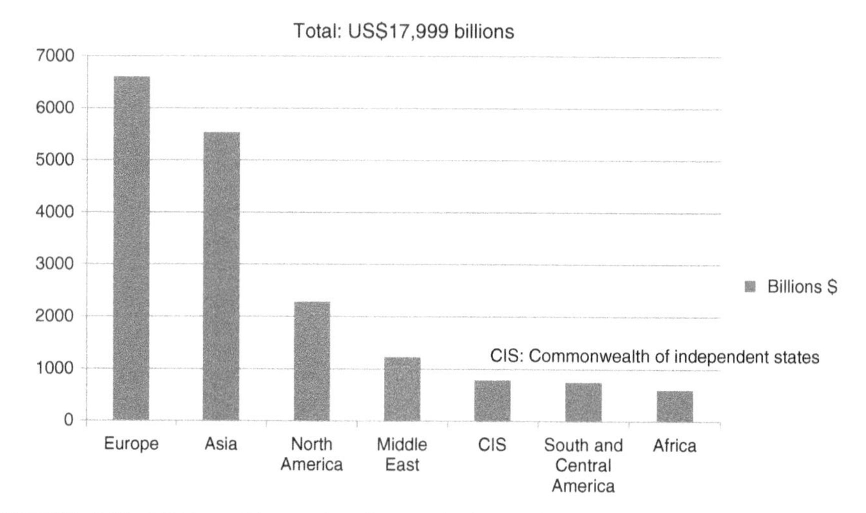

FIGURE 1.12 *2011 world merchandise trade by region: export value. (Source: http://www.wto.org).*

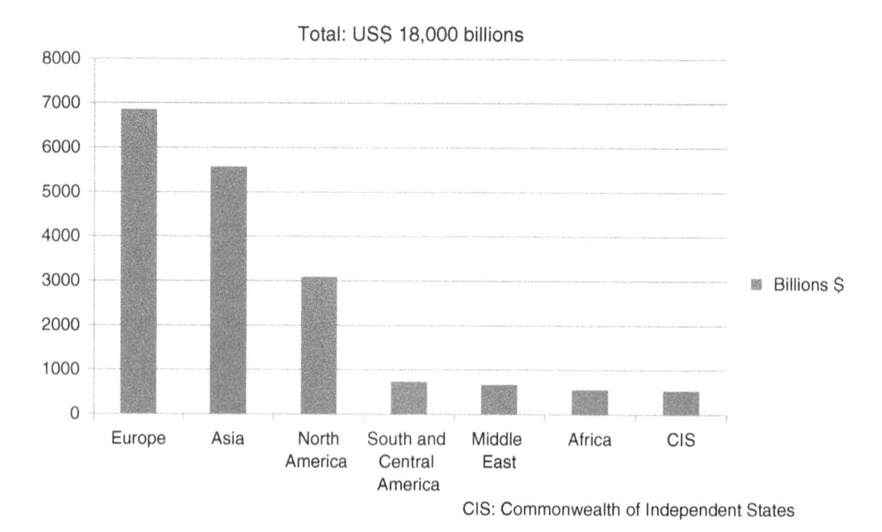

FIGURE 1.13 *2011 world merchandise trade by region: import value.*

respectively. Figures 1.14 and 1.15 show the world commercial service trade in 2011 by region using export value and import value, respectively. Overall, people, who live not only in the developed countries but also in the developing countries, are better off with international trades than without (Deardorff and Stern, 2002).

On the basis of the Bureau of Labor Statistics of the United States, excluding the goods-producing industries—agriculture, mining, construction, and manufacturing, the service industry, in general, spans all other areas from travel, transportations,

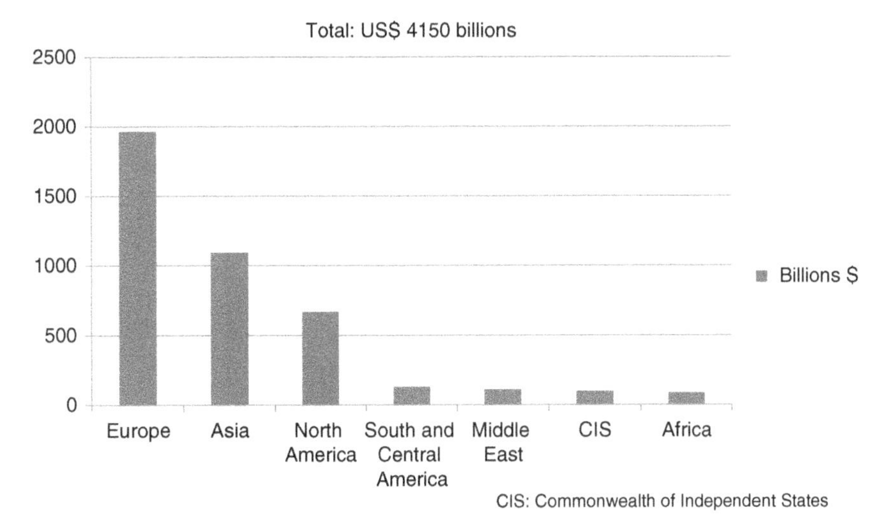

FIGURE 1.14 *2011 world trade in commercial services by region: export value.*

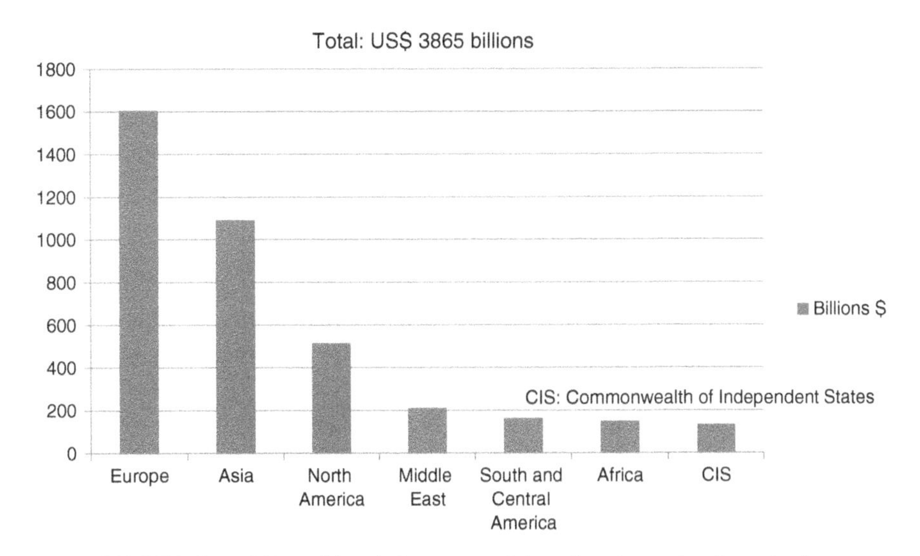

FIGURE 1.15 *2011 world trade in commercial services by region: import value.*

TABLE 1.1 Employment Data of the US Workforce in July 2012

Industries	Employment (In Millions)	Percentage (%)
Trade, transportation and utilities (wholesale trade, retail trade, transportation and warehousing, utilities)	21.483	18.6
Professional and business services	14.824	12.8
Education and health services	17.828	15.4
Leisure and hospitality	11.996	10.4
Government	21.666	18.7
Financial activities	5.954	5.1
Information	2.134	1.8
Other	4.495	3.9
Services sector	100.38	86.7[a]
Manufacturing	8.444	7.3
Construction	4.133	3.6
Agriculture	2.200	1.9
Mining	0.630	0.5
Goods sector	15.407	13.3
Total	115.787	100.0

[a]The percentage is increased from 82.1% in 2006 to 86.7% in 2012.
Source: http://www.bls.gov/ces/.

logistics, communications, utilities, wholesale and retail, trade, education, finance, insurance, real estate, health care, postal operations, governmental supports, to many other public services. Indeed, the service industry has grown to dominate the developed economies while continuing to develop extremely fast in the developing countries. As an illustrative example, Table 1.1 provides the employment data of the US workforce in July 2012.

Table 1.1 clearly shows that it is the service industry that employed the majority of workforce in the United States in 2012. Indeed, the percentage of service employees has kept growing over the years. When compared to the growth of GDPs, the US workforce change is well reflected and matched by the similar change pattern in GDPs over the years. Figures 1.16 and 1.17 provide the changes and comparisons among the agriculture, goods, and service industries. At present, the US economy is surely service-led, so are the other developed countries.

Historically, according to US Department of Commerce (1996), most of the G-7 countries began to see a steady growth in the service industry in the 1960s when the output growth of goods started to slow down. Consequently, the world economy gradually made its structural change. Since the 1980s, the service industry has grown to dominate the developed economies. We have also witnessed that the service industry has been developing extremely fast in the developing countries. Indeed, today's global economy essentially becomes service-led instead of goods-dominant. On the basis of the report published by International Monetary Fund (IMF) (IMF, 2012), the service sector around the world contributed about 63.4% GDP worldwide in 2011. Therefore, the world economy surely became service-led. Along with the economic shift

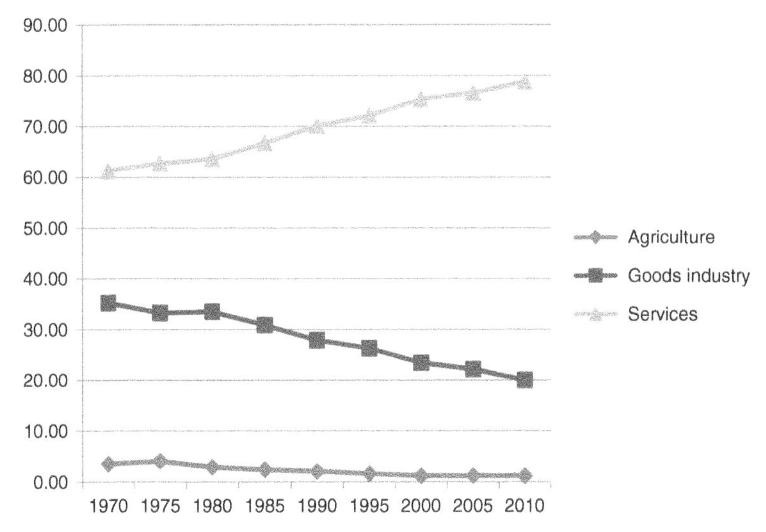

FIGURE 1.16 *US GDP percentage data from 1970 to 2010, where industry includes manufacturing and manufacturing services. (Source: http://www.worldbank.org/).*

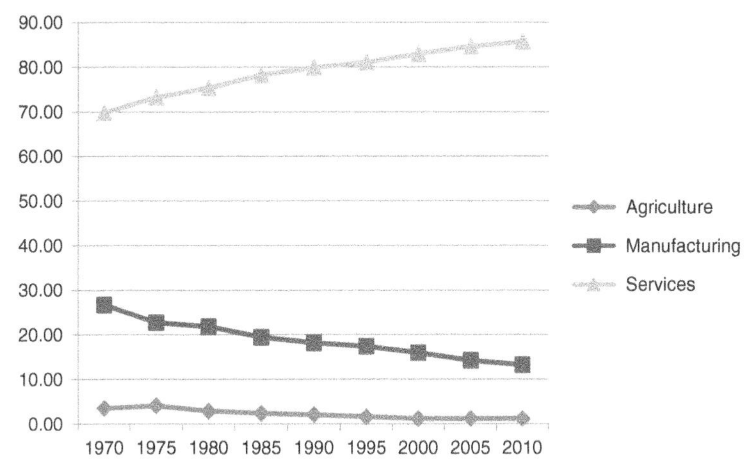

FIGURE 1.17 *US GDP percentage data from 1970 to 2010, where services include commercial services and manufacturing services. (Source: http://www.worldbank.org/).*

from manufacture to service, the changes in business operations and management are significant. Goods-dominant thinking should be replaced with service-dominant thinking in service engineering, operations, and management. With great detail in discussing the shift from manufacturing to service in the developed economy, we will advocate such a mindset change in practice in Chapter 2.

In summary, the customers worldwide are happily enjoying the exuberant markets to fulfill their daily life needs with the support of home and abroad goods and services.

Although the economic globalization is unceasingly accelerated, the world service trade is currently about a quarter of the world goods trade (Figures 1.12–1.15). The world service trade must play a quick catchup as the world economy becomes truly service-led. Therefore, service engineering, operations, and management require new and creative thinking and approaches that can be well applied in practice to help service organizations further leverage the cultural strengths and workforce talents across regions and continents.

1.6 THE EVOLVING AND HOLISTIC VIEW OF SERVICE

At the end of the day, the realized value of delivered products or services lies in their abilities to satisfy the needs of individuals or businesses. Regardless of what type of products or services we are manufacturing or offering, we must always take significant efforts to ensure that our business operations are cost-effective and efficient and our quality products or services are delivered on time. A competitive organizational structure and its corresponding managerial and operational practices should be well defined and executed. Unless it is a small workshop, a service organization is typically developed and organized based on different while necessary business domain functions in pursuit of common business goals and objectives. Although units are separately operated and managed through their well-defined business domains, they must be collaboratively coordinated across the organization in support of daily business operations to accomplish the defined business objectives (Figure 1.18). For a given organization, surely its business models, organizational structures, and accordingly adopted business domain functions, operations, and management all vary with its unique business nature, size, complexity, and regional and global presence.

As illustrated in Figure 1.18, a typical organization would have numerous business domains, most likely including sales and marketing, engineering, logistics, production, finance and accounting, and human resource, in order to fully function in delivering the organization's business promise that has been made to its customers. Figure 1.19 then shows how a typical manufacturing business successfully generates a value (e.g., profit) throughout the strategically synchronized organizational value chain (Porter, 1985; Weske, 2007). The profit margin depends on the efficiency and cost-effectiveness of underlying business operations and management to produce quality products, satisfying the needs of its end users that can be individual and/or business customers.

It is typical that the technical characteristics and physical attributes of manufactured products largely present their brands in the market. As compared to the outcomes of manufactured goods, the highlights of services are not simply and strictly seen in the functions of the services and the physical attributes of the associated products that are included in the services, but the abilities of services to satisfy end users' functional and socioemotional needs (Chase and Erikson, 1989; Dietrich and Harrison, 2006; Chase and Dasu, 2008). The competitiveness of services in the market thus largely relies on the efficacy and quality of service encounters. In addition, the focus shift from supply to customer has confirmed that organizations understand the

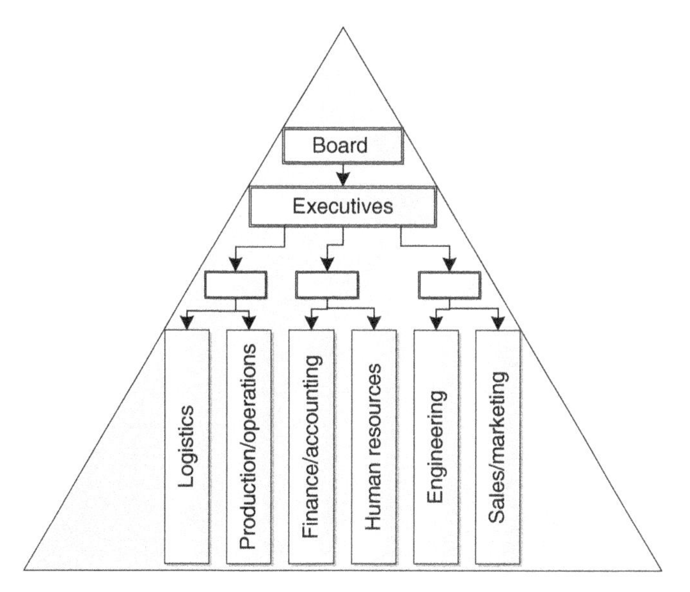

FIGURE 1.18 *A typical organizational structure.*

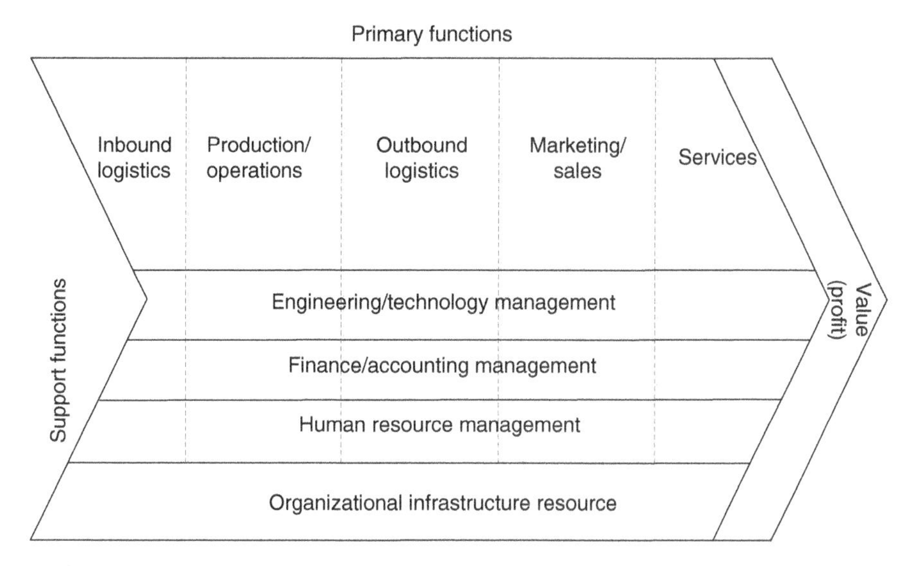

FIGURE 1.19 *The manufacturing organizational value chain (Porter, 1985; Weske, 2007).*

increasing importance of inclusion of customers in business management and operations over the past decade or so. Hence, a service organization cannot well perform services to satisfy customers if service encounters that directly impact service quality and satisfaction are not included and considerably integrated in a gradual and spiral manner on its value chain. Figure 1.20 clearly indicates the substantive change by

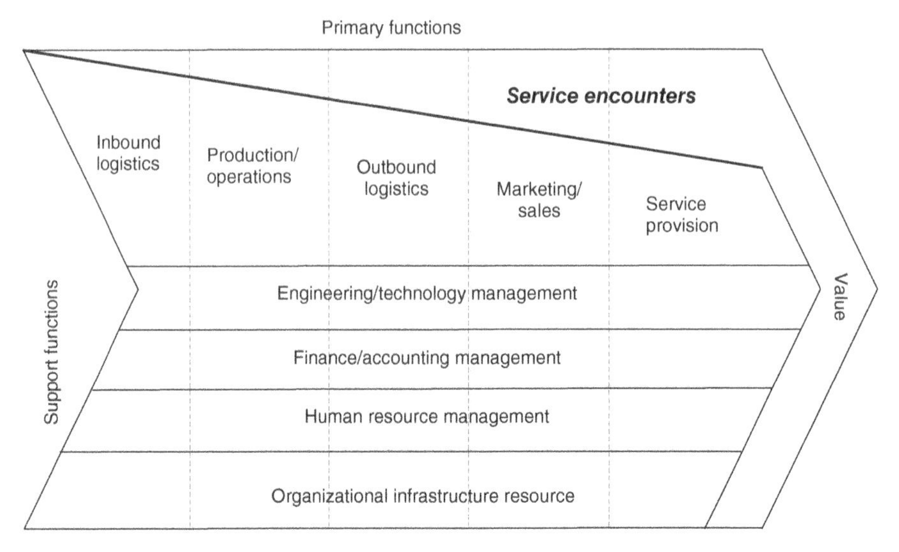

FIGURE 1.20 *The service organizational value chain.*

not only including but emphasizing service encounters on the service organizational value chain. Over the years, the service value chain has essentially evolved from the manufacturing organizational value chain. Nevertheless, the service value chain must continue to evolve by bearing total service encounters in mind to meet the needs of the dynamic marketplaces in the service-led economy (Heskett et al., 1994; Karmarkar, 2004).

The traditional, empirical, or manufacturing-based goods-centric design, development, and delivery of services can be surely applied in practice and might continue to work under certain business circumstances today and in the future. However, we have witnessed that services have evolved substantially and substantively along with the fast development of technologies, societies, and the global economy. It becomes necessary for us to understand how the modern services have evolved from ones not too long ago. Surely, with the comprehensive exploration of service innovation and better understanding of people-centric services in today's information era, we can ensure that our service provision will spiral into the manifests of user experience excellence (IBM, 2004; Cambridge, 2007; Chesbrough, 2011).

Without delving into the details, let us briefly look at a list of core services we rely on at work or primarily in our daily life. Hence, we can capture and abstract the general characteristics of service encounters (Bitner et al., 1994) that are essential for the list of rudimentary services.

- *Restaurant Food Services*. We choose a recipe we like and then go to a restaurant that serves the recipe. We talk to a waiter/waitress and order dishes from a menu. We eat and then pay for the service. If the foods are delicious, the setting is comfortable, and the waiter/waitress is polite and helpful, we will eat there again. Typically, catering services are driven by the quality of foods and customers' perceived pleasure and service satisfaction. Intuitively, we consider that the catering services are act-based as direct and physical service encounters are necessary.

- *Car Services.* For a regular maintenance, we call a car service shop we choose and schedule an appropriate mileage-based service recommended by our car manufacturer. On the scheduled day, we bring the car that is scheduled for its regular maintenance service to the shop. After we confirm with a receptionist on the needed maintenances, we drop the car there and leave for work. A mechanics might call us if there would be something to discuss, such as different problems found during the service, the need for replacing extra parts, the final charge, and/or a different time to pick up. We pick up the car after we pay the due. Again, we intuitively think that car services are generally act-based as direct and physical service encounters mainly occurs throughout the maintenance service process.
- *Residential Gas or Electricity Services.* We call a local office of a gas or an electricity service provider we choose and inform the service provider of the date we move in. When we move out, we simply do the same. We pay a bill based on the monthly usage of gas or electricity. As discussed earlier, unless there would be a problem with power lines, gas pipes, or a discrepancy in a monthly bill statement, we might not physically interact with the service provider at all. At first glance, we think the utilities services are supply-based. Indeed, indirect and virtual interaction types of service encounters dominate across the corresponding service process in the utility industry.
- *Resident Education.* We register a course that can be a required core course or a selective one for a degree or diploma. We go to school to attend instructor-led lectures or lab sessions. We listen to the lectures provided by an instructor. We frequently discuss with the instructor or classmates on a variety of topics related to the lessons. We surely complete assignments and take exams or finish projects in due course. Typically, we think resident instruction-based education services are act-based as direct and physical service encounters dominate in the whole educational service process.
- *Online Training.* We register a training course. No matter where we are, we can log on whenever we have time and an Internet connection. We read lecture notes and watch or listen to recorded lectures via a variety of online social media. We might discuss problems with other trainees who have registered the same training class. The discussions can be done synchronously or asynchronously. By leveraging a variety of online supports, we will complete assignments, take exams, or finish projects as needed. Without question, online training is quite different from resident instruction-based education. As this particular type of online training seems that the offered services are delivered using an on-demand model, it is extremely similar to a utility-type service. Intuitively, we think online training services are supply-based as indirect and virtual service encounters dominate in this type of training process.

- *Federal Bureaus or State Agencies.* We can use a driver license renewal service as a typical example of utilizing state-level governmental services. We fill in a renewal form online. A letter from the Department of Transportation of the state we live in will arrive in a few days, which informs us of the time and location to have our driver licenses renewed. We show up at the designated office on the date indicated in the appointment letter. A staff at the office will assist us in the whole renewal process. A photo will be taken, a signature is then required, and accordingly a new driver license will be issued. Service encounters seem to take a variety of possible social and transactional interactions. We typically perceive that the services provided by both federal and state agencies consist of a series of acts of public services.
- *Global Project Development.* Let us make up a fictional virtual project team first. A software project development team has six small groups of people, populating in six different geographic regions. Each group has certain unique skill sets of from 5 to 15 talent employees, including a software designer, a group architect, programmers, quality assurance staff, business analysts, and a group manager. A top-level management group, managing the entire virtual project team, consists of one team manager, one team architect, and one team business analyst. A project draft specification might be brainstormed when the top-level management group meets with a group of customer representatives. The project specification might be revised and enriched as time goes. Unless the project is completed, it is typical that the specification will keep changing to some extent. Surely each revision will be the outcome of numerous onsite or virtual meetings. Customer representatives could be directly or indirectly contacted by group members if necessary. We surely understand that a project requiring a global virtual team is usually big and complex and its development process is frequently long and complicated. In general, we perceive that global project development services surely are act-based, requiring a series of interactions and coordination, physically and/or virtually. Service encounters throughout the development cycle of global project development services are collaborative in nature.
- *Health care Service Networks.* We use an outpatient, who has a small hand lump removed through a health care service network, to show how a typical US health care service is performed. Figure 1.21 illustrates the process and associated steps that are usually taken by the outpatient to complete his treatment and get fully cured and recovered. Step 1, the patient has to see his family physician (Dr. A) first. He is usually referred to an orthopedic or hand specialist. Step 2, we assume that the patient makes an appointment with the referred orthopedic specialist (Dr. B) and sees the orthopedic specialist accordingly. Dr. B diagnoses the hand lump and then schedules an operation for him. In order for Dr. B to do a hand surgery, Dr. B asks the patient to get his physical examination done by Dr. A before the scheduled operation. Step 3, the patient has to see Dr. A to get his physical exam. Step 4, Dr. A informs Dr. B's office of the result of his physical exam. Step 5, the patient shows up in the hospital where his hand operation will take place. Dr. B has the scheduled operation completed. Step 6, the patient gets

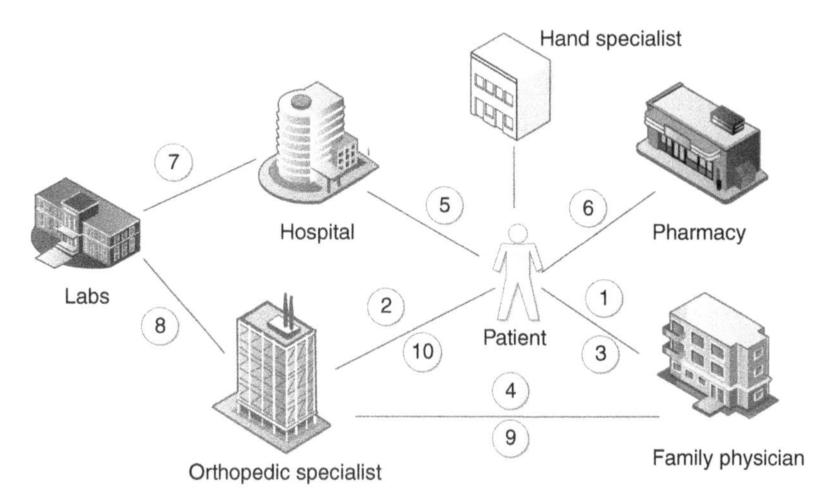

FIGURE 1.21 *A typical US health care service network.*

some prescribed medicines by Dr. B from a pharmacy. Step 7, the removed neoplasm is sent to some labs for further diagnosis. Step 8, the lab result that shows the neoplasm is benign and is delivered to Dr. B's office. Step 9, Dr. B sends a final report of the treatment for the patient to Dr. A's office. Step 10, the patient sees Dr. B., Dr. B checks how the recovery from the surgery goes. The patient gets released once Dr. B determines that the neoplasm is completely removed and patient's hand gets fully cured and recovered from the surgery. Apparently, a health care service is act-based, which mainly requires a series of interactions and coordination, physically and directly. Service encounters throughout the whole process are indeed collaborative in nature. The involved health care personnel should be coordinated in a timely and collaborative manner; the patient must be also well collaborated in order to have the operation service and treatment completed in a quick, successful, and satisfactory manner.

From the above brief discussion on service encounters that were derived from a list of selected core services at work or in our daily life, we can roughly provide a comparison table (Table 1.2) to list the key variations of different services by presenting what customers' general perceptions of these services would be and how a series of service encounters play a pivotal role in the noticeable evolution of the service organizational value chain (Figure 1.20). Banking, online banking, shopping, online shopping, tourism, and transportation services that well serve our daily life needs are also included in Table 1.2. Note that the differences perceived by customers are derived from their perceptions of services throughout the lifecycles of the consumed services.

As the perceptions of services are primarily subjective, the corresponding differences intuitively come from the differences acquired from service encounters by customers during the periods when they receive the offered services. We do not try

TABLE 1.2 Rough Comparisons of Customers' Service Perceptions Across Different Types of Services

Services		Service Encounters				
Type	Product	Physical Versus Virtual	Direct Versus Indirect	Brief Versus Intensive	Process	Customers' General Perceptions of Services
Restaurant foods	Foods	Physical	Direct	Brief	Short and simple	Catering acts
Car services	Fixing or maintenance	Physical	Direct	Brief	Short and simple	Providing fixing acts
Gas or electricity	Utilities	Virtual	Combined	Brief	Long and simple	Supplying acts
Banking	Saving and checking accounts	Physical	Direct	Intensive	Long and simple	Supplying acts
Online banking	Saving and checking accounts	Virtual	Combined	Brief	Long and simple	Supplying acts
Shopping	Merchandises	Physical	Direct	Intensive	Short and simple	Selling acts
Online shopping	Merchandises	Virtual	Combined	Brief	Short and varying	Selling acts
Transportation	Delivery	Physical	Combined	Brief	Short and simple	Supplying acts
Tourism	Tour planning	Varying	Combined	Intensive	Short and complex	Providing tour acts
Health care	Knowledge	Physical	Direct	Intensive	Varying	Providing diagnostic, care, and treatment acts
Resident education	Knowledge	Physical	Direct	Intensive	Long and simple	Providing educational acts
Online Education	Knowledge	Virtual	Combined	Brief	Long and complex	Providing educational Acts
Federal bureaus or state agencies	Polices and, regulations compliances, or law enforcements	Varying	Direct	Brief	Short and simple	Providing public supports acts
Global project development	Knowledge	Varying	Combined	Varying	Varying	Providing knowledge acts

to present perfect and comprehensive understandings of service encounters in this introductory chapter. We use Table 1.2 to simply show a few of examples to provide some clarifications or explanations of the existence of different service definitions mentioned earlier while highlighting the pivotal role of service encounters in service. The readers should also understand that there surely exist other forms of service definitions in academia and practice.

From the approximate comparisons provided in Table 1.2, we can further confirm the three main definitions that have been radically formed from customers' general perceptions of services.

- When a service is performed, if its service encounters are largely physical, intensive, and direct from the customer's perspective, then a social and transactional performance is perceived as the centerpiece of the service. This entails that service is considered as a direct performance of beneficial activities.
- By contrast, when a service is performed, if its service encounters are mainly virtual, brief, and indirect from the customer's perspective, then the usage of a service product or resource is perceived as the centerpiece of the service. Accordingly, service is quite often considered as the supplying of utilities, commodities, or digitalized media.
- In addition, people receive many societal function types of services, such as societal function services that are provided by governmental agencies. People easily view the related service encounters to have a public service nature. Service is thus essentially considered as a performance of supporting the needs for the public.

Overall, we can find that it is the "performance" or "act of performing" in service provision that creates benefits for both service providers and service consumers. As service providers, in addition to the service delivery-based interactions that are mainly perceived by customers, we know that the design, development, and preparation of service encounters must be included and well executed across the service value chain (Figure 1.20). Customers consume and perceive services through a list of service encounters that can be delivered, face-to-face or virtually, directly or indirectly. However, in a systemic perspective, customers are heavily involved in other phases of the service lifecycle including inputs to service design and feedback on consumed services. In other words, the real value of service is the total perceived value of the outcomes developed and accumulated from a series of service encounters that truly cross the lifecycle of service.

Furthermore, the increased degree, magnitude, and/or scope of automation, outsourcing, customization, offshore sourcing, business process transformation, e-business, and self-services continue to evolve. Service provision thus becomes more complicated and challenging. Consequently, service organizations demand a higher efficiency and better cost-effectiveness in service management, engineering, and operations across their service value chains, focusing on further improving their competitiveness in the global service-led economy. We fully understand that the value of service is the total perceived value of the quality outcomes realized through a series of service encounters across the service lifecycle. Hence, we must

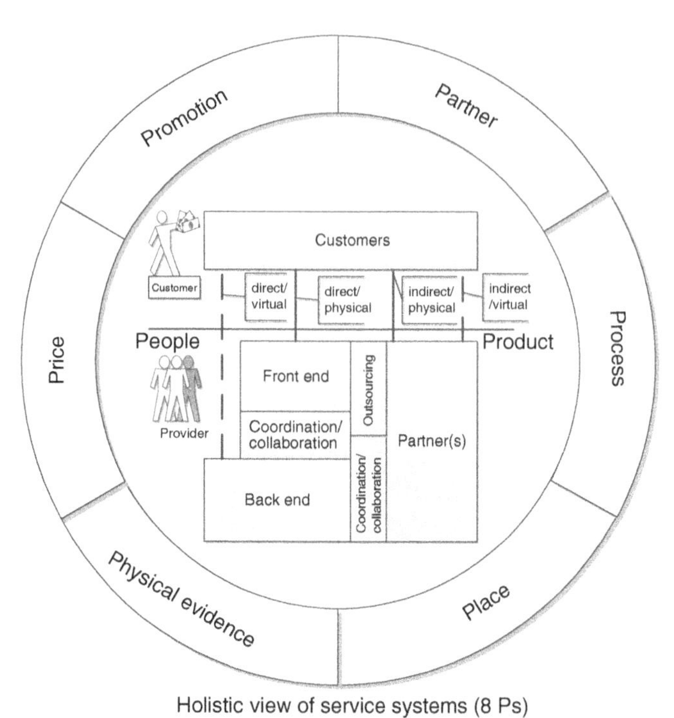

FIGURE 1.22 *A new perspective of service offering and delivering model.*

adopt a new holistic service perspective to study service, aimed at identifying right approaches to help service organizations learn, develop, perform, and improve their offered services. The new holistic perspective of services (Figure 1.22) should simultaneously include the following views:

- *Systems or Systemic View.* A system, focusing on the interdependence of relationships created in an organization, is composed of regularly interacting or interrelating groups of activities within the organization (STWiki, 2012). Thus, a service organization essentially is a service system that consists of a number of interacting and collaborative business domains systems (Qiu, 2007). The systems view is then a perspective of looking at the service organization as a collection of business domain systems that create a whole, allowing us to understand and orchestrate the interacting activities among these business domains systems. Simply put, a corresponding systemic study should focus on the relationships between those systems to determine how they affect the whole on the trajectory of realizing the business goals and objectives of the service organization.
- *People-Centric View.* Both supply side and customer side should be well explored.
- *Global View.* Partnership and cultural aspects should be fully considered. A new 8 Ps should be applied in the service provision model.

- *Lifecycle View*. As discussed earlier, the value of service is the total perceived value of the quality outcomes realized through a series of service encounters across the service lifecycle. Well-designed service encounters thus must span services from beginning to end.

1.7 SUMMARY

In this chapter, we looked into the insights of service from different perspectives. We now understand that customers consume and perceive services through a list of service encounters that occur in the process of service deliveries and beyond. However, service encounters, which are interactions between customers and providers, can occur in different ways, face-to-face or virtually, directly or indirectly. To service providers, we know that customer interactions go beyond the service delivery processes and must include the contacts during the design, development, and preparation of service encounters. Indeed, customers contribute significantly to the design, development, and preparation of service encounters in order to carry out competitive services to prospective customers (Ahlquist and Saagar, 2013). Therefore, the value of service is the total perceived value of the outcomes cocreated by providers and customers from a series of service encounters throughout the service lifecycle.

As a service is largely people-centric, truly cultural and bilateral, the type and nature of a service dictates how a service is performed, which accordingly defines how a series of service encounters could and should occur throughout its service lifecycle. The type, order, frequency, timing, time, efficiency, and effectiveness of the series of service encounters throughout the service lifecycles determine the quality of services perceived by customers who purchase and consume the services (Bitner; 1992; Chase and Dasu, 2008). It is largely true that the perceived service quality by customers substantially impacts the satisfaction and loyalty of the customers. We will fully explore all the aspects of service encounters across the lifecycle of service, chapter by chapter throughout this book.

We never try to craft our definition of service to be more comprehensive and precise than those that have been proposed by many pioneers in the global service research, education, and practice community over the years. However, we do need an appropriate, holistic, and sound definition to lay out the solid foundation for this book. Therefore, regardless of how many versions of service definition exist in academia and practice, one consistent definition is essential for the following chapters of this book. Identifying such an appropriate and sound definition of service surely becomes necessary, which naturally becomes the focus of our next chapter.

REFERENCES

Ahlquist, J., & Saagar, K. (2013). Comprehending the complete customer. *Analytics—INFORMS Analytics Magazine*, May/June, 36–50.

Bitner, M. J. (1990). Evaluating service encounters: the effects of physical surroundings and employee responses. *Journal of Marketing*, 54(2), 69–82.

Bitner, M. J. (1992). Servicescapes: the impact of physical surroundings on customers and employees. *Journal of Marketing*, 56(2), 57–71.

Bitner, M. J., Booms, B. H., & Mohr, L. A. (1994). Critical service encounters: the employee's viewpoint. *The Journal of Marketing*, 58, 95–106.

Bitner, M. J., Brown, S. W., & Meuter, M. L. (2000). Technology infusion in service encounters. *Journal of the Academy of marketing Science*, 28(1), 138–149.

Booms, B. H., & Bitner, M. J. (1981). Marketing strategies and organization structures for service firms. *Marketing of Services*, 47–51.

Cambridge. (2007). Succeeding through service innovation. *Cambridge Service Science, Management and Engineering Symposium*, July 14–15, 2007.

Chase, R. B. (1978). Where does the customer fit in a service operation? *Harvard Business Review*, 56(6), 137–142.

Chase, R. B., & Dasu, S. (2008). Psychology of the experience: the missing link in service science. *Service Science, Management and Engineering Education for the 21st Century*, 35–40, ed. by B. Hefley and W. Murphy. US: Springer.

Chase, R. B., & Erikson, W. (1989). The service factory. *The Academy of Management Executive*, 2(3), 191–196.

Chesbrough, H. W. (2011). *Open Services Innovation: Rethinking Your Business to Grow and Compete in a New Era.* San Francisco, CA: Jossey-Bass.

Czepiel, J. A. (1990). Service encounters and service relationships: implications for research. *Journal of Business Research*, 20(1), 13–21.

Czepiel, J. A., Solomon, M. R., & Surprenant, C. F. (1985). *The Service Encounter: Managing Employee/Customer Interaction in Service Business.* Lexington, MA: Lexington Books.

Deardorff, A., & Stern, R. (2002). What you should know about globalization and the world trade organization. *Review of International Economics*, 10(3), 404–423.

Dietrich, B., & Harrison, T. (2006). Serving the services industry. *OR/MS Today*, 33(3), 42–49.

Flood, R. L., & Carson, E. R. (1993). *Dealing with Complexity: an Introduction to the Theory and Application of Systems Science*, 2nd ed. New York, NY: Springer.

Heskett, J. L., Jones, T. O., Loveman, G. W., Sasser, W. E., & Schlesinger, L. A. (1994). Putting the service-profit chain to work. *Harvard Business Review*, 72(2), 164–174.

IBM. (2004). *Services Science: A New Academic Discipline?* IBM Research.

IMF. (2012). International Monetary Fund. Retrieved Oct. 10, 2012 from http://www.imf.org/.

ITIL. (2011). *Information Technology and Infrastructure Library. ITILv3.* Retrieved Oct. 10, 2012 from http://www.itil-officialsite.com/AboutITIL/WhatisITIL.aspx.

Karmarkar, U. (2004). Will you survive the services revolution? *Harvard Business Review*, 82(6), 100–107.

Lovelock, C. H., & Wirtz, J. (2007). *Service Marketing: People, Technology, Strategy*, 6th ed. Upper Saddle River, NJ: Prentice Hall.

McCarthy, J. (1960). *Basic Marketing – A Managerial Approach.* Homewood, IL: Richard D. Irwin.

Morris, B., & Johnston, R. (1987). Dealing with inherent variability: the difference between manufacturing and service? *International Journal of Operations & Production Management*, 7(4), 13–22.

Porter, M. (1985) *Competitive Advantage: Creating and Sustaining Superior Performance.* New York: Free Press.

Qiu, R. G. (2007). Enterprise Service Computing: From Concept to Deployment. Chapter 1 in *Information Technology as a Service*, 1–24, ed. by Robin Qiu. Hershey, PA: Idea Group Publishing.

Qiu, R. G. (2013). We must rethink service encounters. *Service Science*, 5(1), 1–3.

Qiu, R. G., Fang, Z., Shen, H., & Yu, M. (2007). Editorial: towards service science, engineering and practice. *International Journal of Services Operations and Informatics*, 2(2), 103–113.

Samuelson, P., & Nordhaus, W. (2009). *Economics*, 19th ed. Boston, MA: McGraw-Hill/Irwin.

Svensson, G. (2004). A customized construct of sequential service quality in service encounter chains: time, context, and performance threshold. *Managing Service Quality*, 14(6), 468–475.

STWiki. (2012). Systems Theory. Wikipedia. Retrieved Oct. 10, 2012 from http://en.wikipedia.org/wiki/Systems_theory.

Sullivan, A., Sheffrin, S., & Perez, S. (2011). *Economics: Principles, Applications and Tools*, 7th ed. Upper Saddle River, NJ: Prentice Hall.

Surprenant, C., & Solomon, M. (1987). Predictability and personalization in the service encounter. *Journal of Marketing*, 51(2), 86–96.

Tannenbaum, S. I., & Lauterborn, R. F. (1993). *Integrated Marketing Communications*. Lincolnwood, IL: NTCBusiness Books.

Tax, S., & Brown, S. (1998). Recovering and learning from service failure. *Sloan Management Review*, 40(1), 75–88.

U.S. Department of Commerce. (1996). Service Industries and Economic Performance. Washington, DC: U.S. Department of Commerce. Retrieved on Dec. 10, 2012 from http://www.esa.doc.gov/sites/default/files/reports/documents/serviceindustries_0.pdf.

Vargo, S., & Lusch, R. F. (2004). Evolving to a new dominant logic for marketing. *Journal of Marketing*, 68(1), 1–17.

Weske, M. (2007). *Business Process Management: Concepts, Languages, Architectures*. Berlin: Springer-Verlag.

2

Definition of Service

Today's business environments are characterized with advanced communications, accelerated economic globalization, and increased automation and open source innovations. As we have witnessed, the resultant vibrant but also complex service provision has created higher quality and healthier lives around the world. For a service organization to stay competitive, however, the unceasingly intensified competition demands that the organization must keep improving the efficiency and cost-effectiveness in service management, engineering, and operations across its service organizational value chain.

Over the years, service is typically considered as an application of specialized knowledge, skills, and experiences performed for the benefit of another (Vargo and Lusch, 2004; Spohrer et al., 2007). Quite often, services to customers are regarded as being perishable, heterogeneous, and intangible, commonly provided for either individuals or businesses to create desirable values to satisfy their needs (Sampson and Froehle, 2006; Qiu et al., 2007). Hence, to find an appropriate definition of service in a broad sense to cover a variety of service areas seems difficult and challenging.

Service as a word in economics is mainly defined as an act of helpful activity, the supplying of transportation, communication, and utilities or commodities, or the providing of assistance, accommodation, or leisure activities. Although its meaning might vary with circumstances, a given service substantively implies performing an action or a series of actions. Indeed, no matter what a service product is embedded as part of the offered service, the service is being executed only when the act of a designated service activity is performed. The value of the service thus largely depends on when, where, and how the process of relevant service activities gets executed.

From Chapter 1, we understand that a sound, solid, and holistic definition of service is essential for this book, regardless of the existence of many versions of service definition in academia and practice. Therefore, in order to find a sound, solid, and

Service Science: The Foundations of Service Engineering and Management, First Edition. Robin G. Qiu.

holistic definition of service we need, we must revisit and rethink the process of generating service values by exploring the following aspects of service:

- We must explore how the distinguishing characteristics in service provision differentiate a service execution process from one used in manufacturing.
- We must understand how the globalization of economy impacts the evolution of the service lifecycle.
- We must look into the total service lifecycle, spanning from service concept conceiving, to service phasing out, aimed at capturing the real insights of service-oriented business operations and finding scientific methods and methodologies to help service organizations foster service design, development, delivery, operations, and improvement in a competitive manner.

2.1 FROM MANUFACTURING TO SERVICE: THE ECONOMIC SHIFT

Not long ago when the worldwide economy was dominated by manufacturing, both academics and practitioners paid much attention to the design, development, production, and innovation of physical products. Except for studies that were radically focusing on the employee–customer encounter and service quality in the service marketing research and practice, people's social, physiological, and psychological roles were largely blurred and barely seen in the manufacturing business operations and management (Figure 2.1).

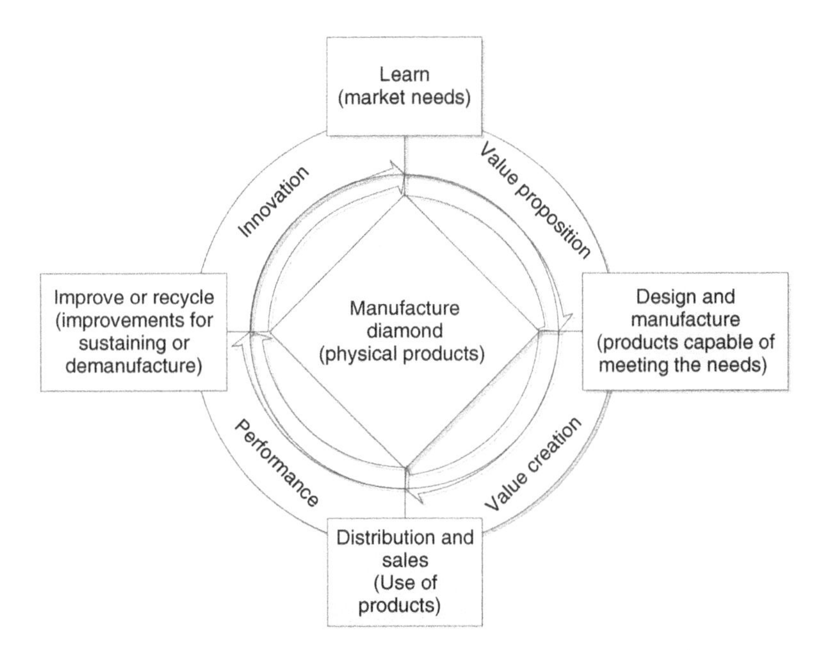

FIGURE 2.1 *Priority shifts in operations and management in manufacturing.*

We can clearly see that Figure 1.4 is directly derived from Figure 2.1. In other words, the priority shifts along with the changing phases defined in the service diamond relationship (Figure 1.4) are similar to the ones defined in the manufacture diamond relationship (Figure 2.1). This is surely not a surprise as manufacturing had played the dominative role in the world economy for over a century or so. Inertial thinking is normal to the majority of human beings, resulting in that many service organizations run their services using manufacturing mindsets.

More specifically, as physical products had been essentially the focus over the last century, manufacture/service organizations had been dominantly considered as technology- or product-driven entities, paying much attention to their product features/functions, applied materials, production/automation means, distribution, and productivity. Hence, it is the product that determines the value of a business which is the mindset of an organizations' executives in business operations and management (Chesbrough, 2011a, 2011b). The physical product surely is the implicit star in a manufacturing organization. A competitive manufacturing business is thus driven by the closed innovation paradigm, focusing on technical breakthroughs from its increased investment in the internal research and development (R&D). Figure 2.2 shows the typical virtuous circle that has been well recognized and adopted in manufacturing (Chesbrough, 2003).

The typical virtuous circle in manufacturing relies heavily on the realized technology breakthroughs in a timely manner to stay competitive (Figure 2.3). Innovations in manufacturing are goods-oriented. In other words, manufacturing

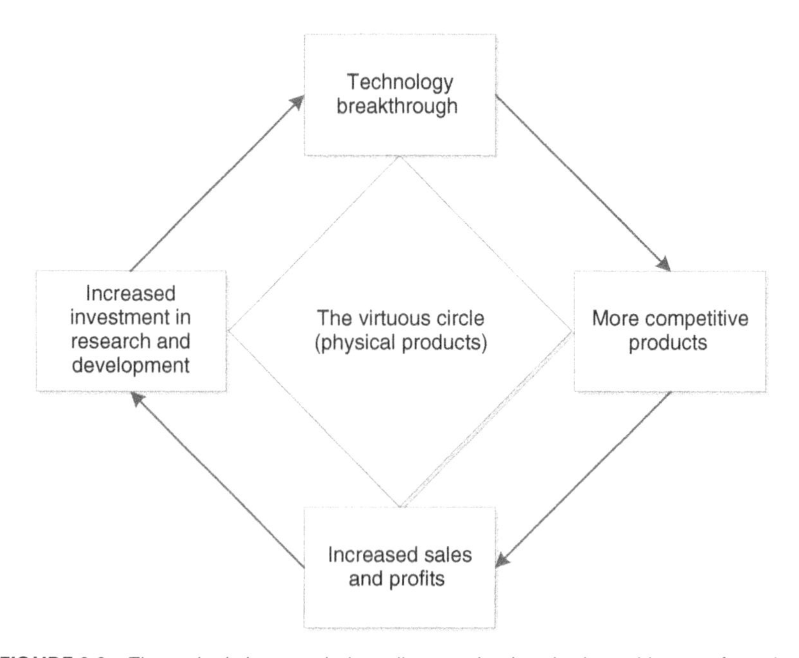

FIGURE 2.2 *The typical virtuous circle well recognized and adopted in manufacturing.*

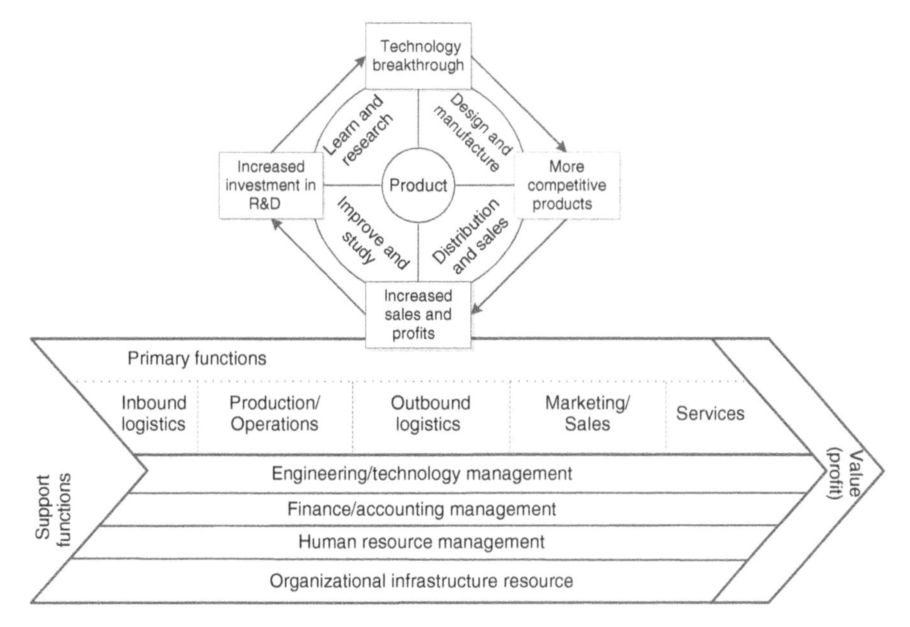

FIGURE 2.3 *Competitive manufacturing driven by goods-oriented innovations.*

organizations primarily invest in the following areas to provide competitive products for mature and emerging markets:

- *Product Features/Functions.* The organizations rely on technology breakthroughs to add newly discovered features/functions to the products, aimed at outperforming competitors considerably in terms of the products' technical capabilities of meeting the needs of customers.
- *Materials.* The organizations increase the performance and reliability of the products using the innovations, resulting in further improved customers' satisfaction from the purchased goods.
- *Production Automation.* The organizations improve the quality of the products and reduce the cost of productions, to improve profit margins and maintain goods quality brands.
- *Supply Chain and Logistics.* The organizations further optimize all transportation, storage, and distribution of raw materials, work-in-process parts, finished goods, and other resources from the point of origin to the point of consumption, focusing on improving the efficiency and cost-effectiveness of the flow of the formed organization networks internally and externally. The lead time and cost of manufactured products in production and on the supply chain thus get cut further.

Without a second thought, the innovation-based virtuous cycle must center and indeed has truly centered at physical products in manufacturing organizations.

In general, the value for an organization is the benefit provided for customers, employees, partners, and investors. The value recognized as a benefit indeed varies with the stakeholders and time. At a different time, the benefit for a different stakeholder might be realized in a different form. In manufacturing, the benefit is usually realized through product-centric business operations that seek to leverage prices over costs by means of organization, policies, management, operations, technology, finance, incentives, and other factors throughout the manufacturing value chain (Figure 2.3).

As the manufacturing productivity and quality of products have been significantly improved, the standard of living has been considerably improved. Although the means that are used in gauging the standard of living vary with political and economic societies and geographic areas, in the last quarter of the twentieth century, the world witnessed significant transformations in many aspects of well-being that were mainly driven by the long-established industrializations and well-improved productivities in the developed economies. As a result, the global economy gradually shifted its focus from manufacturing to services, aimed at further improving the quality of lives around the world. It has been well recognized that the dawn of information era has accelerated the shift.

The quality of life currently takes into account not only the material standard of living but other intangible values of living that are service-oriented and largely subjective. Indeed, people are increasingly demanding supportive, pleasant, and value-added services. The social and perceptive concepts and measures, including success, happiness, satisfaction, and the like, are frequently applied and used to measure the outcomes of performed services. Thus, the measurements used for gauging consumed services are substantively different from the performance measurements such as physical features and technical functions that have been mainly used in manufacturing. Without identifying all the characteristics of manufacture and service, a brief comparison between manufacture and service using the simplified lifecycle phases in Figures 1.4 and 2.1 is provided in Table 2.1.

Many scholars and practitioners have attempted to differentiate service from goods on one or more dimensions ultimately arriving at a continuum (Bell, 1981; Bowen, 1990); goods are arrayed at one end and service on the other end. It is typically true that there is considerable overlap between the two (Solomon et al., 1985). In this book, we incline to have our main discussions by inclining to the service end. However, for a service, we must understand that goods are frequently the conduits of service provision. Therefore, the physical attributes and technical characteristics that specify the goods are surely indispensable to the service.

Let us look into two excellent examples to see how the discussions are reflected in real life. One is a typical car repairing service. The other is a purchase of a popular product, iPad, from the Apple online store.

- *Car Repairing Service.* When we know there is a problem with a car, we call a car service shop that we choose and schedule an appointment. On the scheduled day, we bring the car that is scheduled for a repair service to the shop. After we confirm with a receptionist on the needed repair service, we drop the car there

TABLE 2.1 Highlights of the Operational and Managerial Priorities in the Service Industry

Lifecycle Phases		General Characteristics		Highlights of the Changes in Service Operations and Management
Manufacture	Service	Manufacture	Service	
Learn	Learn	Market analysis: physical features and functions	Market analysis: services to meet the customers' perceived values and their daily needs of service product features and functions	Customers' perceptions of services including service products
Design and manufacture	Develop	Physical product realization: functionality, quality, automation, and cost	Resource development: capability of delivering, quality assurance, and cost	The capacity of resources and capability enabled by the service providers
Distribution and sales	Deliver	Physical product distribution efficiency and price	Satisfactory acts of performing designated service deliveries and price	Service encounters
Improve or Recycle	Improve	New features and functions	New acts to increase customers' perceived values and more capable resources	Enhancements in support of service encounters

and leave for work. A mechanic might call us if there would be something to discuss, including the severity of the identified problem, the misunderstanding or misinterpreting of the stated problem, or other problems found during the conducted diagnosis process, other necessary maintenances recommended by the mechanics, the final charge, and/or a different time to pick up. We pick up the car after we pay the due. Throughout the repair service process, we understand that a series of service encounters must occur. Indeed, the appropriate and timely occurrence of each interacting activity on the service encounter chain ensures user experience excellence in an integrative manner, resulting in that the repair service gets executed in a satisfactory manner.

- *iPad Purchase and Delivery Service*. As a customer, I was amazed by how Apple Inc. can deliver millions of products to customers on time. It is nothing special these days that you can track the delivery process of an ordered product through the Internet, regardless of your choice of a transport organization. However, to deliver millions of products made overseas to customers on the promised dates is exceedingly fascinating. I still remember that the first generation of iPad officially became available for preorder on March 12, 2010. I ordered one on that day. The Apple online store provided several handy and optional shipment alert services that allowed customers to monitor their shipments through Apple's contracted delivery firms. I received my ordered iPad on April 3, 2010. April 3, 2010 was the date that Apple promised on March 12, 2010 that the first generation of iPads would be released and delivered to millions of US customers. Although I cannot get the order history information that is over 18 months old, Figures 2.4–2.6 show good examples of highly coordinated and collaborated processes of purchasing, manufacturing, customs, and shipments across countries that are managed in an effective and satisfactory manner. In addition to calling relevant customer service representatives, we can easily interact with the Apple online store, UPS, or FedEx websites to change orders, monitor the shipments, change how we want the ordered products to be delivered to fit into our busy daily schedules. Customer-centric and satisfaction-focused business operations and management have surely contributed to the success of Apple's business. It is the service that helps sell the product! More discussion on Apple's customer-centric and innovative approach will be further provided in later chapters.

Table 2.1 and the above-discussed two examples surely help us to get a better understanding of how certain priorities have been shifted in business operations and management in the industry. Service encounters play a fundamental role in service offerings, clearly indicating that people's social, physiological, and psychological traits are critical in services (Solomon et al., 1985; Surprenant and Solomon, 1987; Chase and Dasu, 2001). However, these traits are extremely challenging to measure, monitor, and control in service operations and management. Therefore, we understand that, substantively different from traditional manufacturers that have put products in focus, service organizations must put employees and customers at the center

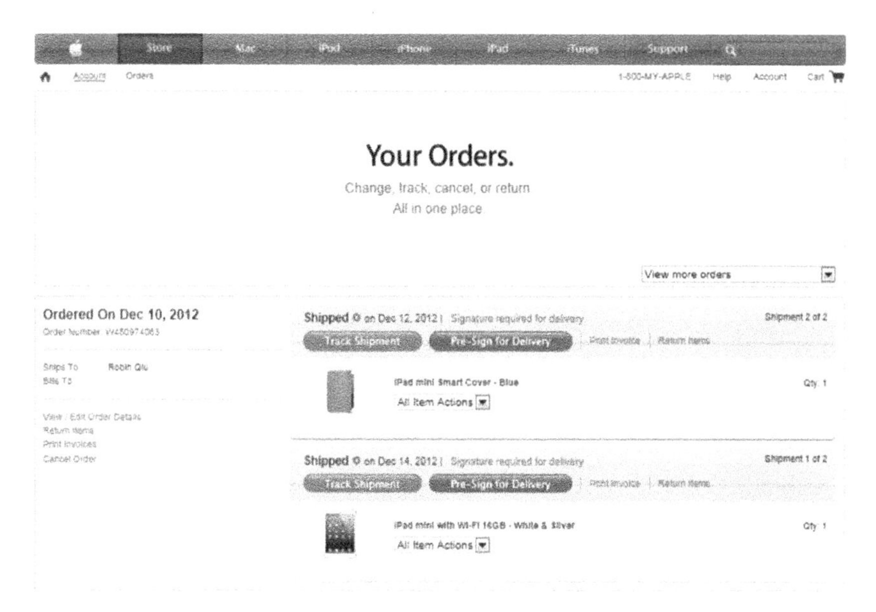

FIGURE 2.4 *Online purchase services provided by Apple online store. (Source: Apple.com).*

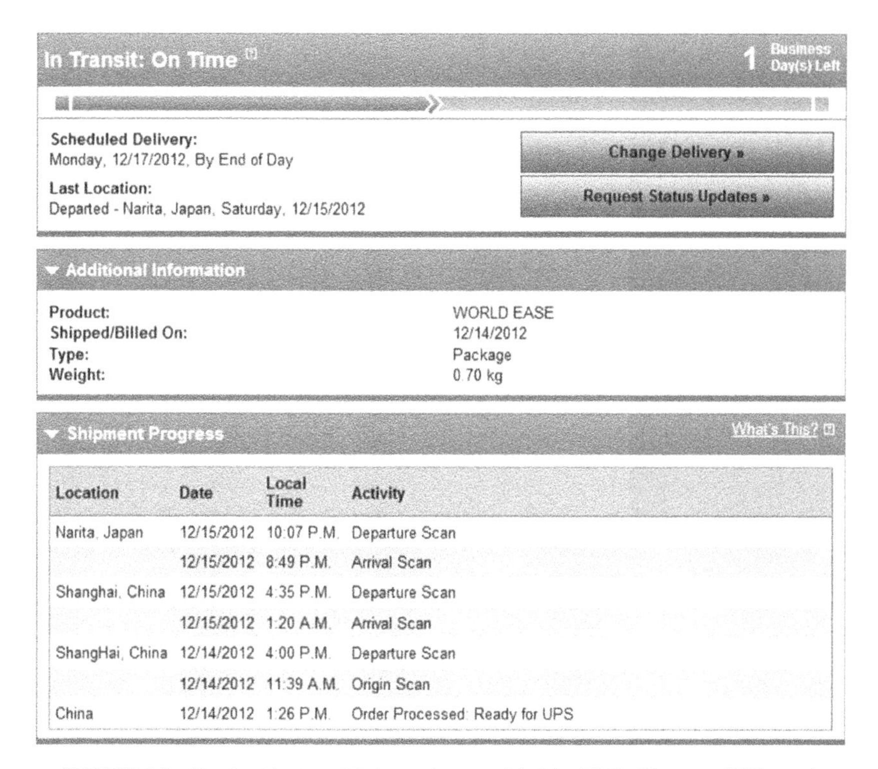

FIGURE 2.5 *Product in-transit information provided by UPS. (Source: UPS.com).*

Shipment Travel History

Select time zone: Local Scan Time

All shipment travel activity is displayed in local time for the location

Date/Time	Activity	Location	Details
Dec 15, 2012 1:07 PM	**Delivered**	COLLEGEVILLE, PA	Left at garage. Package delivered to recipient address - release authorized
Dec 15, 2012 9:49 AM	On FedEx vehicle for delivery	KING OF PRUSSIA, PA	
Dec 15, 2012 8:53 AM	At local FedEx facility	KING OF PRUSSIA, PA	
Dec 15, 2012 6:29 AM	At destination sort facility	PHILADELPHIA, PA	
Dec 15, 2012 3:44 AM	Departed FedEx location	MEMPHIS, TN	
Dec 15, 2012 1:19 AM	Arrived at FedEx location	MEMPHIS, TN	
Dec 14, 2012 4:04 PM	Departed FedEx location	ANCHORAGE, AK	
Dec 13, 2012 6:08 PM	In transit	ANCHORAGE, AK	
Dec 13, 2012 3:31 PM	International shipment release - Import	ANCHORAGE, AK	
Dec 13, 2012 11:08 AM	Arrived at FedEx location	ANCHORAGE, AK	
Dec 13, 2012 10:45 PM	In transit	NARITA-SHI JP	
Dec 13, 2012 10:19 AM	In transit	SENNAN-SHI JP	
Dec 13, 2012 1:12 AM	In transit	TA YUAN HSIANG TW	
Dec 12, 2012 10:28 PM	In transit	CHEK LAP KOK HK	
Dec 12, 2012 2:21 PM	In transit	LANTAU ISLAND HK	
Dec 12, 2012 2:19 PM	In transit	LANTAU ISLAND HK	
Dec 12, 2012 11:04 AM	Left FedEx origin facility	SHENZHEN CN	
Dec 12, 2012 8:51 AM	Picked up	SHENZHEN CN	
Dec 11, 2012 10:06 AM	Shipment information sent to FedEx		

FIGURE 2.6 *Product delivery information provided by FedEx (Source: FedEx.com).*

of concerns in business operations and management (Heskett et al., 1994; Loveman, 1998; Qiu et al., 2007; Schneider and Bowen, 2010).

Even in manufacturing, for farsighted manufacturers in the developed economy, although their product features and functions might lose their competitiveness over time, they recognize that their service components could considerably distinguish themselves from their competitors. Therefore, enterprises are keen on building highly profitable service-oriented businesses by taking advantage of their own unique engineering and service expertise, aimed at shifting gears toward creating superior outcomes to optimally meet their customer needs in order to stay competitive (Rangaswamy and Pal, 2005). General Electric, IBM, Apple, Oracle, HP, and many worldwide bellwethers are great examples in repositioning themselves toward the service-oriented businesses (Qiu et al., 2007).

The economic shift from manufacturing to service makes organizations rethink their business strategies and revamp their organizational structures and operational processes to meet the customers' fluctuating demands on services in a satisfactory manner. World-class enterprises across the board, in general, are eager for seeking new business opportunities by streamlining their business processes, building complex and integrated while more efficient IT-enabled systems, and embracing the worldwide Internet-based marketplace. It is well recognized that business process automation, outsourcing, customization, offshore sourcing, business process transformation, and self-services by leveraging the ubiquitous and pervasive networks and wireless communications became another business wave in today's evolving global service-led economy.

Indeed, the twenty-first century's business environment is considerably enabled by advanced computing, networking, and telecommunications. Business operations are thus significantly impacted by not only the accelerated business globalization but also the increased environmental awareness in societies. By taking advantage of the complex while integrative service interactions involving both providers and customers, emphasis in the service industry has evolved to sources of open innovation, collaboration, integration, and value cocreation, so as to optimally and maximally provide the value (e.g., satisfaction, success, and profitability) for the stakeholders (Figure 2.7).

The discussions we had so far, including the introduction of service encounters throughout the service lifecycle in Chapter 1, clearly show the people-centric emphasis in phases throughout the lifecycle of service. In other words, we now understand that the value of service is the total perceived value of the outcomes cocreated from a series of service encounters by both providers and customers throughout the service lifecycle. This new round economic wave driven by globalization and services seems getting more sophisticated and dynamic than ever before; there is a need for higher efficiency and better cost-effectiveness in business operations and management across the geographically dispersed value chains.

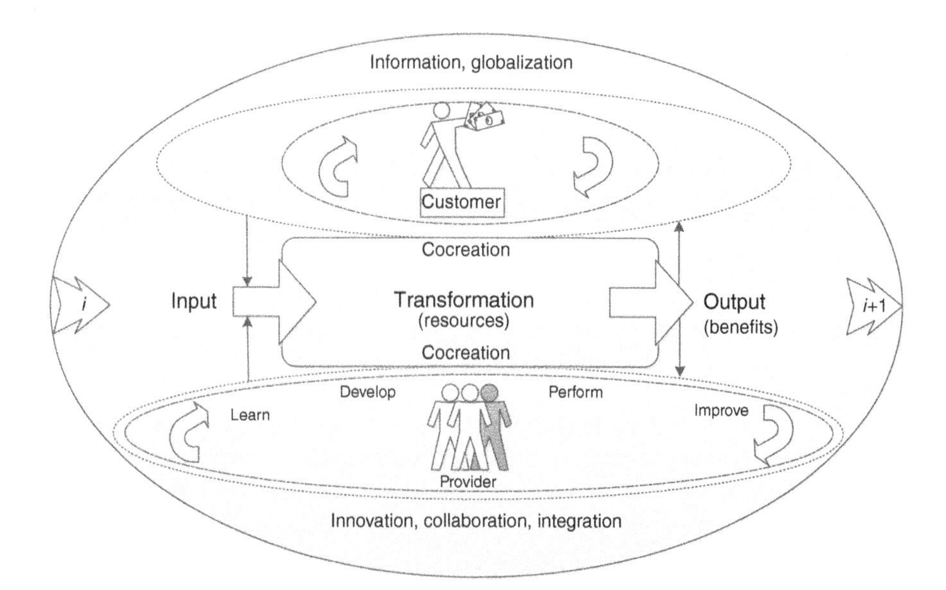

FIGURE 2.7 *Value cocreation in focus in the service industry.*

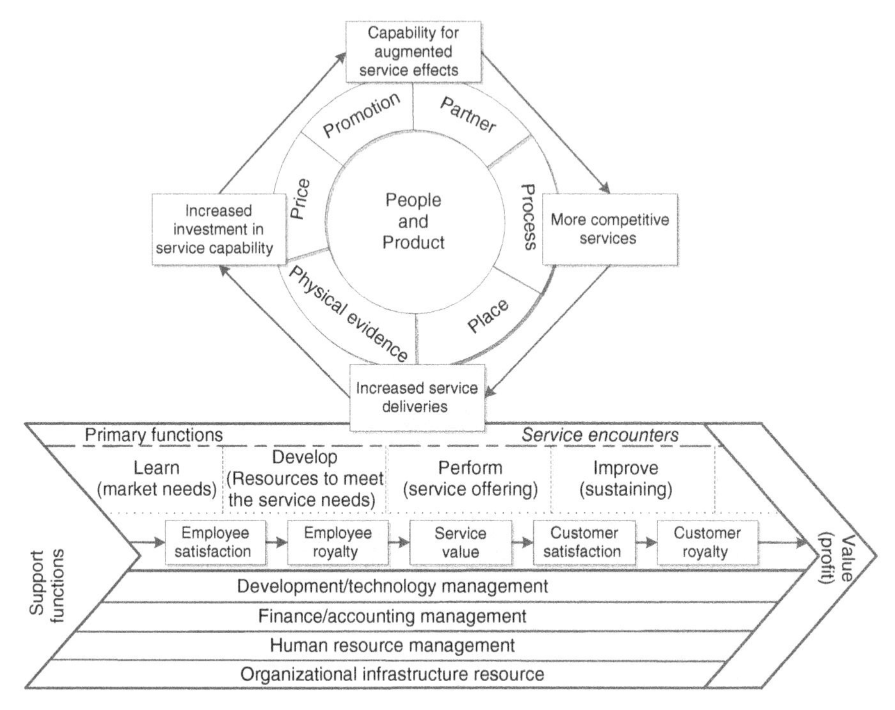

FIGURE 2.8 *Competitive service business driven by service-oriented innovations.*

More specifically, the service value (or profit) chain relies on the creation of lifetime customers' experience excellence from well-crafted and fostered service encounters. Figure 2.8 depicts the complex relationships between employee satisfaction, customer retention, and profitability (Heskett et al., 1994; Lovelock and Wirtz, 2007), emphasizing that we must rethink service encounters and find scientific ways to build and manage people-centric, information-enabled, cocreation-oriented, and innovative service organizations in the service-led economy.

As shown in Figures 2.7 and 2.8, both service providers and customers who are value-creating entities on a service value chain are interwoven in the process of service transformation. The highly correlated value-creating relationships between service providers and customers truly become the general characteristics of the modern services, indispensable for the successful completion of the lifecycle of service. By further examining the operational and managerial priority changes in response to the economic shift from manufacture and service (Table 2.1), we understand that the consistent sensing, interaction, and creativity from customers' feedbacks, participations, or consumptions throughout the lifecycle of service play a pivotal role in satisfactorily performing services that customers want (Ahlquist and Saagar, 2013):

- *Learn.* With the fast development of the Internet and the considerable improvement of living standards and life qualities, our customers have become more knowledgeable and demanding than ever before. For instance, social marketing by leveraging Web 2.0 is crucial for service providers to demonstrate the value of offered services. More importantly, it helps to conceive the concepts of services, know the market trends, engage the prospective customers, and understand customers' changing perceptions in the to-be offered services. In the service industry, hence, discovering and capturing the real and changing needs in a timely manner is what this phase really is about.
- *Develop.* As compared to focusing on the development of main and unique features and technical functions of physical goods in the traditional manufacturing, the development of services in a competitive service organization must frequently involve customers as the customers might have perceived the needs differently and/or changed the needs as time goes. Quite often, in addition to the technical features and functions embodied in services, service providers' soft resources (i.e., operant resources) should be well developed in order to deliver the services successfully. The development of soft resources in the service organization radically relies on the consistent feedbacks from the customers so that the right soft resources can be developed and made readily available for service delivery. In the service industry, thus, developing competitive services cannot be effectively done if customers are not involved in the development of to-be offered services. Customers significantly contribute to the development of service products. In other words, the value of services is indeed cocreated by both service providers and customers.
- *Deliver.* This phase in services is substantively different from the one in manufacturing. As soon as physical goods are sold, customers utilize the provided features and functions supported by the physical goods. However, services are being most likely consumed at the same time when they are being delivered. Service encounters are the key delivery mechanisms in the service industry. Successful and satisfactory deliveries of services significantly depend on efficient and effective service interactions between service providers and customers. Once again, the benefits of services are consistently cocreated through collaborative service delivery processes involving both service providers and customers.
- *Improve.* As discussed earlier, the quality of services is largely influenced and determined by the customers' perceived value, including success, happiness, satisfaction, and the like. The addition of new features and functions mainly used in improving the manufacture of physical goods is insufficient in the service industry. Usually, customers' social, physiological, and psychological roles played throughout the service lifecycle must be analyzed, focusing on the improvement of the resources applied in services and/or the enrichment of service encounters to meet the needs of the customers with continuously increased levels of satisfaction.

2.2 TOTAL SERVICE LIFECYCLE: THE SERVICE PROVIDER'S PERSPECTIVE

Let us briefly recap what we have summarized in Chapter 1 on the general customers' perceptions of services. A service is essentially considered as the "act of performing," which is a mutually beneficial activity for its provider and customer. Quite often, a service evidently manifests itself as a series of service encounters in the marketplace. The resultant value of service is usually the total perceived value of the outcomes generated from the performance of the formed service encounters chain throughout the service lifecycle.

Simply put, this perceived service by customers clearly implies performing actions. No matter what kind of service product (i.e., in a physical, soft, or hybrid form) is offered, a service with its involved service product gets completely executed only after a series of service encounters are successfully conducted. The real value of the service thus largely depends on when, where, and how the process governing all the relevant service encounter activities are performed from beginning to end and particularly how both service providers and customers have participated in the process execution.

Except for sharing the common concept of service, surely, operating a contemporary and sizable food service business compared to the ancient food service example discussed in Chapter 1 is quite different and becomes extremely more challenging. The marketplace is full of competition in many aspects, including a variety of foods, much leisured and cozier catering settings, knowledgeable clients who have a variety of socioeconomic, social, and cultural backgrounds, and different and changing clients' expectations throughout corresponding catering service processes. Accordingly, the value becomes very challenging to measure as it varies with the service providers, consumers, and marketplaces. Thus, experience-based service business operations can hardly survive in the current and competitive marketplace.

Although the understandings of a service from both service providers and service customers should be the same, we must be aware that the lifecycle of services based on the general customers' perceptions of services are substantively different from one that is conceived, developed, and managed from the service provider's point of view. Figure 2.9 schematically compares the perspectives of service lifecycles from a service provider and a service customer. The lifecycle of service in customers' perspective essentially is just a service lifespan. We can clearly see that the life of a given service from a customer point of view is essentially part of the total service lifecycle operated and managed by the service provider. In a service organization, from the operational and managerial perspective, the lifecycle of service spans over all the phases defined in the service diamond relationship. However, as indicated earlier, frequently a customer's service life mainly lies in the service delivery and operations phase.

Let us recap what we just explored. To a typical customer of a service organization, the life of a service offered by the service organization starts when the service is requested and ends when the service is completely performed (Figure 2.9). The corresponding lifespan to the customer is simply a part of the total service lifecycle horizon

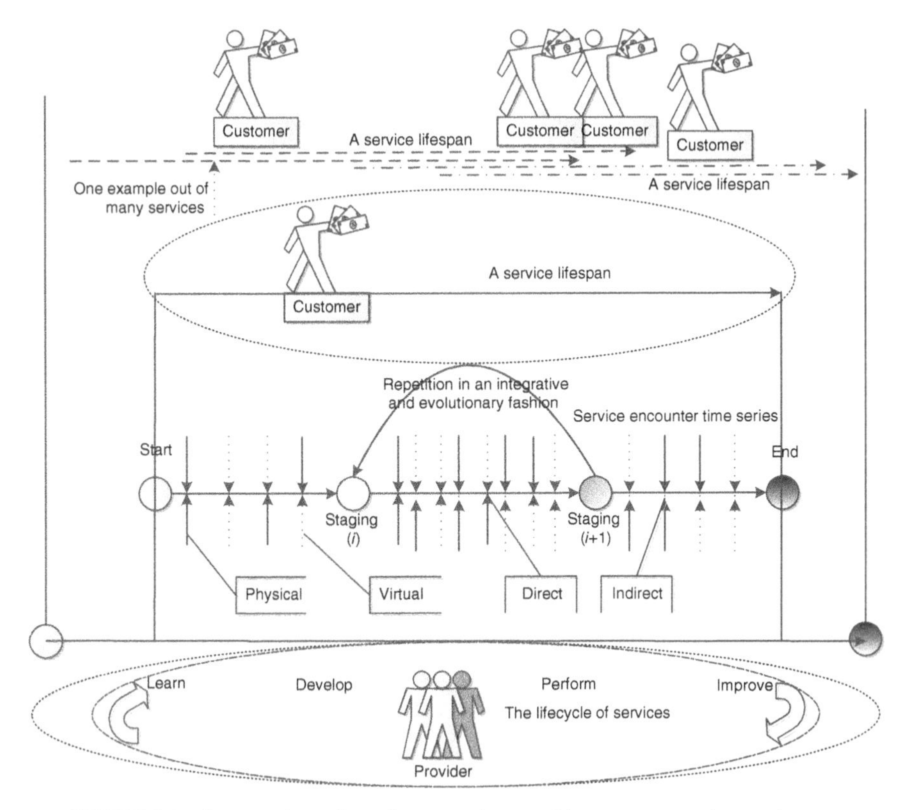

FIGURE 2.9 *Perspectives of services: service providers versus service customers.*

covered by the service organization. Theoretically, a start point can be anywhere while its end point can also be anywhere as long as the corresponding service lifespan is a positive number. In practice, a start point could be a point after such a point at which the marketed service product is requested by a customer after it becomes ready to be offered by the service organization. Then, a corresponding end point will be the time when the service is completely performed and the specified or default contract period expires.

To a customer, the encounter of a service or "moment of truth" frequently is regarded as the service from the customer's perspective (Bitner et al., 1990; Bitner, 1992). However, a systemic view of service encounters throughout the service lifecycle is necessary for a service provider, which can be created by incorporating the organizational view of service encounters (Figure 1.8) into Figure 2.9, which is illustrated in Figure 2.10. Value cocreation-oriented business processes surely are people-centric, involving both providers and customers in pursuit of excellent user experience and high level job satisfaction.

As discussed in Chapter 1, this book promotes a new look of service encounters. Instead of focusing on the interacting activities between providers and customers during the process of service deliveries, we explore all the interactive activities between

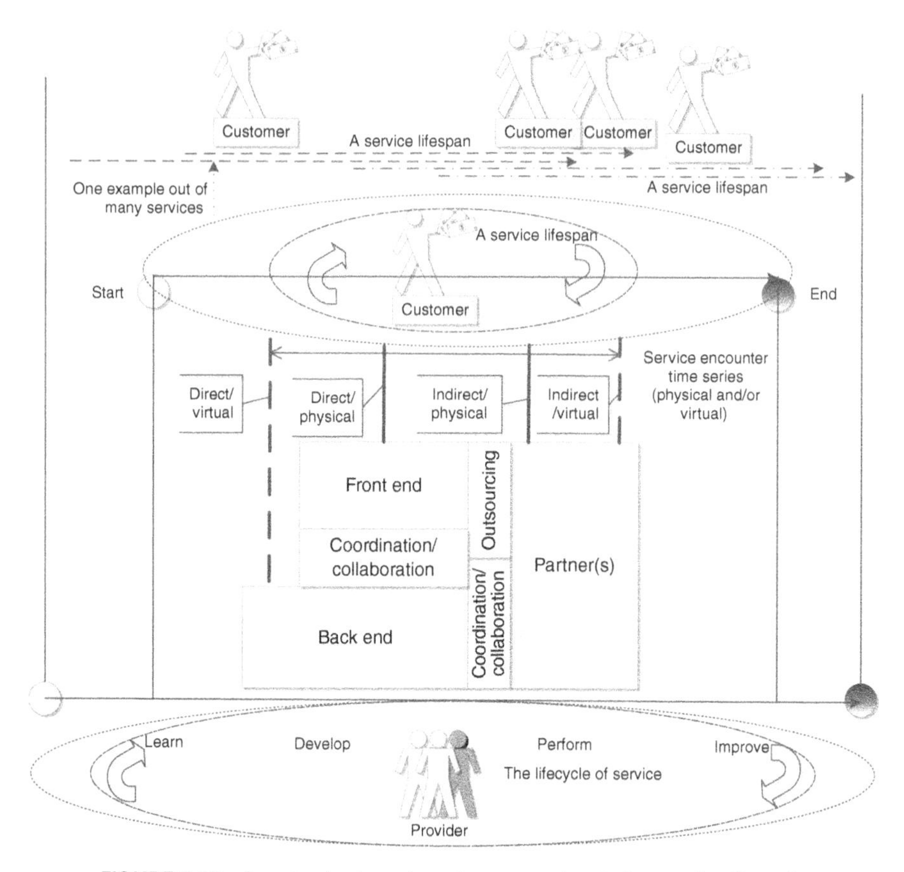

FIGURE 2.10 *A systemic view of service encounters in the service lifecycle.*

providers and customers, including their interactions from the point of service conceiving to the point of service termination. Consecutive service encounters form a service encounter chain (Svensson, 2004; Qiu, 2013), which can be mathematically modeled as an event-based time series. In a service encounter chain, a later encounter is inevitably influenced by the immediately preceding one; employees or customers could also be influenced by other previous encounters that they may have had before if those are somewhat functionally or sociopsychologically related. Apparently, as a series of service encounters entails mutual benefits in a cascading and integrative manner, we can maximize the benefits only if the cocreation-oriented business processes can be totally, cost-effectively, and efficiently executed.

The earlier discussions further confirm that only effective cocreation-oriented business operations throughout the service lifecycle in service organizations can deliver competitive services in the long run. In this book, in analog to 8Ps that have been widely used in the service marketing field, we would like to use 4Ds hereafter, "Discover," "Develop," "Deliver," and "Do Better," to describe the fundamental service diamond relationship, aimed at emphasizing the service-dominant instead

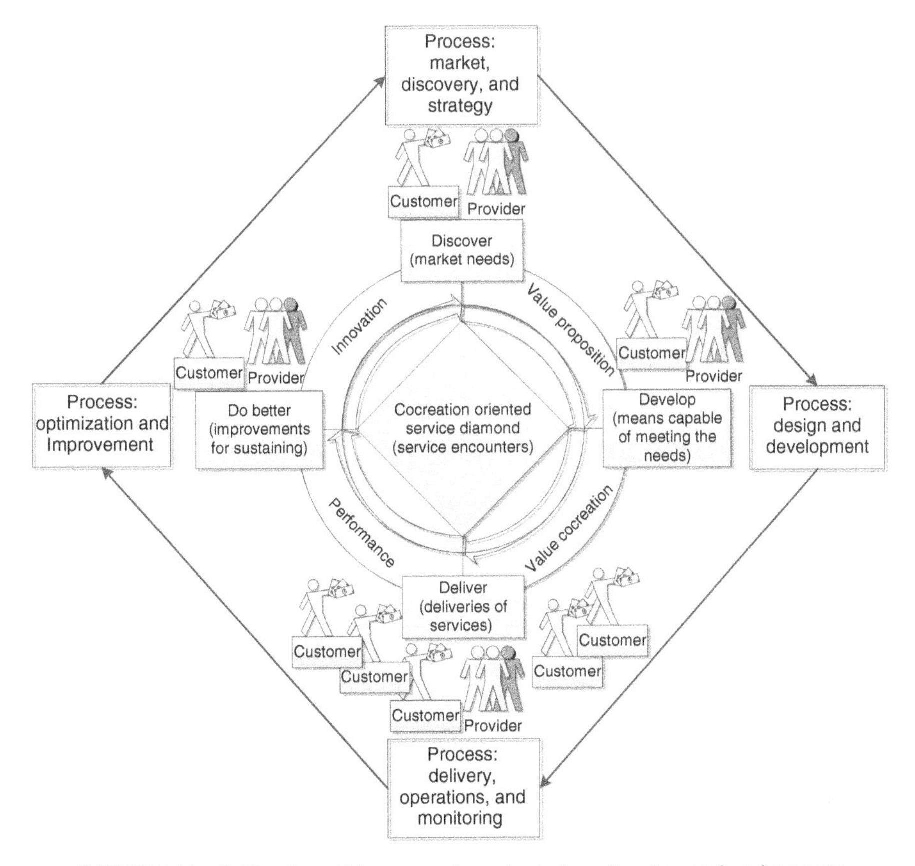

FIGURE 2.11 *A 4Ds view of the cocreation-oriented service diamond and process.*

of goods-dominant logic look of the service diamond relationship in the service industry.

Indeed, the meaning of service lifespan differs when it is viewed from two different perspectives as illustrated in Figures 2.9 and 2.10. This book focuses on the total lifecycle of services from a service organization point of view, aimed at providing a comprehensive understanding of services for service organizations to foster their service business operations, development, and management. When we define the lifecycle of service using the concept of processes, we henceforth adopt four phases or stages to define the milestones of a given service lifecycle. These four stages are "Market, Discovery, and Strategy," "Design and Development," "Delivery, Operations, and Monitoring," and "Optimization and Improvement." Figure 2.11 presents how the 4Ds cocreation-oriented service diamond relationship can be well aligned with the four-stage service lifecycle across the service business operations, development, and management in service organizations.

Let us use the global project development example that was briefly mentioned in Chapter 1 to show what really consists of service encounters throughout the lifecycle of a given project development service and how varieties of service encounters play critical roles in the fulfillment of the needs of service providers and service customers.

Here comes the project background information. An international chemical company called ChemGlobalService has manufacturing facilities in Houston, United States; Beijing, China; and Prague, Czech across three countries in order to serve its customers across different continents. Each facility has its own warehouse. Each warehouse has its own management system application, which was deployed at different times and thus is unique and user-friendly to local employees (Figure 2.12). The products made at each facility are primarily for serving their individual regional markets to ensure that their business operations are responsive and cost-effective. However, different hazard components that are required by all three facilities are separately made by three facilities. This is due to the fact that some pieces of special equipment are extremely expensive, which essentially prohibits from installing the equipment at each manufacturing site. In addition, certain raw materials are extremely risky and prohibitively expensive to transport. As a result, these hazard chemical components must be transported among three warehouses on a weekly basis.

The global project development group (PDGroup) is a software consulting and development service unit that is formed on a project basis in an international bellwether service organization. ChemGlobalService contracted the PDGroup to help integrate three local warehouse system applications to make sure that the effective and timely coordination among three warehouses is conducted in a collaborative manner. The project's main requirements are summarized as follows:

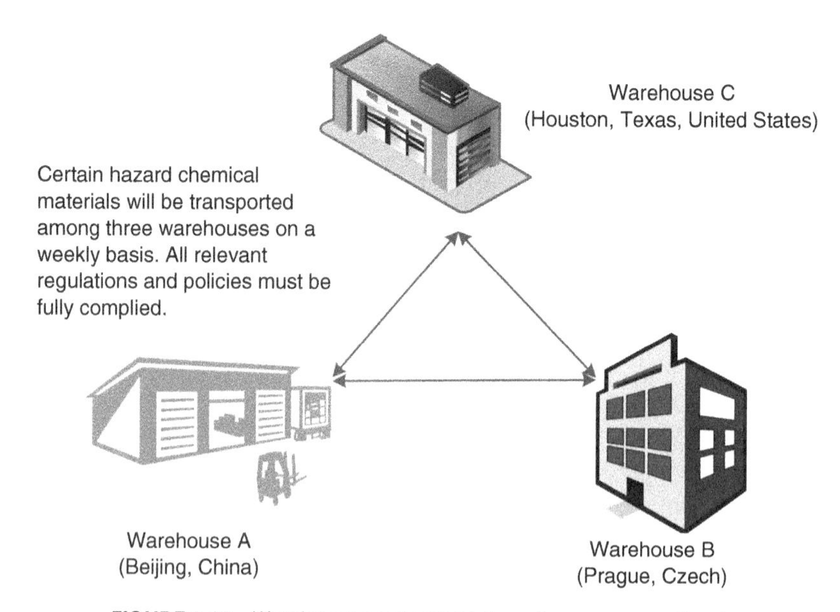

FIGURE 2.12 *Warehouse application integration across continents.*

- Business processes should be defined in support of fully coordinated operational activities within and across warehouses.
- Industry-specific specifications should be supported.
- All environmental protections, treaties, customs, and other related regional and international regulations and policies must be fully complied. Instructions should be provided at the point of need to all the internal and external personnel who are involved in the process of transporting the hazard components.
- User-friendly human interfaces should be provided to warehouse employees. Note that employees in different countries speak different languages and have different educational and cultural backgrounds.

PDGroup as a software consulting and development service unit is hence formed by including six small groups of people. Groups are located in different geographic areas, aimed at leveraging their strengths to meet the project needs. A top-level management unit (i.e., Team A) stays in the New York City (NYC), United States. Team A, consisting of one team manager, one team architect, and one team business analyst, oversees and coordinates the overall project development and deployment across the entire virtual project team. Each of other groups has certain unique skill sets of from 5 to 15 talent employees, including a software designer, a group architect, programmers, quality assurance staff, business analysts, and a group manager. The following list provides the above-mentioned individual group's respective and unique competency (Figure 2.13):

- Team A located at NYC, United States—This team is essentially the administrative team, leading, overseeing, and coordinating the whole project development and deployment across the entire virtual project team.

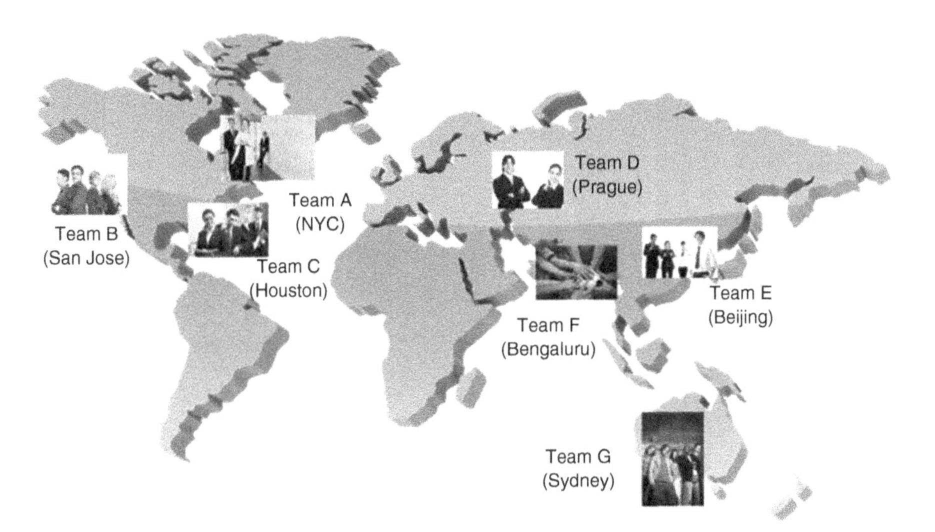

FIGURE 2.13 *Teams with talented people dispersedly populated around the world.*

- Team B located at San Jose, California, United States—This team has a group of persons of talent in human interface design and development.
- Team C located at Houston, Texas, United States—This team has a group of persons of talent in the field of warehouse systems. This team is local and close to the customer facility in Houston, United States. The team will be able to get familiar with the local warehouse system quickly and thus understand local warehouse operations and relevant application and managerial needs.
- Team D located at Prague, The Czech of Republic—Similar to Team C, this team has a group of persons of talent in the field of warehouse systems. This team is local and close to the customer facility in Prague of the Czech of Republic. The team will also be able to get familiar with the warehouse system in Prague quickly and understand local warehouse operations and relevant application and managerial needs.
- Team E located at Beijing, China—Just like Teams C and D, this team has a group of persons of talent in the field of warehouse systems. This team is local and close to the customer facility in Beijing of China. Hence, it will be convenient and easy for the team to get familiar with the customer warehouse system in Beijing and surely understand local warehouse operations and relevant application and managerial needs.
- Team F located at Bengaluru, India—This team is a software outsourcing partner. This team has a group of persons of talent in the field of software design, development, integration, and systems test.
- Team G located at Sydney, Australia—This team has a group of persons of talent in the field of enterprise application integration, business analytics, and international regulation and policy compliance.

We briefly summarize how the project can be completed on time by highlighting service encounters necessary throughout the project development service lifecycle (Figure 2.14):

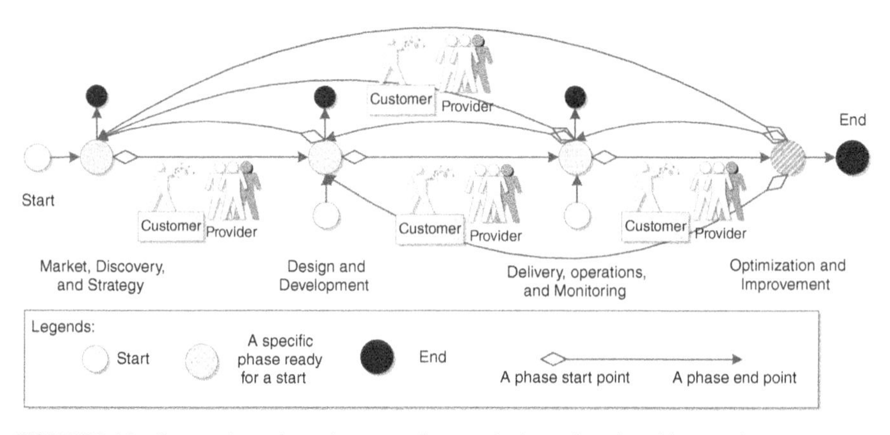

FIGURE 2.14 *Cocreation-oriented process in pursuit of a series of positive service encounters.*

- *At the Market Phase.* A project draft specification might be brainstormed when the top-level management group meets with a group of customer representatives. The participations of managerial and operational personnel from the organization- and unit-level at different facilities of the ChemGlobalService are necessary. Onsite visits might also be needed, aimed at collecting the requirements from daily business operations and end users' sociopsychological constraints by discussing with the end users.
- *At the Design and Development Phase.* The project specification will be revised and enriched as time goes. Unless the project is completed, it is typical that the specification will keep changing to some extent. Surely each revision will be the outcome of numerous onsite or virtual meetings among related representatives from the ChemGlobalService and all teams, that is, Team A to Team G. Customer representatives could be directly or indirectly contacted by group members whenever there is a need.
- *At the Delivery, Operations, and Monitoring Phase.* Most likely, people from Teams A, B, C, D, and E would have to be involved. Before the solution gets fully deployed, PDGroup must make sure that end users will be well trained. Intensive interactions between service providers and customer are necessary during this phase, which ensure that the daily operational needs of end users are fully understood and met. The deployed solution should warrant a twofold success. That is, the daily business operations are well coordinated and monitored among three facilities as expected, while end users' sociopsychological needs are also satisfactorily met.
- *At the Optimization Phase.* All the teams should be involved to some degree. However, Teams A, B, and G would have more interaction with related representatives from the ChemGlobalService, aimed at understanding the weakness of the deployed solution and ensuring that new additions to the changes meet the needs of business operations in the ChemGlobalService.

In general, we understand that this global project development service surely requires a series of interactions and coordination, physically and/or virtually. Varieties of service encounters throughout the lifecycle of global project development service are collaborative in nature. In theory, the lifecycle of service can be completely described using a service encounter graph, which is graphically illustrated in Figure 2.14. The effectiveness of the provided service highly depends on service encounters that should occur in a timely, efficient, and effective manner.

As indicated in Figure 2.14, a series of service encounters for a given end user or customer representative can start at any point and end at a point after his/her start service initiates. To an end user or customer, such an event-based series of service-oriented interactions essentially constitutes the customer's service lifespan, which largely depends on the role of the end user or customer representative with the ChemGlobalService. In other words, individual's service lifespan varies with his/her role at work. By the same token, a member of the PDGroup, including the personnel at the outsourcing group, will also take a role-based trajectory of a series of service encounters. As a value cocreation service interactive activity, each service encounter makes a difference and contributes to the success of the project development service. To the service provider as a whole (i.e., PDGroup's perspective), the sum of all the series of service encounters essentially creates a service encounter network. The efficacy and effectiveness of planning, design, operations, and management of the service encounter network throughout the service lifecycle will directly impact the value of the provided service. Managing and control of service encounter networks in an optimal way become necessary for service organizations to ensure that all the services will be designed, developed, delivered, and operated to meet the needs of both service providers and customers.

2.3 A SERVICE DEFINITION FOR THIS BOOK

Before we formalize our definition of service for this book, let us recap some noticeable definitions of service we have had over the years. In particular, we pay much attention to the service definitions that substantively reflect the status quos of the developed economy in the twenty-first century. As discussed earlier, to most end users, numerous versions of service definitions have been, more or less, intuitively formed from their changing perspectives of consumed services to meet their work and daily life needs under their respective circumstances. On the basis of the discussions in Chapter 1, three quite popular forms of definitions can be recapped as follows:

- When a service is performed, if the service encounters are largely physical, intensive, and direct from the customer's perspective, service is typically defined as an act of beneficial activity. Examples include commonly consumed services that are provided in the bank and finance, hospitality, tourism, resident education, and health care industries.
- By contrast, when a service is performed, if the service encounters are mainly virtual, brief, and indirect from the customer's perspective, then service is quite often defined as the supplying of utilities, commodities, information, or digitalized media. Popular examples, such as online retailing, e-commerce, communications, online social networking, e-banking, digital libraries, online education, and traditional utilities and distributions, surely belong to this category.
- People take many "public" types of services for granted. These types of services can be indeed public or private, profit or nonprofit. Public services are frequently provided by governmental agencies, federal- or state-funded nonprofit organizations, or profit service organizations that subsidized by the governments. Specific examples can include public transportations, varieties of federal and state licenses, post office, security and customs services, etc.

Although the above-revisited definition examples show the viewpoints of end consumers, they indeed capture one of the key components that rudimentarily describe services, which is the performance or act of performing. It is the act of performing that gradually and cumulatively generates the value of service. From the earlier discussions, we also concluded that service encounters are crucial in the service industry although service encounters might be optional for a manufacturer in the manufacturing industry. In addition, no matter how and what kinds of service encounters occur during service business operations and management, people undoubtedly play a central role in the performed services.

We understand that a significant portion of the services provided by the service industry is consumed by individuals, such as medical, education, insurance, legal, financial, transportation, and retailing services. Recently business services that serve different business units or organizations are growing rapidly. For example, international trades, technical support, enterprise resource planning, call center operations, sales management, IT implementation, e-logistics, and business investment and business transformation consulting are well recognized as highly profitable business services in the twenty-first century (Qiu et al., 2007).

For many years, however, when service research and practices were conducted, physical goods-dominant thinking approaches were mainly taken for granted in academia and practice. In addition, due to the prior lack of the necessary means to monitor, capture, and analyze people's dynamics throughout the service lifecycle, the full exploration of the fundamental service theory and principles was prohibitively expensive. The five core elements identified in Figure 1.2 thus were typically studied in a nonintegrated and nonsynergistic manner although the importance of having systems approaches was fully recognized.

To keep abreast of the fast development of the service-led and globalized economy, many academic scholars and professional practitioners have proposed numerous definitions of service since the beginning of this new millennium. The literature shows that many excellent attempts have been well conducted over the last decade or so, aimed at meeting the needs of their focused fields, respectively.

By exploring the marketing shift from the exchange of tangible resources, embedded value, and transaction-based "goods" to the exchange of intangible resources, the cocreation of value, and relationship-based "service", the concept of evolving a service-dominant logic in the field of marketing to replace a goods-dominant logic burgeoned at the very beginning of this new millennium. Many service marketing researchers and pioneers emphasize that we should establish the general concepts, worldview, and small set of fundamental propositions about the services of today and in the future. Vargo and Lusch (2004) have comprehensively reviewed the service marketing research literature in the relevant areas and presented the foundational premises of the emerging service marketing paradigm: "(i) skills and knowledge are the fundamental unit of exchange, (ii) indirect exchange masks the fundamental unit of exchange, (iii) goods are distribution mechanisms for service provision, (iv) knowledge is the fundamental source of competitive advantage, (v) all economies are services economies, (vi) the customer is always a coproducer, (vii) the enterprise can only make value propositions, and (viii) a service-centered view is inherently customer oriented and relational."

Vargo and Lusch (2004) further articulate that the essential concept of "service" should be defined as the application of competences for the benefit of another entity and the term "service" focusing on a process rather than "services" implying "intangible goods" should be used given that the service value is always cocreated during its production. Through further identifying intangibility, heterogeneity, simultaneity, perishability, customer participation, and coproduction (i.e., cocreation) as key commonalities across disparate services businesses, Sampson and Froehle (2006) present the need for a Unifying Services Theory (UST). They particularly argue that the presence of customer dynamic inputs is necessary and sufficient to define a service development process, which is why service processes are typically harder to manage than goods production processes. Their investigation focuses on revealing some principles common to a wide range of services and providing a common ground for further theoretical exploration of capacity and demand management, service quality, service strategy, and so forth.

"A service is the non-material equivalent of a good. Service provision is defined as an economic activity that does not result in ownership, and this is what differentiates it from providing physical goods. It is claimed to be a process that creates benefits by facilitating either a change in customers, a change in their physical possessions, or a change in their intangible assets" (WikiGDPList, 2012). The emergence of the service-dominant logic and cocreation-oriented business process theory makes us rethink service research and practice in general.

Indeed, if we focus on the core difference made due to the economic shift from manufacturing to service, we can surely find that we increasingly emphasize the cocreation-oriented activities between service providers and service consumers throughout the service lifecycle. The disruptive difference compels us to make a considerable change, transforming the way we run service businesses today and in the future. More specifically, people's social, physiological, and psychological capacities identified in Table 2.1, consequently, must be fully understood and incorporated into the lifecycle of service for a service organization to stay competitive. Although physical products might continuously be the core in the manufacturing industry, people-centric service encounters that cocreate service values must become organizational stars in the service industry.

When people-centric and process-driven service encounters are fully incorporated into the five core elements identified in Figure 1.2, we can model a service using the following five core elements (Figure 2.15):

- *Resource*. Traditionally classified resources are natural, human, and/or manufactured or infrastructural. Service products as a fundamental resource in service provision can be in a physical, nonphysical, or hybrid form. For example, an online iPad retailing is a physical service product, a training course is nonphysical product, and a two-year AT&T wireless plan is a hybrid service product. Essentially, with the help of resources the act of performing a transformation task for a customer who asks for it in exchange for acceptable compensation is termed as service provision. Once again, we emphasize that resources are the conduits of service provision to customers.
- *Provider*. Service products are offered by service providers. A service provider as an entity can be an individual, group, organization, institution, or governmental agency.

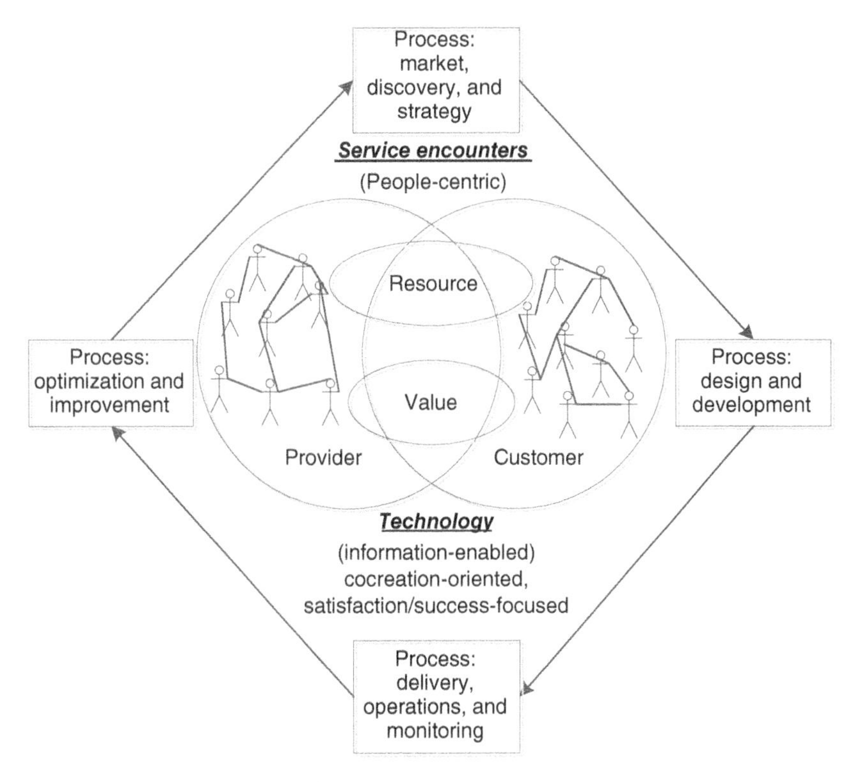

FIGURE 2.15 *A process-driven view of services.*

- *Customer*. Service consumers are human beings who consume, acquire, or utilize the service products offered by their service providers.
- *Value*. Service providers and customers typically have different value propositions. However, their value propositions should be mutually beneficial. The aggregated benefits cocreated from a service are essentially the service value. The service value for a service provider could be profit, satisfaction, and/or competitiveness. The service value for a service customer might be satisfaction, improved competence, possession, and/or productivity. The value of service is typically accumulated by completing a series of service encounters, always involving both the service provider and the service customer.
- *Process*. A typical service process starts from the occurrence of the first interaction between a service provider and a service customer, directly or indirectly. It ends when the offered service product is completely phased out from the engagement agreed by both the provider and the customer. When the lifecycle of service is analyzed, four main stages in a service process can be theoretically identified, market, design and development, delivery and monitoring, and optimization. Value cocreation-oriented service encounters should be well designed and conducted throughout the service lifecycle. In practice, the business

activities at stages are often executed in a concurrent and coordinated manner once a service process completes its initial cycle. Although the priority of a stage might be changed as time goes, the process cycle progresses repeatedly until the offered service gets completely closed out.

No matter what service is designed, developed, and delivered, whether the service need is fully met and the served customer is completely satisfied currently relies on the efficient, effective, and smart operations of its service-oriented delivery network, fully leveraging the advanced resources empowered by the technology, innovation, and information-enabled processes. Being characterized by digitalization and globalization, a competitive service-oriented delivery network essentially is an integrated and process-driven heterogeneous service system (Figure 2.16). As a service system puts people (customers and employees) rather than physical goods in the center of its organizational structure and operations, the service system is a sociotechnical system (Qiu et al., 2007; Spohrer et al., 2007), focusing on service design, development, and delivery using all available means to realize respective values for both service providers and service customers. More detailed discussion on sociotechnical service systems is provided in later chapters.

Since the 1990s, the fast advancement and significant stride in distributed computing and interconnected network has extraordinarily increased the role and power

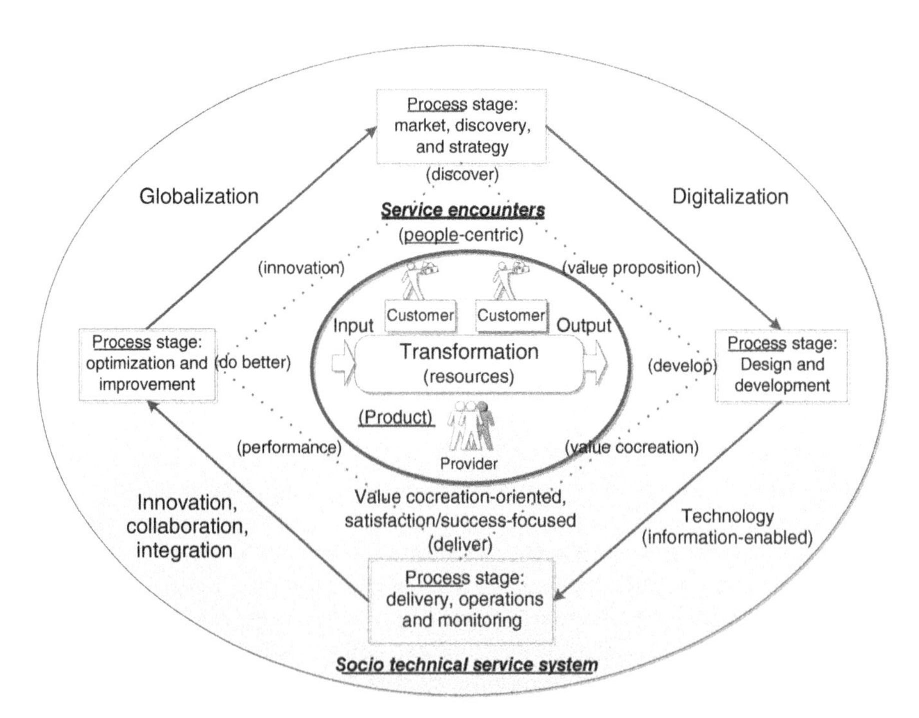

FIGURE 2.16 A sociotechnical process-driven systems view of services.

of IT and communications, transforming the ways how the service industry operates (Schaeffer, 2011; Berman, 2012; Drogseth, 2012; Ahlquist and Saagar, 2013). By using service-dominant thinking while leveraging the increased flexibility, responsiveness, and capability of IT-enabled service business operations and management, service organizations can be realistically operated as effective sociotechnical service systems, improving the business operational productivities and delivering new high levels of job and customer satisfaction (Qiu, 2013).

Figure 2.16 indeed captures and illustrates the marrow of most discussions we have had so far, from which we can try to state a definition of service in a conclusive manner. Here is the definition of service for this book:

> Service is considered as a transformation process in which both provider-side and customer-side people participate in an interactive manner, applying relevant knowledge, skills, and experiences in order to cocreate mutual benefits for the service providers and their customers. Technically and socio-economically the transformation process encompasses a series of service encounters that can be direct or indirect, consecutive or intermittent, physical or virtual, and brief or intensive. The value of the service depends on the socio-technical efficacy and effectiveness of all of the service encounters experienced throughout the service lifecycle.

Please keep in mind, this definition does not aim at replacing the extant definitions that meet the needs of different research disciplines and business environments in academia and practice. This definition simply is an understanding of service that is interpreted in our perspective, focusing on providing the foundation for us to explore service in a systems and holistic manner in this book. The systems and holistic perspective of service includes the following fundamental understandings:

- A service is viewed as a transformation process that creates sociotechnical effects, delivering the values that are, respectively, beneficial for both service providers and customers.
- A service provision through a transformation process is centered at people rather than products, resulting in that service-dominant thinking must replace goods-dominant thinking in service engineering and management.
- A service provision entity is a sociotechnical service system. It is typical that a competitive service system consists of a number of interrelated and interacted domains systems empowered by a variety of operational resources, which are coordinated in a collaborative manner, regionally and/or internationally.
- The realized value of service is the total perceived value of the quality outcomes cocreated by providers and customers through a series of service encounters throughout the service lifecycle.

Here comes a concise version of the definition of service for this book:

> Service is considered as an application of relevant knowledge, skills, and experiences and manifests itself to customers as a service encounter chain that substantively reveals the cocreation of benefits for both service providers and customers.

2.4 FINAL REMARKS

This chapter systematically discussed different perspectives of services, aimed at capturing the marrow of today's services in the information era. We truly understand that service is not just a product. Service is a transformation process in which both provider-side and customer-side people are always involved in an interactive manner. Service is thus an application of relevant knowledge, skills, and experiences, capable of cocreating benefits, respectively, for service providers and customers. As summarized in Chapter 1, a service is people-centric, truly cultural and bilateral. The type and nature of service dictates how a service is performed, which accordingly defines how a series of service encounters could and should occur throughout its service lifecycle. The type, order, frequency, timing, time, efficiency, and effectiveness of the series of service encounters throughout the service lifecycle determine the quality of services perceived by customers who purchase and consume the services (Booms and Bitner, 1981; Bitner; 1992; Chase and Dasu, 2008).

The service business setting has changed substantially. The changes include that (i) more and more data become available, helping to capture the behavior of people and systems dynamics; (ii) service scopes are changing, resulting in that the worldwide competition is essential; (iii) with the help of advanced computing and networking technologies, theories and methodologies can be easily turned into a variety of powerful means that can further empower effective service operations and management to deliver quality services.

In summary, we have defined what service is for this book, laying the foundation for the following chapters of this book. Henceforth in this book, chapter by chapter, different and highly appreciative relationships within service will be further identified and explained in detail along with the discussion of a variety of qualitative and quantitative approaches that should be adopted by service organizations in pursuit of competitive business goals in the service-led economy.

REFERENCES

Ahlquist, J., & Saagar, K. (2013). Comprehending the complete customer. *Analytics—INFORMS Analytics Magazine*, May/June, 36–50.

Bell, M. (1981). A Matrix Approach to the Classification of Marketing Goods and Services in *Marketing of Services*, 208–212, ed. by J. H. Donnelly and W. R. George. Chicago: American Marketing.

Berman, S. (2012). Digital transformation: opportunities to create new business models. *Strategy and Leadership*, 40(2), 16–24.

Bitner, M. J. (1990). Evaluating service encounters: the effects of physical surroundings and employee responses. *Journal of Marketing*, 54(2), 69–82.

Bitner, M. J. (1992). Servicescapes: the impact of physical surroundings on customers and employees. *Journal of Marketing*, 56(2), 57–71.

Bitner, M. J., Booms, B. H., & Tetreault, M. S. (1990). The service encounter: diagnosing favorable and unfavorable incidents. *The Journal of Marketing*, 54, 71–84.

Booms, B. H., & Bitner, M. J. (1981). Marketing strategies and organization structures for service firms. *Marketing of Services*, 47–51.

Bowen, J. (1990). Development of a taxonomy of services to gain strategic marketing insights. *Journal of the Academy of Marketing Science*, 18(1), 43–49.

Chase, R. B., & Dasu, S. (2001). Want to perfect your company's service? Use behavioral science. *Harvard Business Review*, 79(6), 78–85.

Chase, R. B., & Dasu, S. (2008). Psychology of the Experience: The Missing Link in Service Science in *Service Science, Management and Engineering Education for the 21st Century*, 35–40, eds. by B. Hefley and W. Murphy. US: Springer.

Chesbrough, H. W. (2003). *Open Innovation: The New Imperative for Creating and Profiting from Technology*. Boston, MA: Harvard Business School Press.

Chesbrough, H. W. (2011a). *Open Services Innovation: Rethinking Your Business to Grow and Compete in a New Era*. San Francisco, CA: Jossey-Bass.

Chesbrough, H. W. (2011b). Bringing open innovation to services. *MIT Sloan Management Review*, 52(2), 85–90.

Drogseth, D. (2012). User experience management and business impact—a cornerstone for IT transformation. *Enterprise Management Associates Research Report*. Retrieved Oct. 10, 2012 from http://www.enterprisemanagement.com/research/asset.php/2311/.

Heskett, J. L., Jones, T. O., Loveman, G. W., Sasser, W. E., & Schlesinger, L. A. (1994). Putting the service-profit chain to work. *Harvard Business Review*, 72(2), 164–174.

Karmarkar, U. (2004). Will you survive the services revolution? *Harvard Business Review*, 82(6), 100–107.

Lovelock, C. and J. Wirtz, 2007. *Service Marketing: People, Technology, Strategy*, 6th ed. Upper Saddle River, NJ: Prentice Hall.

Loveman, G. W. (1998). Employee satisfaction, customer loyalty, and financial performance—an empirical examination of the service profit chain in retail banking. *Journal of Service Research*, 1(1), 18–31.

Qiu, R. G. (2013). We must rethink service encounters. *Service Science*, 5(1), 1–3.

Qiu, R. G., Fang, Z., Shen, H., & Yu, M. (2007). Editorial: towards service science, engineering and practice. *International Journal of Services Operations and Informatics*, 2(2), 103–113.

Rangaswamy, A., & Pal, N. (2005). Service innovation and new service business models: harnessing e-technology for value co-creation. *An eBRC White Paper*, 2005 Workshop on Service Innovation and New Service Business Models, Penn State.

Sampson, S., & Froehle, C. (2006). Foundation and implication of a proposed unified services theory. *Production and Operations Management*, 15(2), 329–343.

Schaeffer, M. (2011). Capitalizing on the smarter consumer. *IBM Institute for Business Value—IBM Global Business Services Executive Report*. Retrieved on Dec. 6, 2012 from http://www-935.ibm.com/services/us/gbs/thoughtleadership/ibv-capitalizing-on-the-smarter-consumer.html.

Schneider, B., & Bowen, D. (2010). Winning the Service Game in *Handbook of Service Science*, 31–59, eds. by P. Maglio, C. Kieliszewski, and J. Spohrer. US: Springer.

Solomon, M. R., Surprenant, C., Czepiel, J. A., & Gutman, E. G. (1985). A role theory perspective on dyadic interactions: the service encounter. *The Journal of Marketing*, 49(1), 99–111.

Spohrer, J., Maglio, P., Bailey, J., & Gruhl, D. (2007). Steps toward a science of service systems. *Computer*, 40(1), 71–77.

Surprenant, C., & Solomon, M. (1987). Predictability and personalization in the service encounter. *Journal of Marketing*, 51(2), 86–96.

Svensson, G. (2004). A customized construct of sequential service quality in service encounter chains: time, context, and performance threshold. *Managing Service Quality*, 14(6), 468–475.

Vargo, S., & Lusch, R. F. (2004). Evolving to a new dominant logic for marketing. *Journal of Marketing*, 68(1), 1–17.

WikiGDPList. (2012). List of Countries by GDP Sector Composition. Wikipedia. Retrieved on Dec. 10, 2012 from http://en.wikipedia.org/wiki/List_of_countries_by_GDP_sector_composition.

3

The Need for the Science of Service

A wireless communication service provider usually offers a variety of service plans. Wireless service plans, such as a 2-year contract with 600-min monthly usage individual cell phone plan, a 2-year contract with 1000-min monthly usage family cell phone plan, limited or unlimited data and messaging plans, a global positioning system (GPS) navigation plan, and different bundle plans, are popular ones in the United States. A service product does not create any utilitarian and/or sociopsychological benefit until it is consumed. For instance, a customer chooses a competitive 2-year plan with a cellular phone the customer likes. The value cocreation process that defines a service starts at the point when the customer calls a representative, browses the service provider's website, or visits a provider's retailing store to sign up the plan. If the service can meet the needs of customer's daily communications, the customer will be most likely satisfied with the service. Of course, the service provider makes a corresponding profit until the customer terminates the signed service contract.

In the above-discussed example, if the customer experience is outstanding in all aspects with respect to today's highly competitive wireless communication services, the customer most likely becomes a loyal customer and will continue to choose a service product from the service provider in the future. The customer's word of mouth, including posts, blogs, and conversations over varieties of online social media, effectively attracts more customers, intentionally or unintentionally. In business, if the wireless service provider can execute the service plan well, the provider makes a profit from such a service. Loyal and lifetime customers surely help the service provider win the increasingly intensified competition in the marketplace (Heskett et al., 1990; Heskett et al., 1994; Schneider and Bowen, 2010). As the total values of the provided service are the values, respectively, accumulated by the customer and the service

Service Science: The Foundations of Service Engineering and Management, First Edition. Robin G. Qiu.

provider throughout its service lifespan, it is the progression of executing the service plan that truly determines the real values beneficial to the customer and the service provider.

As discussed in Chapter 2, service can be simply defined as an application of relevant knowledge, skills, and experiences and manifest itself as a service encounter chain to cocreate benefits, respectively, for service providers and customers. A valuable, beneficial, and competitive service surely is the operational outcome of a well-operated and managed sociotechnical service system. Given the increasing complexity, dynamics, and scope of services, it becomes essential for a service organization to apply the science of service to the operations and management of its whole service delivery networks (a.k.a. sociotechnical service system) to ensure that the promised services can be performed in a competitive way in the current and future service-led economy.

Science is commonly recognized as knowledge. In a given discipline today, the organized body of knowledge as a disciplinary science is radically derived from systematic observations of focused social or natural phenomena. It is well known that the systematic observations help us to discover and organize knowledge in the form of laws and principles about the observed social or natural phenomena in the universe. Indeed, the formulated laws and principles allow us to test the explanations and make predictions for further investigation and exploration. In this chapter, by taking a holistic and systems perspective we discuss an approach on how we can take steps with scientific rigor to study services and service systems.

3.1 A BRIEF REVIEW OF THE EVOLUTION OF SERVICE RESEARCH

Service research in several focused areas, including service operations, marketing, and organizational structure and behavior, and economic transformation, has been conducted worldwide for many decades. Reviewing all very influential service research literature that contributed to the service research, development, and practice at time when the work was done is not the purpose of this chapter. However, it is worthy of briefly highlighting many pioneer research work that continues to substantially impact current service research. Note that this brief highlight could be quite limited as it simply reflects the author's viewpoint (Qiu, 2012).

As early as in the 1970s, a rational approach deviating from the traditional physical-product-based rationalizations was explored by Chase (1978), aimed at identifying a new course to help organizations understand and manage service business operations by addressing the newly confronted service-oriented challenges in business back in the 1970s. Larson (1989) has been a longtime proponent of applying operations research and management science to improve services business operations. Recently, applying optimization and queuing theory in solving a variety of managerial and operational problems in service systems is comprehensively discussed by Daskin (2011).

Considerable research efforts have significantly contributed to the development of service marketing and economics. Lovelock (1983) pioneered the education and exploration of modern service marketing and leadership with a focus on the synergistic effects by fully leveraging people, technology, and strategy in service organizations. Grönroos (1994) has been leading the study in service management by applying service logic to market-oriented management in service and manufacturing firms. By comprehensively analyzing the profitability drivers throughout the service-profit chain, Heskett et al. (1994) argue that putting employee and customers first would radically make a shift in the way service organizations manage and measure success. How service quality can be financially accountable has been specifically investigated by Rust et al. (1995), giving rise to further exploration of marketing investments, customer equity, and relationships in services.

As discussed in Chapter 2, because the developed economies further moved away from producing goods to providing services after the turn of this new millennium, Vargo and Lusch (2004) strongly propose a new service dominant logic focusing on intangible resource, the cocreation of value, and service-for-service exchange relationships in the service marketing field. The service-dominant logic thinking has been well incorporated into the research in the fields of organizational structures, employee behavior, and economic transformation (Vargo and Akaka, 2009; Hilton and Hughes, 2013; Löbler, 2013).

Karmarkar (2004) then emphasizes the industrialization of services and argues that service organizations can survive the ever-changing business environments because of the digitalization and globalization only if they can effectively reorganize strategies, processes, and people within and across organizations for the unprecedented challenge ahead. Hsu (2009) shows how a theory of service scaling and transformation through leveraging digital connections could contribute to the development of Service Science for the knowledge economy. In particular, he advocates that digital connection scaling plays a key role in value cocreation for providers and customer.

When the worldwide economy was dominated by goods, both academics and practitioners paid much attention to the development, production, and innovation of physical products (Chesbrough, 2003). Because physical products were essentially the focus, production/service organizations were more or less considered as technology-driven organizations or systems. Consequently, when service research was conducted, physical goods-dominant thinking was frequently taken for granted.

However, as discussed in Chapter 2, the emphasis in the developed economy has been shifted away from manufacturing to service since the 1960s. As compared to goods-dominant thinking, service-dominant thinking must have people clearly identified and centered across the lifecycle of service. The interactions between service providers and service consumers play a crucial role in the process of transformation of the customer's needs utilizing the operations' resources. The value of service lies along with the process trajectory throughout the lifecycle of service. Therefore, the behavior of systems of a sociotechnical service system should be well designed, managed, and operated with the full support of Service Science (Spohrer and Riechen, 2006; Qiu et al., 2007; Qiu, 2009), so that an effective and satisfactory service path toward the realization of business objectives can be formed in an optimal manner. A service path is nothing but a sequence of relevant service business activities, which typically manifest themselves to a customer as an event-based series of service encounters.

Despite the recognition of the importance of service research, the shift to focus on disparate and global-scale services (Karmarkar, 2004) and the servitization of products (Chase and Erikson, 1989) to compete in the service-led global market has created an education and research gap (IBM, 2004; IBM Palisade Summit Report, 2006; Dietrich and Harrison, 2006). The gap has not been fully filled largely because of the granted physical goods-dominant thinking and the prior lack of the means that allowed us to fully explore the people-centric systemic interactions and their sociotechnical impacts on service system dynamics in the service research. According to Spohrer et al. (2007), "the role of people, technology, shared information, as well as the role of customer input in production processes and the application of competence to benefit others must be described and defined."

In summary, the science of service, or service science, must be explored, clearly defined, and well developed. When the discovered service theories, laws, and principles are applied in practice, practitioners can effectively manage and control systemic behavior and leverage sociotechnical effects in a service system, so that the system can be scientifically and wisely guided to maneuver throughout the process-driven service lifecycle to create, develop, and deliver valuable, beneficial, and/or competitive services.

3.2 SERVICE AS A PROCESS OF TRANSFORMATION

From the preceding chapters, we understand that the word of "service" has many connotations, which varies with business domains and settings. A very good paper from Morris and Johnston (1987) provides a great discussion on the inherent variability between manufacturing and service operations management. They classify three types of general production operations, characterized by the inputs that are processed rather than the outputs of the processing operations: material processing operations (MPOs) (Figure 3.1a), customer processing operations (CPOs) (Figure 3.1b), and information processing operations (IPOs).

Morris and Johnston further discuss the differences among the three types of general production operations. They differentiate service from manufacturing by the nature of the thing that is processed. The CPO acts upon a customer to create sociotechnical and economic effects on the customer. The IPO processes information to convert it into a desirable form. The MPO processes input materials and then produces goods. Manufacturing organizations are largely MPO-based entities. Service firms are then mainly CPO-based. They argue that issues such as capacity planning, operations planning and control, inventory or queue management, and quality control must be considered in each of the three types of operations. However, as service is inherently different from manufacturing, the inherent difference must be well considered in service operations and management by service organizations to ensure their successes in business.

Figure 3.1a uses the simple and traditional three categories of resources: natural, human, and manufactured or infrastructural resources to highlight the inherent nature of goods-dominant production operations in which materials are the input.

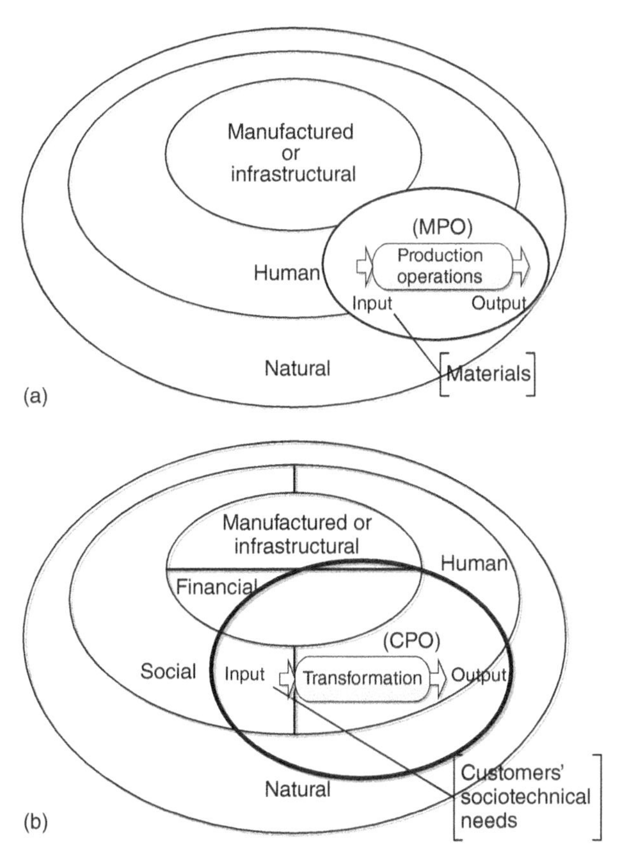

FIGURE 3.1 *Goods-dominant production operations versus service-oriented transformation. (a) Goods-dominant production operations in the resource model and (b) service-oriented transformation in the five capitals model.*

Natural resources essentially are the source of raw materials. Human resources consist of human efforts provided in the transformation of the materials into physical products. Surely, manufactured or infrastructural resources, consisting of man-made goods or means of production (e.g., machinery, buildings, computers, networks, and instruments), must be utilized in the processing operations to make the transformation cost-effective and efficient (Morris and Johnston, 1987; Samuelson and Nordhaus, 2009; Sullivan et al., 2011).

As discussed in Chapter 2, the economic shift from manufacturing to service entails a disruptive change in business, transforming the way business would operate in the service-led economy. A process of transformation that focuses on people-centric service encounters surely becomes the organizational and business core in a service organization. Table 3.1 highlights the disruptive change and shows the focus shift in service business operations and management. The intangibility, heterogeneity, simultaneity, perishability, customer participation, and cocreation are

TABLE 3.1 Main Characteristics Comparison Between Service and Goods

Focus	Service	Goods
Production	Cocreated	Produced
Variability	Heterogeneous	Homogeneous
Physicality	Intangible	Tangible
Product	Perishable	Imperishable
Satisfaction	Expectation-related	Utility-related

the key commonalities across disparate services businesses (Sampson and Froehle, 2006).

In response to the social, political, economic, and environmental issues in today's globalized economy, the Forum for the Future (FftF) as a nonprofit organization proposes the five capitals model, a framework for sustainability. In addition to the traditional three categories of resources, the five capitals model further includes social and financial capitals as shown in Figure 3.1b. The five capitals model provides a basis for organizations to consider the impact of its business activities on each of the capitals in an integrated manner. As a result, this resource model allows organizations to implement a responsible and balanced business model to ensure their sustainable outcomes in the long run (FftF, 2012).

On the basis of the definition of service concluded in Chapter 2, service is essentially an application of relevant knowledge, skills, and experiences and manifests itself to customers as a service encounter chain that substantively reveals the cocreation of benefits, respectively, for service providers and customers. As the service encounter chain is essentially created and managed through a process of transformation as illustrated in Figure 3.1b, a good understanding of social and human capitals in an organization becomes important (Lepak and Snell, 2002). According to the FftF (2012), human and social capitals are well defined as follows:

- "Human capital incorporates the health, knowledge, skills, intellectual outputs, motivation and capacity for relationships of the individual. Human Capital is also about joy, passion, empathy and spirituality."
- "Social capital is any value added to the activities and economic outputs of an organization by human relationships, partnerships and co-operation. For example networks, communication channels, families, communities, businesses, trade unions, schools and voluntary organizations as well as social norms, values and trust."

Simply put, we must emphasize cocreation-oriented business activities between service providers and service consumers throughout the service lifecycle. More specifically, people's social, physiological, and psychological traits must be fully and explicitly explored, understood, and incorporated into the process of transformation of customers' sociotechnical needs by service organizations for competitive and sustainable outcomes.

From the early discussion, we understand that because CPO is the rudimentary business operational paradigm in service organizations, service-oriented operations must focus on transforming "customers" instead of materials that are used in MPO. In other words, the shift from materials to "customers" as the input to CPO indicates that customers' sociotechnical needs should be the main concerns in a service-oriented transformation. In service business, a service thus is a process of transformation of customers' sociotechnical needs with the support of operations resources, fostering and operating positive service encounters to meet the needs of customers and providers. Therefore, capturing customer's needs and leveraging customers' participations in the process of transformation of customers' sociotechnical needs truly play a key role in service operations and management.

3.3 FORMATION OF SERVICE ENCOUNTERS NETWORKS

Now it is well understood that service is a transformation process that takes "customer" as its input. Both provider-side and customer-side people must be involved in an interactive manner, applying relevant knowledge, skills, and experiences to cocreate benefits, respectively, for service providers and customers. For a given service, the interactions, service encounters, essentially function as the delivery mechanism of rendering the promised service. Therefore, we must have a full understanding of service encounters in order to grope for a new approach to a creative and comprehensive study of service science.

Let us briefly review what we discussed about a service encounter in the preceding chapters. A service encounter essentially is a social and transactional interaction in which a service provider performs a service activity beneficial to its corresponding service customer. To a service customer, a service encounter is a moment of truth for the wanted service with which the customer interacts. To a service provider, a service encounter is an act of communicating and rendering the promise.

The PDGroup project service example in Chapter 2 is a good source for us to revisit how service encounters are crucial in delivering a successful and satisfactory service. Throughout the project cycle, we briefly discussed some indispensable interactions between different groups of consultants from the PDGroup and employees from the ChemGlobalService who are located across different continents. Here, we would like to emphasize the challenges and discuss what might impact the outcomes of the PDGroup project service in both a short term and the long run.

Assume that ChemGlobalService initiated the first interaction by consulting with the PDGroup for a possible project service. A memorandum of understanding or preliminary service agreement might be written before the project got started. A project draft specification would then be brainstormed when the top-level management group from the PDGroup met with a group of customer representatives from ChemGlobalService. The customer-side participants from the organization- and unit-level at different facilities of the ChemGlobalService would be critical for identifying the challenges at the systems level. Onsite visits might also be needed, focusing on collecting the detailed requirements at the operations level. A final service agreement

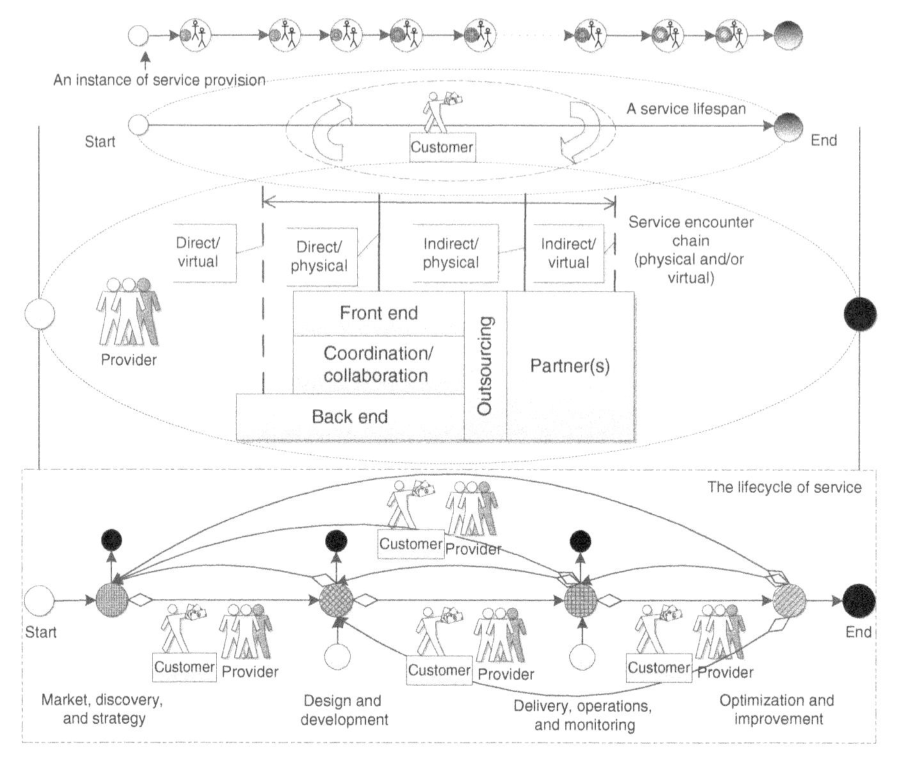

FIGURE 3.2 *An instance of service derived from a service encounter graph.*

is typically signed off during this stage. A series of service encounters that occurred along with the progression of the project development form a service encounter chain (Figure 3.2), which is essentially described as an instance of a service encounter graph in theory.

At the design and development phase in the PDGroup project service example, numerous direct and productive interactions would be essential to ensure that the technical and nontechnical requirements would be fully considered. For instance, the project specification would be revised and further enriched after this phase was kicked off. Unless the project is completed, it is typical that the specification would keep changing to some extent. Surely each revision would be the outcome of many onsite or virtual meetings among related representatives from the ChemGlobalService and all teams, that is, Team A to Team G, from the PDGroup. Effective communication means should be established so that customer representatives could be directly or indirectly contacted by PDGroup project group members whenever additional end users' inputs are needed.

The most intensive and productive interactions throughout the service lifecycle in the PDGroup project service example should occur at its delivery, operations, and monitoring phase. Most likely, people from Teams A, B, C, D, and E would have to be involved. The PDGroup must make sure that end users will be well trained; the daily business operations would thus be well coordinated and monitored among three facilities. Frequently, service encounters at this phase would be direct and physical. In fact, the study of service quality in this phase by the literature has been voluminous as scholars and practitioners around the world have paid exceeding attention to these direct interactions that must occur in delivering services and to how these service encounters impact the perceived service quality by both the service providers and the service customers (Czepiel et al., 1985; Parasuraman et al., 1988; Czepiel, 1990; Bitner et al., 1990; Bitner et al., 1997; Bradley et al., 2010).

At the optimization phase in the PDGroup project service example, many productive interactions should also be required to ensure that the weakness of the delivered integration project and occurred service encounters would be well and promptly identified. In particular, Teams A and G would have more interaction with related representatives from ChemGlobalService, aimed at understanding the weakness of the deployed solution and identifying new additions in support of the ongoing changes of business operations and management.

For any given phase in the service lifecycle, the literature has shown many outstanding works. For instance, the technical requirements within services are most likely materialized in service products with the support of operations resources. How service encounters substantially impact perceived service quality in a variety of dimensions from the perspectives of the service providers, customers, or both has been well studied (Taylor, 1977; Parasuraman et al., 1988; Bitner, 1990; Bitner, 1992; Chase and Dasu, 2001; Svensson, 2002; Bradley et al., 2010). However, these human interactions, in general, have not well explored as they have not been considered throughout the service lifecycle in an integrative and collaborative manner. In other words, how these intensive and broad interactions at one phase impact other phases of service provision have been largely ignored so far in service research. We must rethink service encounters by integrating these human interactions into a service encounter chain, from beginning to end across the service lifecycle, so that we can look into services defined in this book using a holistic, systems, and integrative approach (Qiu, 2013b).

To extend the popular cross-section service quality studies in the literature, Svensson (2004) proposes a framework for exploring sequential service quality in service encounter chains, that is, examining the consecutive service performances in a series of service encounters during the service delivery processes. Conceptually, Svensson provides a customized six-dimensional construct of sequential service quality to highlight the importance of time, context, and performance threshold in service encounter chains. Although Svensson's conceptual work has not well validated empirically or theoretically, the service encounter chain concept certainly sheds some light on our groping for an alternative direction of further developing service science.

"At the heart of every service is the service encounter" (Heskett et al., 1990, p. 2). To develop a holistic, systemic, and integrative approach to the scientific study of service, the concept of service encounters should be further extended from the old one focusing on the study of service quality and satisfaction at the delivery, operations, and monitoring phase to a new one spanning the total service lifecycle. In other words,

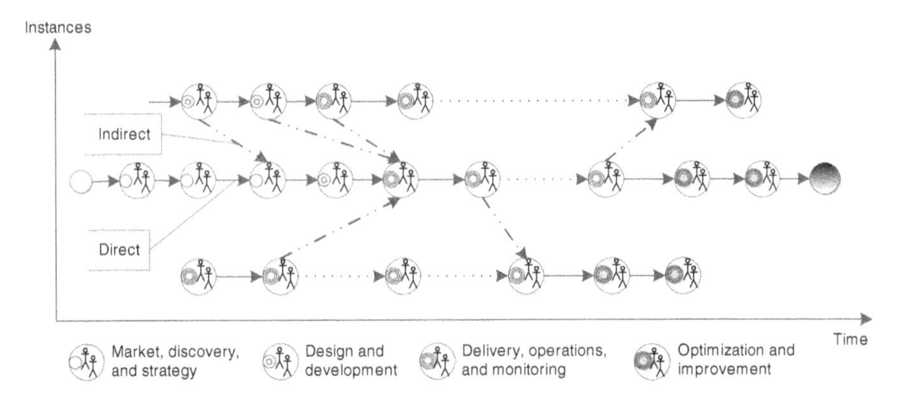

FIGURE 3.3 *Service encounter chains to form a service encounter analytic network.*

all interactions between the service providers and the customers as a whole across the service lifecycle should be explored and analyzed.

The first three service encounters in the middle instance shown in Figure 3.3 provides a graphic view of the occurrence of these types of service encounters at the market, discovery, and strategy phase in the PDGroup project service example. They will surely, sequentially, and substantially impact the service quality and value perceived by customers in the later phases, directly or indirectly. Put in a broader context, for a given service encounter chain, essentially, the preceding service encounters impact the current service encounter and the succeeding ones, psychologically, socially, and economically.

The effectiveness of the deployed project solution in the above-mentioned example largely depends on a series of positive and productive service encounters that must occur in a timely, efficient, and effective manner. At each point of customer interaction, customer experience is the perceived effect of the fulfillments in both functional and socioemotional dimensions (Durvasula et al., 2005; Chase and Dasu, 2008; Qiu, 2013b). Functional needs are met by performing desired service functions that are specified in a signed service agreement. Meeting the psychological needs of customers and employees becomes extremely challenging as socioemotional needs manifested through an array of psychological needs vary with time, duration, and servicescape, and individual's expectation and competency (Bitner, 1990; Bitner, 1992; Bitner et al., 1994; Bitner et al., 2000; Svensson, 2004; Meyer and Schwager, 2007; Bradley et al., 2010). Moreover, these psychological needs at a service encounter are frequently influenced directly and indirectly by the outcomes of the preceding service encounters as illustrated in Figure 3.3.

As discussed earlier, varieties of service encounters throughout the lifecycle of the global project development service are collaborative in nature. An individual's service lifespan varies with his/her role at work. As a value cocreation and service-oriented interactive activity, each service encounter certainly makes a difference, positively or negatively impacting the final outcome of the rendered service (Lloyd and Luk, 2009; Svensson, 2006). To the service provider as a whole (i.e., PDGroup's perspective),

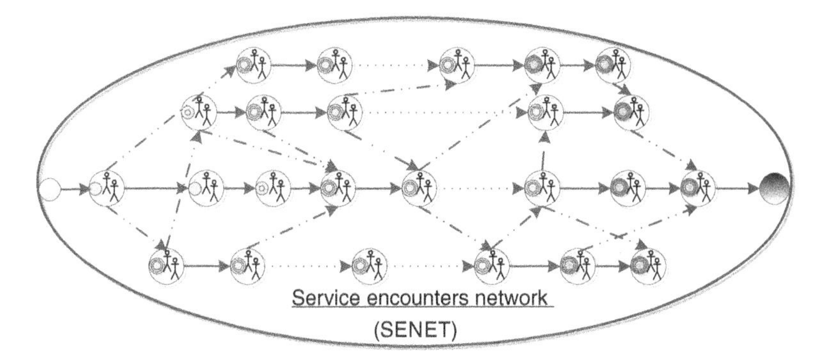

FIGURE 3.4 *An illustration of SENET.*

the sum of all the series of service encounters spanning the service lifecycle indeed creates a service encounters network (SENET) (Figure 3.4). Hence, the values of the provided service for the provider and the customers largely depend on the efficacy and effectiveness of planning, operations, and management of the SENET throughout the lifecycle of service.

3.4 INHERENT NATURE OF SOCIOTECHNICAL SERVICE SYSTEMS

As the world is now all connected economically, technically, and socially, service organizations must integrate products and services into solutions that are desirable, amicable, and environmentally sustainable. Because services dominate the developed economy and radically drive the growth of the world economy, service firms must continuously improve their service business competitiveness that is clearly characterized by customization, integration, intelligence, and globalization in order to serve their diversified customers across the continents. Indeed, over the last decade or so we have witnessed that this new service-oriented social business wave, through leveraging the advancement of IT and the diversity of cultures and societies, provides end users better satisfaction and quality of life—the ultimate prosperity goal of human being (Palmisano, 2008; Qiu, 2009).

A system, focusing on the interdependence of relationships created in an organization, is composed of regularly interacting or interrelating groups of activities within the organization (STWiki, 2012). From the systems' perspective, a service system essentially consists of a number of interacting business domains entities that must be well coordinated (Qiu, 2007; Qiu, 2013a). A service system can simply be a software application, or a business unit within an organization, from a project team, business department, to a global division; it can be a firm, institution, governmental agency, town, city or nation; it can also be a composition of numerous collaboratively connected service entities within and/or across organizations. No matter what a service system is, small or large, individual or composed, and intra- or interconnected, it must radically consist of people, technologies, infrastructures, and processes of service operations and management (Spohrer et al., 2007; Spohrer, 2009).

Generally speaking, a competitive service organization is a well-built, controlled, and managed service system. The systems view of a service organization is then a perspective of looking at the service organization as a collection of business domain systems that create a whole, allowing us to understand and orchestrate the interacting activities among these business domains systems. As discussed earlier, today's competitive service organizations must put people (customers and employees) rather than physical goods in the center of their organizational structures and operations. Moreover, real-time explorations of human behaviors and sociopsychological dynamics within services become essential for service organizations to stay competitive in the information era.

Because service firms must focus on engineering and delivering services using all available means to meet both technical functional and socioeconomic needs and accordingly realize respective values for both providers and consumers, service firms essentially are social-technical service systems. The fast advancement in distributed computing and interconnected network has significantly increased the role and power of IT and communications, transforming the ways how the service industry operates. We are sure that the science of service can be fully developed in the near future. We believe that service firms can be realistically operated as effective socio-technical service systems using service-dominant thinking while leveraging the increased flexibility, responsiveness, and capability of service business operations and management (Qiu, 2013a), resulting in improving business operational productivities and delivering new high levels of job and customer satisfaction (Qiu, 2013b).

As the world is becoming better instrumented and interconnected, and more intelligent, a service system must be people-centric, information-enabled, service-oriented, and satisfaction-focused; it should encourage and cultivate people to collaborate and innovate (Qiu, 2009). The incorporation of the five-capital model into the process of transformation or service provision provides a clear and conceptual illustration of a sustainable socio-technical service system (Figure 3.5).

No matter what service is conceived, designed, developed, and delivered, whether the functional and socioeconomical needs are fully met and the served customer is completely satisfied rely on the efficient, effective, and smart operations of its service-oriented value delivery network, that is, an integrated and collaborated heterogeneous service system (please refer to Figure 3.2). It is well known that competitive systems are not always at equilibrium as time goes; they are very dynamic and adaptive. A service organization as a service system or ecosystem surely becomes more integrated and capable while more dynamic, complicated, and challenging than ever before (Qiu, 2009).

"Indeed, almost anything from people, object, to process, for any organization, large or small—can become digitally aware and networked" (Palmisano, 2008). On one hand, the world becomes smaller, flatter, and smarter, which creates more opportunities and enormous promise; on the other hand, more challenges and issues appear in many aspects from business strategy, marketing, modeling, innovations, design, engineering, to operations and management in order for businesses to stay competitive in a globally integrated economy. Consequently, an enterprise has to rethink its operational and organizational structure by focusing on people (e.g., implementing a

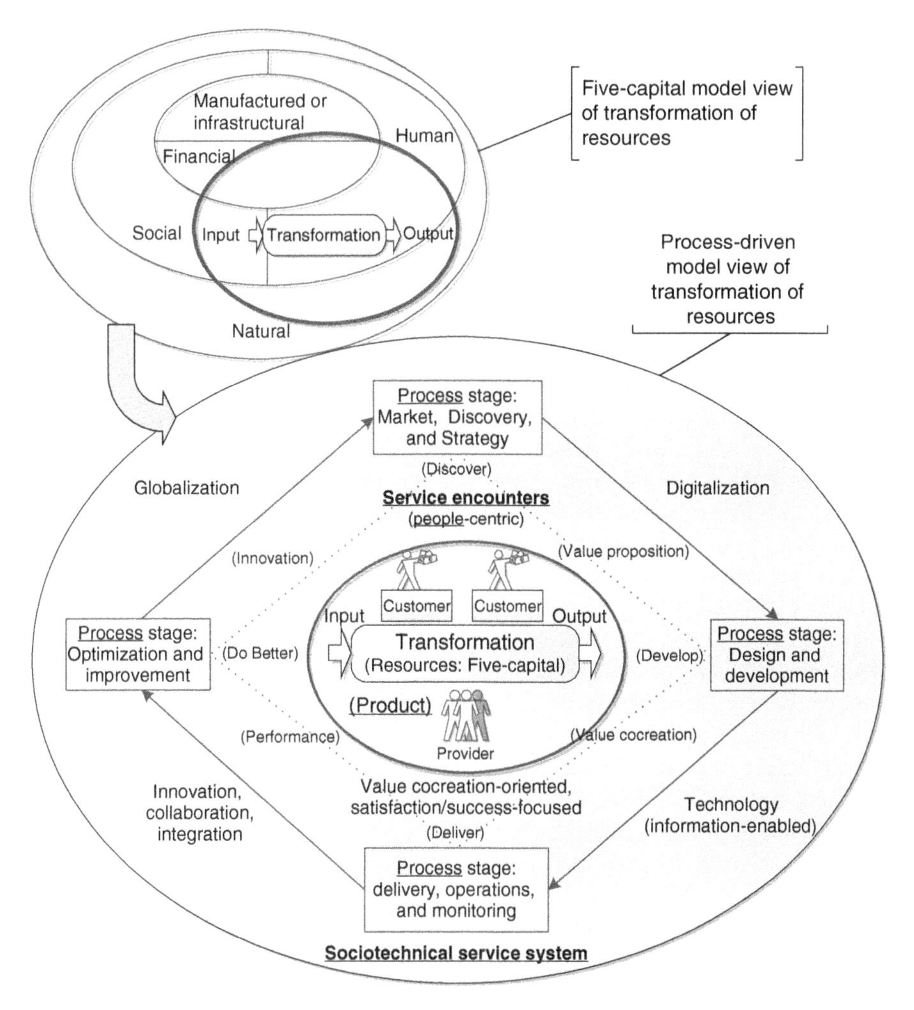

FIGURE 3.5 *A sustainable socio-technical process-driven service system.*

novel approach to overcoming social and cultural barriers to cultivate and enhance the cultures of cocreation, collaboration, and innovations), so as to ensure the prompt and cost-effective development and delivery of competitive and satisfactory services for customers throughout its geographically dispersed while digitally integrated dynamic service systems (Qiu, 2009). More detailed discussions of digitalization of service systems will be provided in the next section.

In summary, regardless of the complexity and type of service provision that is enabled by a service system, it is typical that a service consists of a series of social and transactional interactions. We understand that the series of service encounters can be direct or indirect, consecutive or intermittent, physical or virtual, and brief or intensive. Regardless of the occurring time and servicescape of a service encounter, each service encounter that constitutes the offered service plays its unique role in contributing to the final service outcomes. The value of service indeed relies on the sociotechnical efficacy and effectiveness of the total occurred service encounters throughout the service lifecycle. The successful operation of a service system thus largely depends on whether its formed SENET-oriented operations can be well performed, monitored, and controlled. Therefore, service systems must be efficiently and cost-effectively managed, realizing their business objectives and goals tactically and strategically.

3.5 DIGITALIZATION OF SERVICE SYSTEMS

Enterprises have benefited from building collaborative partnerships with geographically dispersed partners. Hence, businesses are frequently operated under the umbrella of global virtual enterprises. This becomes a common practice as enterprises can fully leverage the best-of-breed goods and service components at a more competitive price while meeting the changing needs of today's on-demand business environment. As the competition in the global economy unceasingly intensifies, without exception, service organizations must leverage their digital connections to scale and transform, internally and externally, so that they can meet their consumers' fluctuating demand for innovation, flexibility, and shorter lead time of their provided services (Chesbrough, 2011a).

Internally, business domain or enterprise-wide information systems in support of all aspects of business operations and management in an organization are essential (Figure 3.6). It is the information technology (IT) that enables real-time information flow. Consequently, the right data and information in the right context can be delivered to the right user (e.g., people, machine, device, component, etc.) at

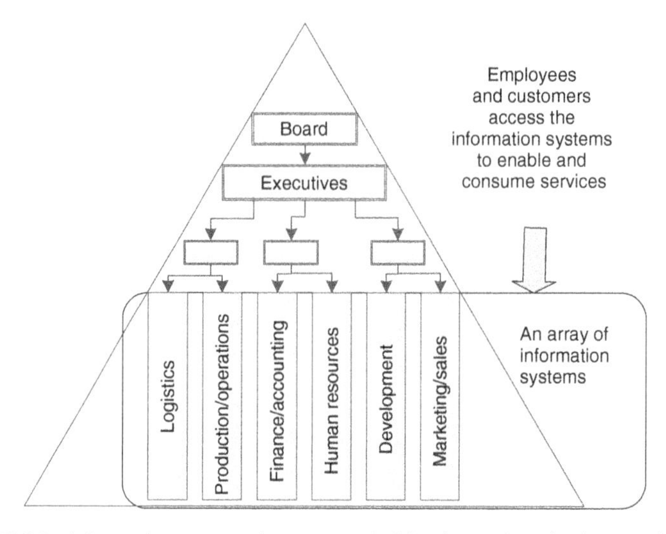

FIGURE 3.6 *Information systems in support of all business domains in organizations.*

the right place and right time, facilitating efficient and effective coordination across business domains within the organization. For instance, the top management can pinpoint business weakness areas and make the informed decisions accordingly with the support of the real-time information on sales, finances, production, and resource utilization of the organization. In general, the enabled and fostered coordination among business domains results in the substantial increase of the degree of business process automation, the continual increment of production productivity and services quality, the reduction of service lead time, and the improvement of job and customer satisfaction (Qiu, 2007; Berman, 2012; Qiu, 2013a).

Because a variety of devices, hardware, and software become network aware, almost everything is currently capable of being handled through the networks. Services can be completed onsite if necessary; while many tasks or functional e-service components can also be done remotely or even self-performed over the Internet. Indeed, at the end of the day, end users do not care about how and where the products and services were made or engineered, by whom, and how they were delivered; what the end users or consumers essentially care about is that their functional and sociopsychological needs are met in a satisfactory manner. In today's globalized and service-led economy, it is the total customer satisfaction and loyalty that drives further and more sales.

Under the unceasingly increased pressure of market competitions, organizations have to be capable of offering and delivering services fast and cost-effectively. In an organization, the employed business operations are essentially derived from its adopted corporate best practices. Managerially, daily business operations are largely reflected and driven by varieties of domain-based business activities that are logically grouped as an array of business processes. Operationally, these business processes are mainly executed by its employees with the support of the deployed enterprise information systems across the organization. A significant portion of business operations might be fully or partially automated by complying with an array of predefined business processes. However, it is also normal for an organization to have a number of *ad hoc* business processes in operation to meet the uncertain or changing needs of employees and customers. Therefore, the enterprise information systems in use in the organization must be efficient, adaptable, ready for integration, and easy to make changes in order for the organization to stay competitive in business from time to time (Qiu, 2013a).

In a service organization, business processes are fully involved with people, tools, and information. Information technology (IT) makes possible the digitalization of the whole organization or socio-technical service system from the systems' perspective. As shown in Figure 3.7, the realization of dynamic, collaborative, and connected ways of business operations with the support of well-integrated enterprise information systems defines smarter working practices, resulting in greater agility than ever before in business (Pearson et al., 2010; Qiu, 2013a):

- *Dynamic*. Instead of retaining static and rigid ways of executing business operations, organizations full of processes, people, and information should be capable of being adjusted rapidly to the changing needs of employees and customers.

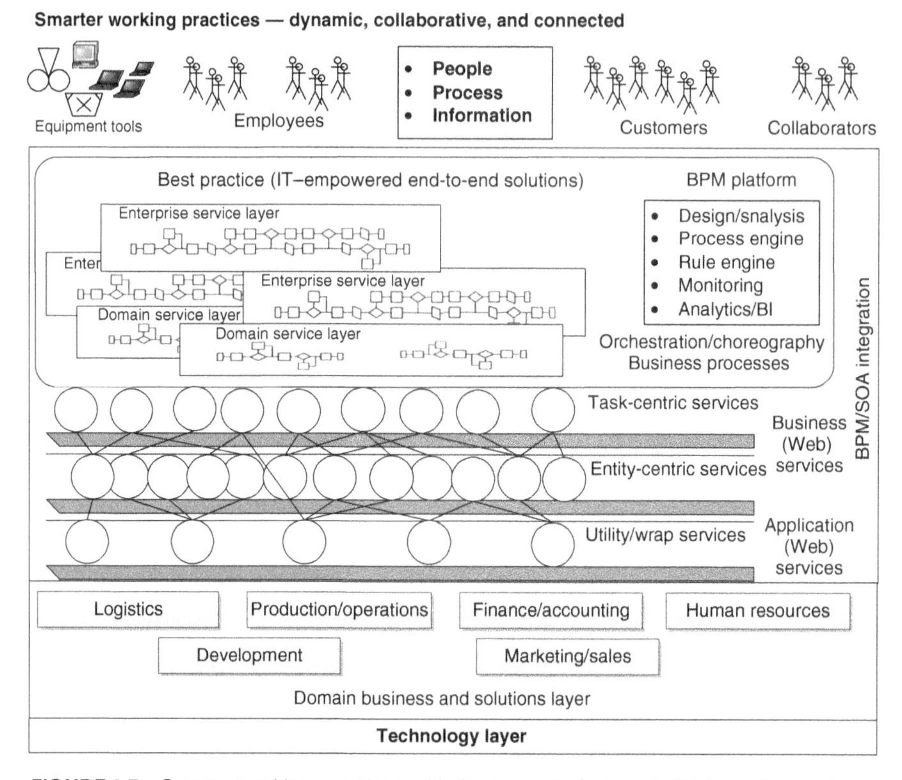

FIGURE 3.7 *Smarter working practices with the support of integrated information systems.*

- *Collaborative.* Instead of relying on a monolithic system (or so-called a monopoly business model), organizations apply best-of-breed service models in practice, focusing on the capability of fully leveraging resources (including people, tools, and information) to share insights, solve problems, and cooperate business operations, internally and externally.
- *Connected.* Telecommunications and networked computers make possible the delivery of the right data and information to the right users at the point of need, regardless of time and location. The world becomes flattened. People and communities are more connected than ever before, so are the business operations across organizations nationally and/or internationally, which surely make today's business operations naturally collaborative across organizations, nationally and/or internationally.

Digitalization of service systems makes possible for us to capture and understand customer experience in service provision in a comprehensive manner (Drogseth, 2012; Ahlquist and Saagar, 2013). Over the years, we have seen that data are overwhelming everywhere and data volumes are exponentially growing year by year. However, the majority of data coming from disparate and heterogeneous sources are either semistructured or unstructured; thus, conventional approaches and tools that were designed to work with large structured data sets simply cannot handle this big data. New analytic methodologies and frameworks must be explored and introduced to the market to help service organizations bring order to the big data from diverse sources and thus harness the power of the big data. By doing so, service organizations can glean the insights and values and also grope for new opportunities that were previously unattainable. This is particularly critical for service organizations to implement SENET-oriented service systems because of the fact that the people behavioral data are typically overwhelming and utterly unstructured (Figure 3.8).

Indeed, the economic globalization has been accelerated with the advancement of networking and computing technologies over the past two decades or so. IT currently plays a more and more critical role in enabling and supporting business and societal development collaborations across the world. Service organizations can leverage the five capitals in an effective and sustainable manner. Particularly, world-class service organizations must take advantage of human capitals including the cultural diversity and focused-area talents to deliver satisfactory services across the continents. Nowadays, not only service organizations but also competitive manufacturers eagerly embrace service-led business models to build and sustain highly profitable service-oriented businesses. They take advantage of their own unique ways of marketing, engineering, and application expertise and shift gears toward creating superior outcomes to best meet their customers' functional and sociopsychological needs in order to outperform their competitors (Rangaswamy and Pal, 2005; Qiu, 2007).

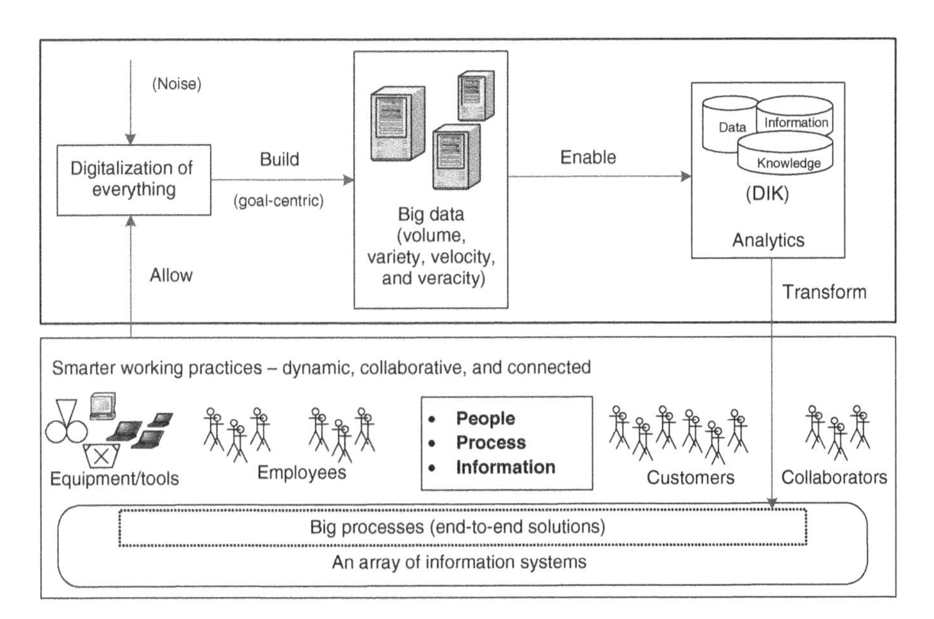

FIGURE 3.8 *Big data and processes enabling SENET-oriented service system.*

Distributed computing and interconnected networks have made IT ubiquitous and pervasive over the years, and thus have significantly increased the capacity and capability of IT in service organizations. Information-enabled systems have continuously increased the flexibility, responsiveness, and capability of business operations. As a result, competitive service organizations have unceasingly improved their productivities and job/customer satisfactions. Regardless of the involved service complexity and nature, a service organization makes every effort to make the organization as a sustainable sociotechnical process-driven service system (Figure 3.9), so as to join and stay in the world-class business club.

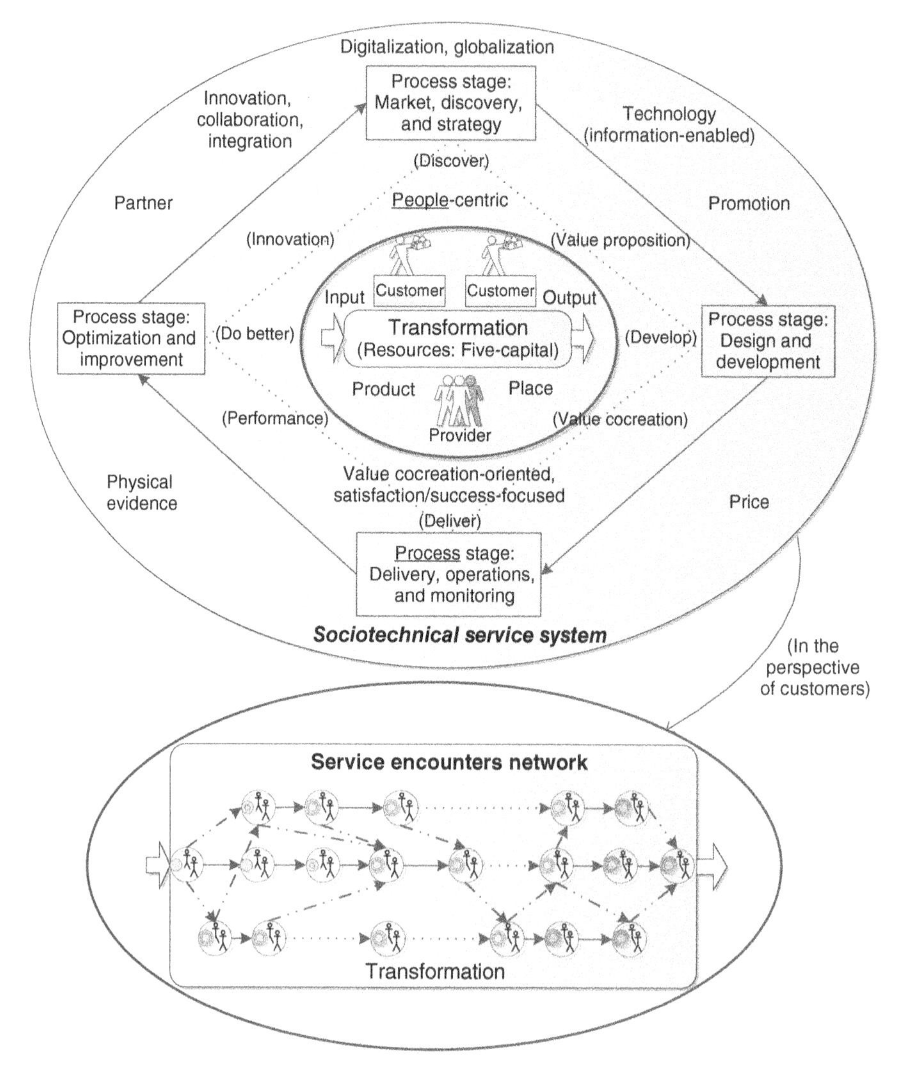

FIGURE 3.9 Sustainable sociotechnical process-driven system.

As discussed earlier, with the push of ongoing "industrialization" of the information technologies, digitalizing information across all business domains in organizations allows the information on service provision to be fully preserved, accessed, and shared within or across organizations. By creating integrated, scalable, and adaptable value networks through collaborating with geographically dispersed business partners, service organizations are capable of delivering their service-led total solutions to customers efficiently and cost-effectively. Note that the value of a delivered service lies in its ability to satisfy an end user's functional and sociopsychological need, which apparently is not strictly seen in the physical attributes and technical characteristics of the provided product or the technical functions of the delivered service.

3.6 AN INNOVATIVE APPROACH TO DEVELOPING SERVICE SCIENCE

As discussed earlier, today's service concept has dramatically evolved beyond the traditional nonagricultural and/or nonmanufacturing performance in delivering customers' benefits. For example, many new emerging high value areas, such as IT outsourcing, after-sales training, on-demand innovations consulting (e.g., consulting services that help customers innovate and improve their product designs, business processes, goods and services delivery operations, and IT systems' efficiency and effectiveness), are well recognized as services, drawing substantial attention from many industrial bellwethers. As a result, the service sector nowadays covers from commercial transportation, logistics and distribution, health care delivery, training and education, financial engineering, e-commerce, retailing, hospitality and entertainment, issuance, supply chain, enterprise knowledge discovery, transformation and delivery, to a variety of high tech and high value consulting services across different industries.

Regardless of service business types, "service drives sale" is not a secret in the current service-led economy. On one hand, unique and satisfactory services differentiate an organization from its competitors; on the other hand, delivery of highly satisfactory services frequently drives more product or service sales for the organization to outperform in its marketplace. As the shift from manufacturing to services becomes inescapable from the developed countries to the developing countries, organizations are gradually embracing for service-oriented business models by defining and selling anything as a service. Note that the service-led economy not only conceptualizes a quantitative increase in the percentage of GDP, more importantly, also indicates a substantive shift in which the service sector must become a main driving engine for future economy growth and innovation.

3.6.1 Service Value Chains in the Service Encounter Perspective

As discussed in the preceding chapters, regardless of many definitions of service existing in the literature, all definitions are more or less based on the same fundamental concept. That is, service is considered as an application of relevant knowledge, skills, and experiences to cocreate benefits, respectively, for service providers and customers. To a customer, the encounter of a service or "moment of truth" frequently is the service in the customer's perspective (Bitner et al., 1990). However, from the perspective of a service system, service is a process of transformation of the customer's needs utilizing the operations' resources, in which dimensions of customer experience manifest themselves in the themes of a service encounter or service encounter chain.

Centered at both provider-side and consumer-side people in services, service encounters are mainly interaction-focused and inherently dyadic and collaborative both socially and psychologically (Shostack, 1985; Solomon et al., 1985; Lu et al., 2009; Schneider and Bowen, 2010). Surely it is a service encounter that enables the necessary manifest function that engages the providers and the customers in order to show the "truth" of service. For example, consumers are the customers in the retailing service sector; students are the customers in educational service systems; patients then are the customers in health care delivery systems. Service is a process of transformation of customer needs; service takes time to complete, resulting in reciprocal influence between service providers and customers. In the extant literature, service encounters within a service largely considered all interacting activities involved in its corresponding service delivery process. In this book, we advocate that a service organization should explore all service encounters across the lifecycle of service, groping for the optimal opportunities and outcomes in services to stay competitive in its marketplace.

The quality of a provider's services is the overall perception that results from comparing the provider's actual performance with the customers' general expectations of how providers in that industry should perform. More specifically, customers' satisfaction is mainly determined by their experience with the service provider; users' experience in turn is their perception based on their experienced service encounters. Empirical studies affirm the fundamental role of the service encounters in evaluating the overall quality and satisfaction with services (Parasuraman et al., 1985, 1988; Bitner et al., 1990). Significant research on the service profit chain has indeed revealed the changing, complex, cascading, and highly correlated relationships among job satisfaction, customer satisfaction and loyalty, and business profitability in service firms (Bolton and Drew, 1991; Heskett et al., 1994; Zeithaml, 2000; Lovelock and Wirtz, 2007; Gracia et al., 2010). In other words, the service profit (or value) chain for a service provider relies on the creation of loyal customers through experiencing excellent service encounters (Figure 3.10).

This new and emerging field is truly interdisciplinary in nature, and explores new frontiers of research in the service research arena. As this book's attempt is to establish the foundation for understanding of future competitiveness in services, we must have a new viable approach by taking a new and innovative path to study services. In manufacturing, products are central to both manufacturers and customers. When we trace the lifecycle of products, we can essentially analyze and understand how the manufacturing businesses have been operated (Qiu and Joshi, 1999). By the same token, service encounters are central to providers and customers in service businesses. Therefore, if we can trace the lifecycle of services, we can also analyze and understand how the service businesses have been operated. Ultimately, if we can track

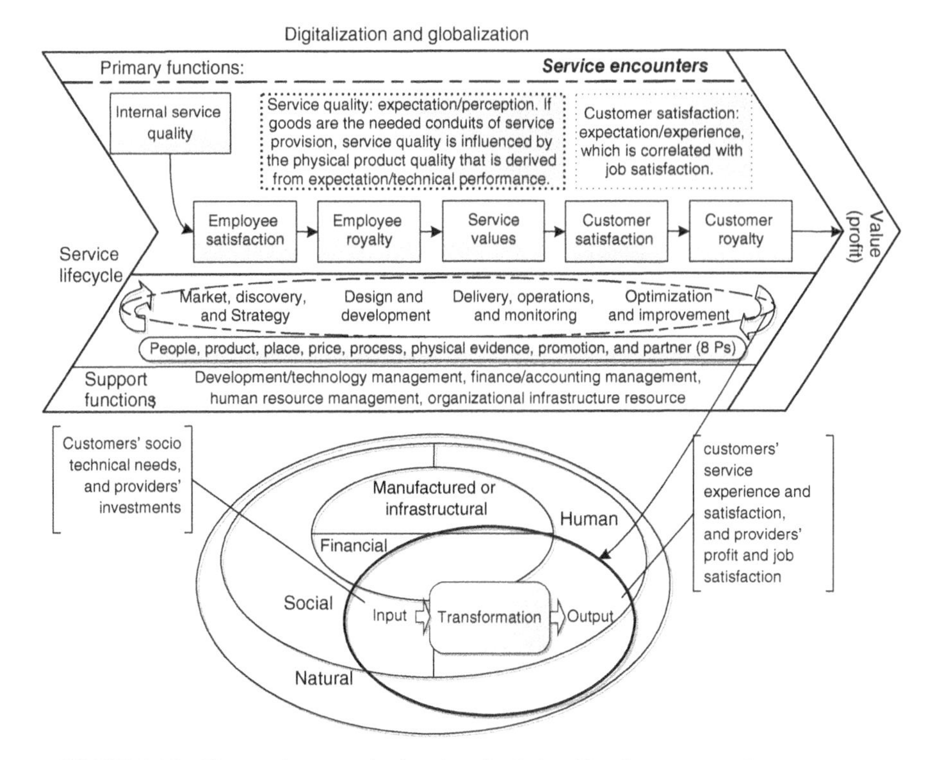

FIGURE 3.10 *The service organizational profit chain with a focus on service encounters.*

and understand service encounters well, we then have opportunities to develop and manage service businesses as desired.

To identify and develop a new approach to explore the science of service, surely it is wise for us to recapture what we have discussed and concluded so far in this chapter. By doing so, we can ensure that the direction of our proposed approach can be articulated in a profound and comprehensive manner. Here comes a list of our understandings of services and service systems in today and the future's globalized service-led economy:

- *Sociotechnical Perspective.* A service organization that offers and delivers services is essentially and holistically functioning as a service system. Truly, physical goods are the conduits of service provision. We understand that human and social capitals play an essential role in service provision. Therefore, we have to develop and manage efficient, effective, and smart operations of an integrated and collaborated sociotechnical service system that human and social capitals can be fully leveraged in service provision (Figure 3.9).
- *Interactive–cocreative Perspective.* Indeed, service provider-side and customer-side people must interact and hence cocreate the value of service. In other words, service encounters are central to providers and customers

in the process of service provision; the dynamics of service offering and delivering processes largely represents the systemic behaviors of service systems in service business (Figure 3.9).

- *Service Encounter Network Perspective.* When we look into service provision over its offering and delivering time horizon, we find that the inherently interactive and collaborative nature among service providers and customers highlights the dynamics of the service value delivery networks. Hence, the formed service encounter networks in service provision can be modeled to describe the processes of transformation of customer needs in service operations and management across the service value delivery networks. As a result, by analyzing the dynamics of service encounter networks, we can interpret and understand the dynamics of corresponding service systems (Figure 3.9).
- *Holistic or Systemic Perspective.* Across the lifecycle of service, the dynamics of service encounter networks substantively depends on how 8 Ps are managed from stage to stage, meeting changing needs under different business circumstances, such as marketing, operations and management, and delivery. In other words, the priority of an individual P within 8 Ps might shift from stage to stage throughout the service lifecycle. When a service is performed, its customer and provider interact with each other, directly or indirectly, consecutively or intermittently, physically or virtually, briefly or intensively, during the process of performing the service. Therefore, 8 Ps must be fully leveraged for optimal service outcomes. That is to say, it must be explored in a holistic or systemic perspective (Figure 3.10).
- *Computational Thinking and Analytics Perspective.* The intelligent connection of people, processes, data, and things makes possible capturing and abstracting of the behavioral dynamics of service encounter networks in a meaningful way so that the changing needs of service customers can be well analyzed and optimally aligned with the business objectives of service providers (Figure 3.10).

With the above-mentioned understandings, it becomes essential for us to adopt service-dominant thinking and technically leverage digitalization and intelligent connections to explore services and accordingly service systems. When we investigate a systemic approach to explore service science, it now becomes clear that we must take the following key concepts into our considerations:

- It must be process-driven and people-centric. Once again, service is a process of transformation of the customer's needs utilizing the operations' resources, in which dimensions of customer experience manifest themselves in the themes of a service encounter or service encounter chain. As compared to manufacturing, service is people-centric, which must be cocreated by customers and providers.
- It must be holistic. The holistic or systemic viewpoint focuses on the "big picture" and the long-range view of systems dynamics, looking at a service organization as a collection of domain systems that constitute a whole. The systematic view allows us to see how each and every service activities is operated. We analyze the efficiency and effectiveness of each activity and accordingly control and

manage them in a decisive manner. By paying attention to the whole, a holistic perspective thus allows us to understand and orchestrate service encounters among business domains across the service lifecycle.

- It must utilize computational thinking. Computational thinking that fully leverages today's ubiquitous digitalized information, computing capability and computational power has evolved as an optimal way of solving problems, designing systems, and understanding human behavior. Computational thinking promotes qualitative and quantitative thinking in terms of abstractions, modeling, algorithms, and understanding the consequences of scale and adaptation, not only for reasons of efficiency and effectiveness but also for economic and social reasons.

3.6.2 A Systemic and Lifecycle Approach to Exploring Service

The prior lack of means to monitor and capture people's dynamics throughout the service lifecycle has prohibited us from gaining insights into the service encounter chains or networks. Promisingly, the rapid development of digitization and networking technologies has made possible the needed means and methods to change this. From all the previous discussions, we conclude that one of the approaches to the explorations of services can be oriented to service encounters. We also understand that the adopted approach should be process-driven, people-centric, and holistic, while leveraging computational thinking.

Now is surely the time for us to explore service science by rethinking service encounters and study those social and transactional interactions in a deeper and more sophisticated manner than ever before. Let us briefly discuss what we could and should explore to optimize the service profit chain from the perspective of a service lifecycle (Figure 3.10).

It is worth pointing out that the Agile Project Management (APM) Group Limited in United Kingdom has pioneered a lot of IT service management studies with a focus on quality, integrity, and the cultivation of international best practice for knowledge-based workers. By focusing on IT service management, the APM Group publishes Information Technology and Infrastructure Library (ITIL) that defines five phases of IT service lifecycle, service strategy, design, transition, operation, and continual service improvement (ITIL, 2011). ITIL essentially provides comprehensive guidelines throughout the phases for aligning IT services with the needs of business. Note that the defined phases and the provided corresponding guidelines, based on goods-dominant thinking, essentially are product-oriented. Surely they can be well applied to the development and management of information-enabled solution products through fully leveraging the fast advancement of computing and networking technologies. In the following discussion, we present a high level guideline for executing service business by referring to the defined phases and the provided corresponding guidelines in ITIL v3 (ITIL, 2011). However, evolving from goods-dominant thinking, we use service-dominant thinking to define the new guideline with a focus of the means and methods to trace, control, and manage service encounters throughout the service lifecycles.

At the market, discovery, and strategy process stage, a service provider must identify and discover the current needs and ongoing trends in its serving marketplaces (Rust et al., 2004), and correspondingly develop and prepare its strategic and operational resources for business execution. It is critical for the provider to understand its operational capability and how to utilize and develop its capability, and hence determine its innovative and competitive service portfolios. Innovation is the highest priority (Chesbrough, 2011a). The provider should leverage all the potential interactions with current or prospective customers to capture their utilitarian needs and understand what would impact their sociopsychological needs. In particular, a service provider must now take advantage of the big data that are amassed on online social networks because online social networks have given rise to a new breed of monitoring and analytical means and methods. By digging into while parsing the cacophony of voices and conversations to uncover actionable information relevant to the provider's service offerings, social-media-based analytics can be well applied at this stage to help reveal needs and patterns and identify trends in its designated and potential service marketplaces (Schaeffer, 2011).

To make sure that a provider can well execute the market, discovery, and strategy process stage of a service, the provider should have done at least the following tasks:

- *Defining Service Value Propositions.* The values of services to the provider and customers must be simultaneously defined using the appropriate marketing mindset; service utilities and warranties shall be clearly identified in a deliverable manner.
- *Planning Supportive Service Resources.* Resources and capabilities in support of service provision across all business domains shall be identified and planned, including internal business units and external service collaborators.
- *Determining Value-Added Service Structures and Corresponding Operational Trajectories.* the dynamics of service systems shall be analyzed and determined through designing service structures or delivery networks in support of desirable, viable, and competitive service value chains.
- *Creating a Contingency Plan.* This should cover a variety of areas, from resources, operations, to recoveries, across the service value chains.

To accomplish those fundamental tasks, the provider should look into all 8 Ps at this stage. However, "People," "Product," "Place," and "Price" are the critical ones; they must be well understood, identified, and planned. "Partner" could be included as a critical P in the service mix if the third party would be involved in the service. "People," including providers and customers, are central to this stage. It is critical to meld assurance and empathy items into the service ideas during a service conceiving process. Therefore, service encounters do occur at this stage. Positive customers' involvements and contributions to this stage foster the development of customer experience excellence, ultimately leading to a great success of the provider's service business.

At the design and development stage, a service provider focuses on the design and development of both the identified innovative service products and the means and capabilities of delivering the services. Goods as the necessary conduit of service provision might be identified or purchased. Both operant and operand resources (Vargo and Lusch, 2004) should be developed and made ready for service provision. Processes, policies, and documentation to meet current and future agreed service requirements must be well designed and developed. The utilitarian needs based on agreed service requirements vary with services. The specification for a given service should clearly, completely, and consistently describe its customers' benefit and warranty, how the service will be delivered and consumed, and what responsibilities the provider and customers have during its service provision.

To make sure that a provider can well execute the design and development process stage of a service, the provider should have completed at least the following tasks:

- *Defining and Validating Service Specification.* On the basis of the identified value proposition, detailed service specifications should be defined and validated. Service-level agreements in great detail are typically finalized at this stage. Service contracts with partners shall also be finalized if partners are essential to the service provision.
- *Developing and Preparing the Resources.* This is critical for the realization of service value proposition defined at the market, discovery, and strategy process stage. All needed resources, in particular, the needed social and human capitals (Lepak and Snell, 1999), should be developed and prepared. As compared to goods, those resources must be well prepared and developed for not only delivering the specified technical and functional attributes in the service specifications but also meeting the customers' dynamic behavioral needs in their sociopsychological dimensions during the period of service consumption.
- *Defining Measurements and Metrics and Developing Corresponding Means for Collecting the Necessary Performance Data.* Unless appropriate measurements and metrics are clearly defined, the performance of service provision can be hardly evaluated. Truly, essential tools to collect the necessary performance data can be hardly developed or acquired without well-defined measurements and metrics.

To have those fundamental tasks done at this stage, the provider should certainly look into all 8 Ps as usual. However, "People," "Product," and "Process" are the critical ones; operant resources, products, and processes must be well analyzed and developed. It is similar to the preceding stage; "Partner" could be included as a critical P in the service mix if the third party would be involved in the service. Validating customers' requirements is the key to having the next stage of a service rolled out successfully. Without question, employees are central to this stage. Therefore, service encounters must occur at this stage. Customers' involvements and contributions to this stage truly lays out a sound and solid foundation that helps deliver customer experience excellence.

At the delivery, operations, and monitoring stage, service providers understand that this stage of a service significantly manifests itself as service encounters (Chiba, 2012). To many customers, the encounter of a service or "moment of truth" frequently is the service in the customer's perspective (Bitner et al., 1990). In general, service providers and customers substantively cocreate the values to the customers during the stage of service delivery and operations. Service operations responsible for all aspects of delivering and managing the customer service needs over the specified service period. The delivery and operations should be well monitored and controlled to ensure that the promised service levels are fully met.

To make sure that a provider can well execute the delivery, operations, and monitoring stage of a service, the provider should have completed at least the following tasks:

- *Educating the Customers.* Customers play a critical role in the creation of service values. The provider should educate the customers by leveraging all the means available in today's information era. The effectiveness of service encounters highly depends on how much the customers know about the service and their competencies to consume the service when the service is offered.
- *Delivering the Service.* This is the interactive point at which the provider and the customers cocreate the service values. Customer experience with the service is mainly assessed through both the outcomes and processes of service deliveries in light of meeting customers' utilitarian and sociopsychological needs over the service lifespans in the customers' perspective.
- *Managing and Facilitating the Service Consumption Process.* As customer experience with a service is the perception of meeting customers' utilitarian and sociopsychological needs over the service lifespans, the service provider should make sure that all involved service encounters during this stage of the service are well managed and facilitated. Service encounters vary with the 8 Ps (i.e., service provision mix). Technically, socially, and psychologically, a personal, optimal, while viable service consumption process for a customer should be carried out.
- *Monitoring the Service and Detecting Ongoing and Potential Service Problems.* The process of cocreating the service should be fully monitored and all the relevant data are collected, so any issue with the delivery and operations can be analyzed. In particular, if ongoing unsatisfactory issues can be detected in a timely manner, promptly, recoverable actions could be taken to compensate the customers.

To have those fundamental tasks done at this stage, the provider should surely look into all 8 Ps once again. However, "People," "Product," "Place," "Process," and "Physical Evidence" are the critical ones; they must be well understood, identified, and managed and operated. "Partner" could be included as a critical P in the service provision mix if the third party would be involved in the service provision. As the service must be cocreated by both providers and customers, "People" are central to this stage of delivering the "moments of truth." The ultimate goal of this stage is to turn one-time customer into a lifetime customer through managing and delivering excellent user experience.

At the optimization and improvement stage, essentially service providers focus on the continual alignment, adjustment, and innovation of the offered services to meet the changing customers' needs. The service providers must keep challenging themselves in improving the service quality on continual basis, enhancing customer experience through continual innovation, and increasing the values that customers and providers can cocreate with the services. Continual service optimization and improvement is the outcomes of improvement actions that are identified through pre-defined analysis and optimization modules. Performance data must be collected in a comprehensive and timely manner. For instance, leveraging service encounter networks and/or relevant social networks to uncover complains and weaknesses can help recover service failures more effective, increase customer satisfaction, improve the effectiveness of market research efforts, and foster open service innovations (Chesbrough, 2003; Chesbrough, 2011b). The ultimate business award for us as service providers will surely be that customers become our service innovators and advocates over time.

To make sure that a provider can well execute the optimization and improvement process stage of a service, the provider should have completed at least the following tasks:

- *Understanding and Embracing a High Level Business Vision.* Continuing to be fully engaged with both current and prospective customers is the key to understand the ongoing changes of the business environment in which the provider is. The provider should be vigilant and readily embrace for the changes and accordingly define and adjust its high level business vision. The business vision should be timely and well communicated across the organization.
- *Verifying that Measurements and Metrics are Working So that the Current Situation Can be Truly Assessed.* This is a critical step to ensure that the needs of customers were fully understood and defined measurements and metrics are appropriate for the needed assessments. Adjustments and changes should be made if needed.
- *Determining the Priorities and Corresponding Plan for Improvement.* On the basis of the assessments of provided services, the priorities for improvement can be identified. Furthermore, the identified changes from the updated high level business vision should be well considered in determining the priorities and corresponding plan for improvement.
- *Ensuring that Actions for Improvement are Embedded into the Organization.* As discussed earlier, a systemic approach is a must in delivering competitive services in today and the future's service-led economy. Therefore, the identified actions for improvement must be embedded into the organization to ensure an optimal outcome from the adopted improvement measures.

To have those fundamental tasks done at this stage, the provider should unexceptionally look into all 8 Ps as usual. However, "People," "Place," "Process," and "Physical Evidence" are the critical ones; they must be well understood and analyzed. "Partner" could also be included as a critical P in the service mix if the third party would significantly contribute to the service. Without question, "People," including providers and customers, are continuously centered at this stage. Customers' involvements and contributions to this stage foster the continual alignment, adjustment, and innovation of the offered services, ultimately leading to a great success of the provider's service business.

REFERENCES

Ahlquist, J., & Saagar, K. (2013). Comprehending the complete customer. *Analytics—INFORMS Analytics Magazine*, May/June, 36–50.

Berman, S. (2012). Digital transformation: opportunities to create new business models. *Strategy and Leadership*, 40(2), 16–24.

Bitner, M. J. (1990). Evaluating service encounters: the effects of physical surroundings and employee responses. *Journal of Marketing*, 54(2), 69–82.

Bitner, M. J. (1992). Servicescapes: the impact of physical surroundings on customers and employees. *Journal of Marketing*, 56(2), 57–71.

Bitner, M. J., Booms, B. H., & Mohr, L. A. (1994). Critical service encounters: the employee's viewpoint. *The Journal of Marketing*, 58, 95–106.

Bitner, M. J., Booms, B. H., & Tetreault, M. S. (1990). The service encounter: diagnosing favorable and unfavorable incidents. *The Journal of Marketing*, 54, 71–84.

Bitner, M. J., Brown, S. W., & Meuter, M. L. (2000). Technology infusion in service encounters. *Journal of the Academy of marketing Science*, 28(1), 138–149.

Bitner, M. J., Faranda, W. T., Hubbert, A. R., & Zeithaml, V. A. (1997). Customer contributions and roles in service delivery. *International Journal of Service Industry Management*, 8(3), 193–205.

Bolton, R. N., & Drew, J. H. (1991). A multistage model of customers' assessments of service quality and value. *Journal of Consumer Research*, 17(4), 375–384.

Bradley, G. L., McColl-Kennedy, J. R., Sparks, B. A., Jimmieson, N. L., & Zapf, D. (2010). Service encounter needs theory: a dyadic, psychosocial approach to understanding service encounters. *Research on Emotion in Organizations*, 6, 221–258.

Chase, R. B. (1978). Where does the customer fit in a service operation? *Harvard Business Review*, 56(6), 137–142.

Chase, R. B., & Dasu, S. (2001). Want to perfect your company's service? Use behavioral science. *Harvard Business Review*, 79(6), 78–85.

Chase, R. B., & Dasu, S. (2008). Psychology of the Experience: The Missing link in Service Science. *Service Science, Management and Engineering Education for the 21st Century*, 35–40, eds. by B. Hefley and W. Murphy. US: Springer.

Chase, R. B., & Erikson, W. (1989). The service factory. *The Academy of Management Executive*, 2(3), 191–196.

Chesbrough, H. W. (2003). *Open Innovation: The New Imperative for Creating and Profiting from Technology*. Boston, MA: Harvard Business School Press.

Chesbrough, H. W. (2011a). *Open Services Innovation: Rethinking Your Business to Grow and Compete in a New Era*. San Francisco, CA: Jossey-Bass.

Chesbrough, H. W. (2011b). Bringing open innovation to services. *MIT Sloan Management Review*, 52(2), 85–90.

Chiba, T. (2012). Service encounter model focused on customer benefits and satisfaction: reconsideration of psychological model (No. 2011-034). Keio/Kyoto Joint Global COE Program.

Czepiel, J. A. (1990). Service encounters and service relationships: implications for research. *Journal of Business Research*, 20(1), 13–21.

Czepiel, J. A., Solomon, M. R., & Surprenant, C. F. (1985). *The Service Encounter: Managing Employee/Customer Interaction in Service Business*. Lexington, MA: Lexington Books.

Daskin, M. S. (2011). *Service Science*. Hoboken, NJ: Wiley.

Dietrich, B., & Harrison, T. (2006). Serving the services industry. *OR/MS Today*, 33(3), 42–49.

Drogseth, D. (2012). User experience management and business impact—a cornerstone for IT transformation. *Enterprise Management Associates Research Report*. Retrieved Oct. 10, 2012 from http://www.enterprisemanagement.com/research/asset.php/2311/.

Durvasula, S., Lysonski, S., & Mehta, S. C. (2005). Service encounters: the missing link between service quality perceptions and satisfaction. *Journal of Applied Business Research*, 21(3), 15–25.

FftF. (2012). The five capitals model—a framework for sustainability. *White Paper from Forum for the Future*. Retrieved Oct. 10, 2012 from http://www.forumforthefuture.org/.

Gracia, E., Cifre, E., & Grau, R. (2010). Service quality: the key role of service climate and service behavior of boundary employee units. *Group & Organization Management*, 35(3), 276–298.

Grönroos, C. (1994). From scientific management to service management: a management perspective for the age of service competition. *International Journal of Service Industry Management*, 5(1), 5–20.

Heskett, J. L., Jones, T. O., Loveman, G. W., Sasser, W. E., & Schlesinger, L. A. (1994). Putting the service-profit chain to work. *Harvard Business Review*, 72(2), 164–174.

Heskett, J. L., Sasser, W. E., & Hart, C. W. (1990). *Service Breakthroughs: Changing the Rules of the Game*. New York, NY: The Free Press.

Hilton, T., & Hughes, T. (2013). Co-production and self-service: the application of service-dominant logic. *Journal of Marketing Management*, 29(7/8), 861–881.

Hsu, C. (2009). *Service Science: Design for Scaling and Transformation*. Singapore: World Scientific and Imperial College Press.

IBM. (2004). *Services Science: A New Academic Discipline?* IBM Research.

IBM Palisade Summit Report. (2006). *Service Science Education for 21st Century*. Palisades, NY: IBM.

ITIL. (2011). *Information Technology and Infrastructure Library*. ITILv3. Retrieved Oct. 10, 2012 from http://www.itil-officialsite.com/AboutITIL/WhatisITIL.aspx.

Karmarkar, U. (2004). Will you survive the services revolution? *Harvard Business Review*, 82(6), 100–107.

Larson, R. (1989). OR/MS and the services industries. *OR/MS Today*, April, 12–18.

Lepak, D. P., & Snell, S. A. (1999). The human resource architecture: towards a theory of human capital allocation and development. *Academy of Management Review*, 24(1), 31–48.

Lepak, D. P., & Snell, S. A. (2002). Examining the human resource architecture: the relationships among human capital, employment, and human resource configurations. *Journal of Management*, 28(4), 517–543.

Lloyd, A. E., & Luk, S. T. (2009). Interaction behaviors leading to comfort in the service encounter. *Journal of Services Marketing*, 25(3), 176–189.

Löbler, H. (2013). Service-dominant networks: an evolution from the service-dominant logic perspective. *Journal of Service Management*, 24(4), 4–4.

Lovelock, C. H. (1983). Classifying services to gain strategic marketing insights. *The Journal of Marketing*, 47, 9-20.

Lovelock, C. H., & Wirtz, J. (2007). *Service Marketing: People, Technology, Strategy*, 6th ed. Upper Saddle River, NJ: Prentice Hall.

Lu, I. Y., Yang, C. Y., & Tseng, C. J. (2009). Push-pull interactive model of service innovation cycle-under the service encounter framework. *African Journal of Business Management*, 3(9), 433–442.

Meyer, C., & Schwager, A. (2007). Understanding customer experience. *Harvard Business Review*, 85(2), 116–126.

Morris, B., & Johnston, R. (1987). Dealing with inherent variability: the difference between manufacturing and service?. *International Journal of Operations & Production Management*, 7(4), 13–22.

Palmisano, S. (2008). A Smarter planet: instrumented, interconnected, intelligent. Retrieved on Dec. 17, 2008 at http://www.ibm.com/ibm/ideasfromibm/us/smartplanet/20081117/sjp_speech.shtml.

Parasuraman, A., Zeithaml, V. A., & Berry, L. L. (1985). A conceptual model of service quality and its implications for future research. *The Journal of Marketing*, 49(Fall), 41–50.

Parasuraman, A., Zeithaml, V. A., & Berry, L. L. (1988). SERQUAL: a multi-item scale for measuring customer perceptions of service quality. *Journal of Retailing*, 64(1), 12–40.

Pearson, N., Lesser, E., & Sapp, J. (2010) A new way of working: insights from global leaders. *IBM Institute of Business Value, Executive Report (IBM Global Business Services).*

Qiu, R. G. (2007). Information Technology as a Service. Chapter 1 in *Enterprise Service Computing: From Concept to Deployment*, 1–24, ed. by Robin Qiu. Hershey, PA: Idea Group Publishing.

Qiu, R. G. (2009). Computational thinking of service systems: dynamics and adaptiveness modeling. *Service Science*, 1(1), 42–55.

Qiu, R. G. (2012). Editorial column—launching service science. *Service Science*, 4(1), 1–3.

Qiu, R. G. (2013a). *Business-Oriented Enterprise Integration for Organizational Agility*. Hershey, PA: IGI Global.

Qiu, R. G. (2013b). Rethinking service encounters. *Submitted to Harvard Business Review*.

Qiu, R. G., Fang, Z., Shen, H., & Yu, M. (2007). Editorial: towards service science, engineering and practice. *International Journal of Services Operations and Informatics*, 2(2), 103–113.

Qiu, R. G., & Joshi, S. B. (1999). A structured adaptive supervisory control methodology for modeling the control of a discrete event manufacturing system. *IEEE Transactions on Systems, Man and Cybernetics, Part A: Systems and Humans*, 29(6), 573–586.

Rangaswamy, A., & Pal, N. (2005). Service innovation and new service business models: harnessing e-technology for value co-creation. *An eBRC White Paper*, 2005 Workshop on Service Innovation and New Service Business Models, Penn State.

Rust, R., Ambler, T., Carpenter, G., Kumar, V., & Srivastava, R. (2004). Measuring marketing productivity: current knowledge and future directions. *Journal of Marketing*, 68(4), 76–89.

Rust, R., Zahorik, A., & Keiningham, T. (1995). Return on quality (ROQ): making service quality financially accountable. *Journal of Marketing*, 59(2), 58–70.

Sampson, S., & Froehle, C. (2006). Foundation and implication of a proposed unified services theory. *Production and Operations Management*, 15(2), 329–343.

Samuelson, P., & Nordhaus, W. (2009). *Economics*, 19th ed. Boston, MA: McGraw-Hill/Irwin.

Schaeffer, M. (2011). Capitalizing on the smarter consumer. *IBM Institute for Business Value—IBM Global Business Services Executive Report*, Retrieved on Dec. 6, 2012 from http://www-935.ibm.com/services/us/gbs/thoughtleadership/ibv-capitalizing-on-the-smarter-consumer.html.

Schneider, B., & Bowen, D. (2010). Winning the service game. *Handbook of Service Science*, 31–59.

Shostack, G. L. (1985). Planning the Service Encounter. In *The Service Encounter: Managing Employee/Customer Interaction in Service Businesses*, 243–254, ed. by J. Czepiel, M. Solomon, and C. Suprenant. Lexington, MA: Lexington.

Solomon, M. R., Surprenant, C., Czepiel, J. A., & Gutman, E. G. (1985). A role theory perspective on dyadic interactions: the service encounter. *The Journal of Marketing*, 49(1), 99–111.

Spohrer, J. (2009). Welcome to our declaration of interdependence. *Service Science*, 1(1), i–ii.

Spohrer, J., Maglio, P., Bailey, J., & Gruhl, D. (2007). Steps toward a science of service systems. *Computer*, 40(1), 71–77.

Spohrer, J., & Riechen, D. (2006). Services science. *Communications of the ACM*, 49(7), 30–34.

STWiki. (2012). *Systems Theory*. Wikipedia. Retrieved Oct. 10, 2012 from http://en.wikipedia.org/wiki/Systems_theory.

Sullivan, A., Sheffrin, S., & Perez, S. (2011). *Economics: Principles, Applications and Tools*, 7th ed. Upper Saddle River, NJ: Prentice Hall.

Svensson, G. (2002). A triadic network approach to service quality. *Journal of Services Marketing*, 16(2), 158–179.

Svensson, G. (2004). A customized construct of sequential service quality in service encounter chains: time, context, and performance threshold. *Managing Service Quality*, 14(6), 468–475.

Svensson, G. (2006). New aspects of research into service encounters and service quality. *International Journal of Service Industry Management*, 17(3), 245–257.

Taylor, J. C. (1977). Job satisfaction and quality of working life: a reassessment. *Journal of Occupational Psychology*, 50(4), 243–252.

Vargo, S. L., & Akaka, M. A. (2009). Service-dominant logic as a foundation for service science: clarifications. *Service Science*, 1(1), 32–41.

Vargo, S., & Lusch, R. F. (2004), Evolving to a new dominant logic for marketing. *Journal of Marketing*, 68(1), 1–17.

Zeithaml, V. A. (2000). Service quality, profitability, and the economic worth of customers: what we know and what we need to learn. *Journal of the Academy of Marketing Science*, 28(1), 67–85.

4

Service Science Fundamentals

At the end of the day, the cocreated total value of a service lies in its ability to satisfy the needs of its provider-side and customer-side people. Hence, the resource, operations, and management models of service systems should be centered on the end users. From the discussions in the preceding chapters, it is understood that both service providers and customers are the core elements that constitute a service system, cocreating services by transforming the customers' needs with the support of infrastructural, financial, social, and natural resources. Even though in a solely self-service system, we are frequently personifying its serving units and processes to improve our service effectiveness and customer satisfaction. For instance, personifying allows a service provider to have empathic and pleasing considerations in service provision to enrich personal touches. As a result, the service provider can avoid creating apathy and negativity that people might feel when physical machines are only present in carrying on certain service encounters from the service provider side at a given time.

As compared to manufacturing that has been mainly centered with physical matters, services are people-centered. Because the resources in service systems, largely people, cannot be held. It becomes extremely challenging for us to model the dynamics of service systems. We understand that people participating in service production and consumption have physiological and psychological characteristics, cognitive ability, and sociological constraints (Dietrich and Harrison, 2006). People's behaviors are extremely difficult to model in general. As a result, when we compare studies between services and manufacturing in the literature, we can easily find that there generally lacks quantitative modeling of the dynamics of service systems although the literature has a long history of publishing empirical studies or qualitative research

Service Science: The Foundations of Service Engineering and Management, First Edition. Robin G. Qiu.

of services or service systems in focused areas, including service market, operations, and management.

As discussed in the previous chapters, we must use service-dominant thinking to explore the people-centric and systemic interactions and their impacts on the dynamics of service systems. Bearing in mind that real-time quantitative explorations of service systems are the ultimate goal in the service research community, we know that an applicable approach must be process-driven, people-centric, holistic, and computational. Undoubtedly, Service Science as metascience of services must build on predecessors' excellent work from many traditional disciplines. Like scores of other constituent parts of the study of services, Service Science must follow the scientific method and must be rigorous and scholarly. Service Science must be built upon combining the best of a variety of perspectives into an integrated and interdisciplinary discovery and analytics of service systems (Larson, 2011; Qiu, 2012).

Because the laws of service radically are a set of rules and guidelines, we can apply them to founding and fostering further scientific explorations in the field of service engineering and management. In this chapter, we first explore Service Science fundamentals, focusing on finding the laws of service in general. By referring to Newton's law of motion that explains and investigates the dynamics of physical objects and systems, we articulate that a similar set of principles can be deducted from the dynamics of service. Secondly, we discuss that service-encounter-based sociophysics wins a focus on the formed service encounter networks to explain and investigate the dynamics of service systems. Then, we present an overview of Service Science in a sociophysics perspective. Finally, a brief conclusion will be provided, highlighting the potential of Service Science in general.

4.1 THE FUNDAMENTAL LAWS OF SERVICE: A SYSTEMIC VIEWPOINT

"If we don't know what the laws of service are, or we think they don't matter, and thus continually break them, we will pay the consequences." "Without knowing the laws of service and how they impact our business, we will most surely fall into chaos, lose our competitive edge and cease to be profitable" (Meany, 2012). Many scholars and practitioners (Wyckoff and Maister, 2005) have attempted to put together a system of rules and guidelines as service laws, aimed at identifying managerial and operational guidance with scientific rigor for service organizations so as to help them achieve effective service marketing, management, and operations in practice.

For instance, Meany (2012) compiles the following seven real, concrete service laws that could assist service organizations to establish managerial guidelines in cultivating their business operations:

- *The Law of Customers.* "Treat everyone around you like a customer, or someone else will."
- *The Law of Consistency.* "Don't just make it a priority .... keep it a priority."

- *The Law of Expectation.* "If you are going to assume anything, assume customer loyalty."
- *The Law of Challenge.* "Good customer communication means bridging service gaps, not falling into them."
- *The Law of Control.* "If your house is in order, your customer's house is in order."
- *The Law of Image.* "Nobody knows where the beef is without the sizzle."
- *The Law of The Basics.* "First things first, second things not at all."

In theory, the provided effective guidance helps service organizations improve their competitiveness and profitability so as to enjoy long-term success in their respective industries.

In general, law is a system of rules and guidelines that are enforced to control and govern social behavior. There are varieties of laws and regulations. The discussions on the distinctions of public and private laws, nationally and internationally, are certainly beyond the scope of this book. As we are interested in the rules and guidelines that can be applied to service engineering and management in general, we essentially focus on finding the principles that can scientifically guide us to enable and govern service offerings in a repeatable, competitive, and profitable manner. According to the Oxford English dictionary (Oxford, 2001), a physical law or scientific law in the physical world is "a theoretical principle deduced from particular facts, applicable to a defined group or class of phenomena, and expressible by the statement that a particular phenomenon always occurs if certain conditions be present."

Before we fully delve into the compilation of the fundamental laws of service, we can briefly illustrate the similarity and dissimilarity between the motion of a physical object and the user experience perceived from a service. In physics, the action and reaction between two objects define how these two objects interact. Similarly, the action (or request) and reaction (or response) in cocreating the benefits of the service then define how service provider-side and customer-side people execute the essential social and transactional interaction within the process of transformation of the needs of customers. Furthermore, the motion state change of a physical object is due to the fact that there is an external force applied to the object. Similarly, the change of user experience perceived from a service is due to the fact that there is an extra effort applied to the service. However, an exact relationship between force, mass, and acceleration can be defined for the motion of a physical object. The perceived user experience holds a similar measure but is mainly subjective in nature. Hence, we must look into these challenges next in a comprehensive manner.

4.1.1 Newton's Three Laws of Motion

An excellent example of physical laws is Newton's three laws of motion. According to Wikipedia (NewtonWiki, 2013), the three laws of motion have been used to explain and investigate the motion of many physical objects and systems over three centuries. The three laws of motion are fundamental in Physics, which were first compiled by Sir Isaac Newton in his book *Philosophiæ Naturalis Principia Mathematica*. The book was written in Latin; its first edition was published in 1687. Newton's three laws are useful approximations at the scales and speeds of daily life, in which physical objects move at the much slower speed than light. It is well recognized that the combination of Newton's laws of motion, his universal law of gravitation, and calculus provided a unified quantitative explanation for a wide range of physical phenomena over three centuries.

Newton's first law is well known as the law of inertia. "If there is no net force on an object, then its velocity is constant. The object is either at rest (if its velocity is equal to zero), or it moves with constant speed in a single direction" (NewtonWiki, 2013). This simply means that there is a natural tendency of an object to keep on doing what the object is doing. An object resists a change in its state of motion, at rest or moving at a constant velocity. "Changes in motion must be imposed against the tendency of an object to retain its state of motion. In the absence of net forces, a moving object tends to move along a straight line path indefinitely" (NewtonWiki, 2013).

Newton's second law states an exact relationship between force, mass, and acceleration. "The acceleration A of a body is parallel and directly proportional to the net force F acting on the body, is in the direction of the net force, and is inversely proportional to the mass M of the body, i.e., $F = MA$" (NewtonWiki, 2013). Simply put, if an object is accelerating, then there is a force on it. "Consistent with the first law, the time derivative of the momentum is non-zero when the momentum changes direction, even if there is no change in its magnitude; such is the case with uniform circular motion. The relationship also implies the conservation of momentum: when the net force on the body is zero, the momentum of the body is constant. Any net force is equal to the rate of change of the momentum" (NewtonWiki, 2013).

Newton's third law is well known as the law of action–reaction. "When a first body exerts a force F_1 on a second body, the second body simultaneously exerts a force $F_2 = -F_1$ on the first body. This means that F_1 and F_2 are equal in magnitude and opposite in direction" (NewtonWiki, 2013). The third law essentially defines how different objects interact. "The action and the reaction are simultaneous, and it does not matter which is called the action and which is called reaction; both forces are part of a single interaction, and neither force exists without the other" (NewtonWiki, 2013).

We understand that the three laws of motion have been used to explain and investigate the motion of physical objects and systems over three centuries although they have never been proved in a scientific way. We adopt the same approach in this book. By referring to Newton's law of motion that explain and investigate the dynamics of physical objects and systems, we radically advocate that a similar set of principles can be compiled for exploring the dynamics of service encounter networks. With the help of derived theoretical foundations, ultimately we can apply approximations of systems behavior to explain and investigate the overall dynamics of service systems.

4.1.2 The Three Fundamental Laws of Service: The Newtonian Approach

Before we discuss the three fundamental laws of service, let us review the definition of service we had in Chapter 2. Service is considered as a transformation process in which both provider-side and customer-side people participate in an interactive manner, applying relevant knowledge, skills, and experiences in order to cocreate mutual benefits for the service providers and their customers. Technically and socioeconomically the transformation process encompasses a series of service encounters that can be direct or indirect, consecutive or intermittent, physical or virtual, and brief or intensive. The value of service depends on the sociotechnical efficacy and effectiveness of all of the service encounters experienced throughout the service lifecycle.

Without question, the interactions between service providers and customers are the key to the services performed by the service providers and their customers. In a simple service interaction, two entities, a provider and a customer, are exactly the same as the physical objects in the above-discussed Newton's interpretation. Each interaction requires an effort from each of the entities and creates an impact on the entities. Thus, it is the interaction that generates a service experience perceived by the two entities in our sociotechnical service world. In other words, the effort generating the service experience in Service Science is our "force" that is similar to the force that changes the motion of an object in Physics.

In the physical world, mass, acceleration, momentum, and force are assumed to be externally defined quantities. Newton's three laws of motion describe how force, mass, acceleration, and momentum are related in a quantitative manner. If a service system's systemic mass, acceleration, momentum, and effort in our sociotechnical service world can also be assumed to be externally defined quantities, we can try to form a similar interpretation describing the laws of service.

As discussed earlier, Newton's three laws are fundamental to physical objects, essentially describing basic rules about how the motion of physical objects changes. The question we ask here is what the fundamental laws of service are. As a force applied to an object changes the motion of the object, effort that is viewed as a sociotechnical force changes the experience perceived by a participant or entity. As an analogy to Newton's three laws, the fundamental laws of service must be the ones that can be applied to describe the basic rules about how the experience of a service perceived by a participant changes. Here is our first attempt to compile the laws of service from this discussion:

- *Service's First Law.* If there is no changing effort on a service, then the service experience perceived by an entity remains the same.
- *Service's Second Law.* The acceleration A of an entity is directly proportional to the changed effort E applying to the entity, is in the direction of the changed effort, and is inversely proportional to the systemic mass M of the entity, that is, $A = E/M$.
- *Service's Third Law.* When a first entity applies an effort E_1 on a second entity, the second entity simultaneously applies an effort $E_2 = -E_1$ on the

first entity. This means that E_1 and E_2 are equal in magnitude and opposite in direction.

Just like Newton's first law, the first law of service aims to establish a general frame of reference for which the other service laws are applicable. In the sociotechnical service world, services are always cocreated by service providers and customers. An entity that is defined in the laws of service thus is either a human being or personalized system. At each point of customer interaction, a customer gains experience that essentially is the perceived effect of the fulfillments in both the utilitarian and sociopsychological dimensions. Therefore, to an end user, the experience perceived by the end user is what a service is all about.

Here are some real-life examples of services, which help us interpret the practical meaning of the above-compiled laws of service. A guest who stayed in a hotel in a city for 3 days had his/her unique experience. Assume that the hotel has applied no changing effort to the provided service. If the guest would stay there for another 3 days, he/she realistically should perceive the same experience he/she perceived before. Similarly, a group of students as customers took a course that was offered by a professor in Lion's College. They surely had their learning experience. Assume that Lion's College made no changing effort on offering the course, including classroom equipment, instructions, and other related learning supports. If another group of students with a similar background that the first group of students had would take the course in a new semester, the second group students should also perceive the same experience that the first group of students perceived earlier. We understand that each service might have its unique value measured in certain utilitarian and sociopsychological dimensions. However, the established reference frame for services remains the same.

Service's second law aims to define how a changing effort on a service makes a difference in a quantitative manner. In reality, we know that "the only constant in the universe is change" is well applied to today's business world. Indeed, during the period of time when a customer consumes a service, the service provider can make a variety of changes in order to meet the customer's changing needs in both the utilitarian and sociopsychological dimensions.

Let us continue to use the above-discussed hotel and learning services as examples. The guest who stayed in the hotel for 3 days at the second time actually had a morning meeting in an organization that is located 10 miles away from the hotel. Assume that he/she was not familiar with the city and the hotel does not provide shuttle services. Because the hotel cannot provide shuttle services to its customers, the hotel would like to help him/her arrange a taxi. His/her experience on his/her second stay in the hotel might be positively impacted because of the additional help provided by the hotel. As for the Lion's College case, changing efforts could include enhanced lectures from the professor, a newly equipped classroom, and/or more experienced teaching assistants. As a result, the second group students could perceive much better experience than what the first group of students perceived earlier.

With respect to the established frame of reference, the acceleration $\overrightarrow{A}_t$ of an entity defined in the second law of service is directly proportional to the changed effort $\overrightarrow{E}_t$

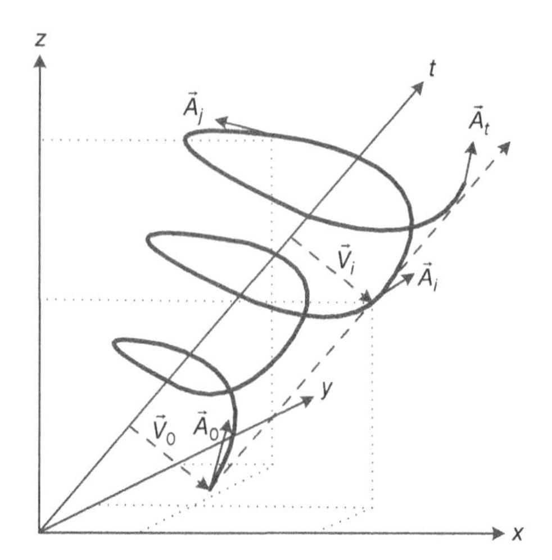

FIGURE 4.1 *An illustration of the momentum of service based on the laws of service.*

applied to the entity. An illustration of the momentum of service based on the second law of service is shown in Figure 4.1. In reality, when a given changing effort $\vec{E}_t^{\prime}$ is applied to a small and simple entity with a systemic mass of $\vec{M}_t^{1}$, we assume that the corresponding acceleration of service dynamics can be quantified as $\vec{A}_t^{1}$; when the same changing effort $\vec{E}_t^{\prime}$ is applied to a large and complex entity with a systemic mass of $\vec{M}_t^{2}$, we assume that the corresponding acceleration of service dynamics can be quantified as $\vec{A}_t^{2}$. We know that $|\vec{A}_t^{2}| < |\vec{A}_t^{1}|$ holds for sure. In other words, the smaller and less complex an entity to which a changing effort is applied, the more significant the resultant impact is. However, in the sociotechnical service world, it is unnecessary that $\vec{A}_t^{1}$ is more beneficial than $\vec{A}_t^{2}$ in light of realizing the business goal.

Service's third law can also be called the mutuality law, defining the equality of efforts made, respectively, by interacting entities during service provision. A good and classical example can be "a bushel of wheat exchanged for a barrel of oil" between two entities that was discussed in Chapter 1. The first entity applies an effort E_1 on growing a bushel of wheat, while the second entity must apply the same amount of effort E_2 on making a barrel of oil. It is easy to understand the real meaning of the mutuality law if we apply modern economics here. To the first entity, an effort of E_1 is made to exchange for an equal effort E_2 made by the second entity. Because of an effort made by the interacting entity, its direction must be opposite if its direction is applied. Therefore, $\vec{E}_1 = -\vec{E}_2$ holds. In Economics, we can convert an effort made by an entity into a value $\vec{V}$. A customer pays a service provider to exchange for an equivalent value of service that the customer asks for. Therefore, we have $|\vec{V}_1| = |\vec{V}_2|$ to represent the mutuality law in the sociotechnical service world.

Newton's laws of motion can be verified through experiments and observations. Indeed, scientists have positively verified and demonstrated them for over three centuries. As compared to the physical quantities that are well defined and physically measurable, a service system's systemic mass and effort in the sociotechnical service world are largely subjective. However, we can always find scientific ways to have them to be externally defined quantities when a service setting is given. Consequently, by applying the three fundamental laws of service, scientific findings, and enabling technologies in service businesses, we can make business operations competitive and profitable and enjoy long-term successes in our respective service industries.

4.1.3 A Systemic View of the Fundamental Laws of Service

In the previous section, we proposed three laws of service analogous to Newton's three laws of motion. To make the analogous discussion of the three laws of service easily comprehendible and fully understood, we use Table 4.1 to show the analogous highlights between the laws of motion and service. Indeed, the current socioeconomic service activities largely obey the underlying principles that are implied in the laws of service. In other words, the developed three fundamental laws of service seem to make sense in reality. However, we understand that variables must be truly and externally defined to be quantities in satisfying the canons of scientific rigor. By doing so, the above-mentioned three fundamental laws can be truly applied to describe basic rules about how the experience of a service changes. The three fundamental laws of service can then be the three essential pillars in support of Service Science (Figure 4.2).

As discussed earlier, we are interested in the rules and guidelines that can be applied to service engineering and management in general. Thus, we must focus on identifying managerial and operational guidance with scientific rigor for service organizations to achieve effective service marketing, management, and operations in practice. In the physical world, by relying on Newton's three laws of motion we can truly describe the motion dynamics of physical objects in a scientific manner with respect to an inertial frame of reference. In the sociotechnical service world, we lack scientific methods of measuring the systemic mass of a first entity and effort made by a second entity or the interacting entity of the first entity. Therefore, the first compiled three laws of service can be hardly applied to guiding service organizations to enable and govern service offerings in a repeatable, competitive, and profitable manner.

The mass of an object in Physics is well defined and can be instrumentally measured. A scientific definition of mass is given as follows: the mass of an object is the

TABLE 4.1 Analogous Highlights Between the Laws of Motion and Service

Law	Motion	Service Experience
The first law	Inertia due to zero net force ($\Sigma F = 0$)	Inertia due to no changing effort ($\Delta E = 0$)
The second law	$F = MA$	$E = MA$
The third law	$F_2 = -F_1$	$E_2 = -E_1$

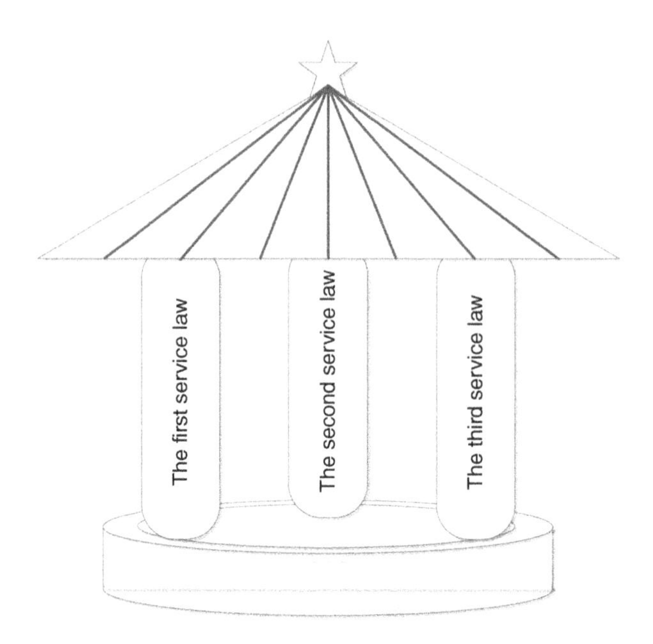

FIGURE 4.2 *The three fundamental laws of service.*

quantity of matter in the object body regardless of its volume or of any forces acting on it. However, different from a physical object, the systemic mass of an entity in the sociotechnical service world is totally subjective. For example, the service experience perceived by a customer will surely be enriched if an additional personal touch has been developed in the service provided for the customer.

Let us use a specific study to show the challenges of defining a systemic mass. In addition to a customer's competence and expectation, Johnston (1995) reveals that satisfactory determinants of a banking service can be radically categorized into two types. One is the instrumental determinants that define the performance of technical functions of the service; the other is the expressive determinants that then define the psychological performance of the service. Johnston identifies some determinants of quality predominate over others in the banking industry. The main sources of satisfaction are attentiveness, responsiveness, care, and friendliness, while the main sources of dissatisfaction are integrity, reliability, responsiveness, availability, and functionality.

The inertia of an object is a property of matter in the object by which it remains in its current state unless acted upon by some external force. As soon as there is an external force acted on the object, the rate of change of its current state is directly proportional to the external force but inversely proportional to its mass. On the basis of logical thinking and inductive reasoning, we know that the service experience perceived by a customer has a similar property. We can call it the inertia of service experience. From the earlier discussion, we also understand that the experience of service perceived by a customer remains the same unless acted upon by changed external "force" or effort. The earlier example clearly argues that the attentiveness, responsiveness, care, friendliness, integrity, reliability, responsiveness, availability, and functionality of a banking service are largely measured across the whole banking service system in a subjective manner. The quantity of "matter and mind" in the customer is surely of a systemic nature. We know that the second law of service holds, but we can hardly know how to measure the systemic mass of a customer.

In general, a service satisfies both the utilitarian and sociopsychological needs of a customer; the quantity of "matter" in the customer mind is subjective in nature. It is not difficult to find out that defining a systemic mass for an entity is almost impossible. As an individual entity cannot be precisely measured in a quantitative manner, we should take a different path by looking into its collective nature.

When we look at the services in a collective manner, we find that the approach taken in social sciences could shed us a promising light in finding appropriate methods of studying services in a quantitative manner (Figure 4.3). In many aspects of social sciences, such as psychology, economics, sociology, marketing, drug discovery, and political science, quantitative research as a scientific method has been widely used for many decades. Statistics is the fundamental mathematics that is widely used by social scientists, researchers, and practitioners in conducting quantitative research. For example, based on a big sample of data that are collected over a given study period, we conduct quantitative research using statistical methods to confirm or decipher the causal relationships of certain social phenomena in which we have great interest.

"In the social sciences, quantitative research refers to the systematic empirical investigation of social phenomena via statistical, mathematical or computational techniques. The objective of quantitative research is to develop and employ mathematical

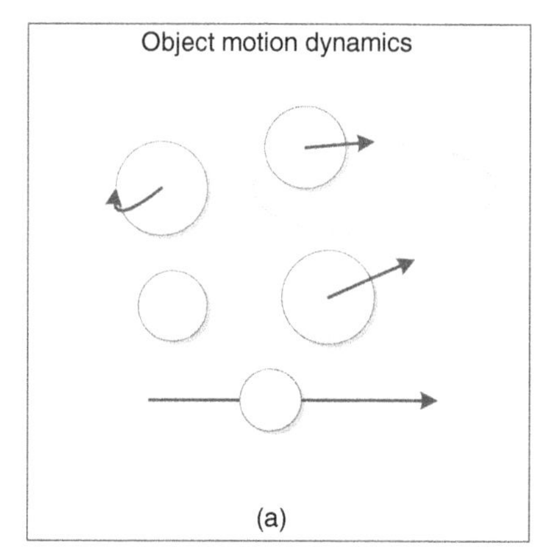

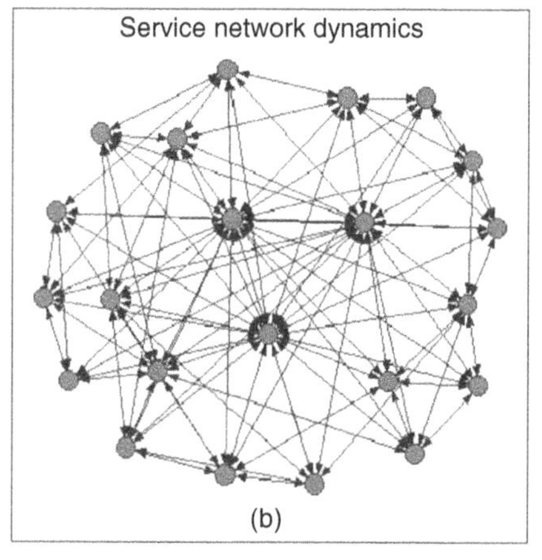

FIGURE 4.3 *The service network dynamics versus the object motion dynamics. (a) Individual object approach and (b) holistic and systemic approach.*

models, theories and/or hypotheses pertaining to phenomena. The process of measurement is central to quantitative research because it provides the fundamental connection between empirical observation and mathematical expression of quantitative relationships. Quantitative data is any data that is in numerical form such as statistics, percentages, etc." "The researcher analyzes the data with the help of statistics. The researcher is hoping the numbers will yield an unbiased result that can be generalized to some larger population. Qualitative research, on the other hand, asks broad questions and collects word data from participants. The researcher looks for themes and describes the information in themes and patterns exclusive to that set of participants" (QRWiki, 2012).

Further, Figure 4.3 shows the fundamental difference between the dynamics of service network and object motion. The motion of an object can be well described; the collectives of minds and technical functions that mainly constitute services and considerably drive corresponding sociotechnical phenomena are highly interactive and subjective. Obviously, the approach to studying individual physical objects cannot be directly applied to studying service experiences.

We know that service is considered as a transformation process in which both provider-side and customer-side people participate in an interactive manner, applying relevant knowledge, skills, and experiences in order to cocreate mutual benefits for the service providers and their customers. Thus, the interactions between service providers and customers are the key to the services performed by the service providers and their customers. Because it is the collection of minds and technical functions in services that mainly constitutes the services, the corresponding sociotechnical services are highly social and interactive. The measured behavior and outcomes are largely subjective in nature.

Therefore, in an effort to further develop the laws of service that are practically applicable for the service industry to meet the needs of today's service-led economy, we must take all the service characteristics we discussed in the previous chapters into full consideration. In other words, we must rethink the laws of service using a systemic or holistic approach, aimed at overcoming the challenges of finding the scientific method to define the systemic mass of a service system or an alternative rule in a similar nature of the second law of service in Service Science. In summary, the following considerations become necessary:

- Services are cocreated by the service providers and customers; the service providers and customers as a whole constitutes a sociotechnical service system. Scientifically assessing services should be done in a collective manner across the service system. In other words, we might evaluate a service when needed; however, only the collection of services provides scientific insights into the service system under study. As a result, instead of studying the service experience perceived by an individual entity using the systemic mass of the studied entity, we should focus on the behaviors of the group of entities under study or the total service experiences perceived by the group using statistic methods that are well adopted in quantitative research.

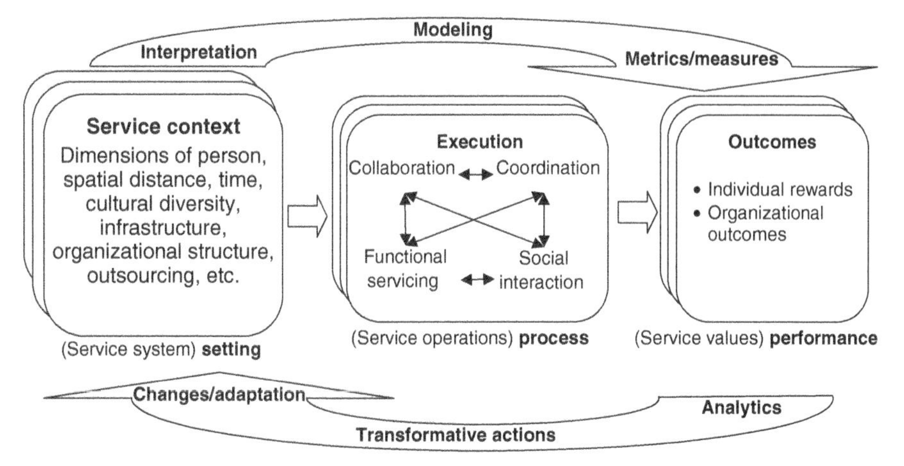

FIGURE 4.4 A qualitative and quantitative approach to interpret a service system throughout its lifecycle. Adapted and revised from Andriessen and Verburg (2004).

- Because services are people-centric, truly cultural and bilateral, holistic or systemic viewpoints are essential. The holistic or systemic viewpoint focuses on the "big picture," interacting relationships, and long-range view of systems dynamics. By paying attention to the whole, a holistic perspective helps us understand and orchestrate service interactions across business domains and throughout the service lifecycle. Figure 4.4 shows our envisioned qualitative and quantitative approach, which must be people-centric, systemic, evolutionary, and iterative (Andriessen and Verburg, 2004; Qiu, 2009). More detailed discussions of this proposed approach will be provided in Chapters 5 and 6.

By relying on quantitative approaches, we can then recompile the laws of service from the earlier discussion:

- *Service's First Law.* If there is no changing effort on a collection of services, then the value V of the collection of services perceived by customers remains the same. V is proportional to or a function of E, that is, $\vec{V} \propto \vec{E}$, or $\vec{V} = f(\vec{E})$.
- *Service's Second Law.* The service efficacy A of a service system is directly proportional to the changing effort E that is applied to the service system, is in the direction of the changing effort, but is inversely proportional to the systemic resistance R of the entity, that is, $A = E/R$. The systemic resistance essentially is the inertia of a service system that can be approximately measured by its relative systemic complexity and/or inefficiency due to the bureaucracy embedded in the service system. Therefore, $\vec{V} = f(\vec{E})$, $\vec{E} = \vec{R} \times \vec{A}$, that is, $\vec{V} \propto \vec{R} \times \vec{A}$, or $\vec{V} = f(\vec{R} \times \vec{A})$.
- *Service's Third Law.* When customers apply an effort E_1 on providers in a service system, the providers simultaneously apply an effort $E_2 = -E_1$ on the customers. This means that E_1 and E_2 are equal in magnitude and opposite in

TABLE 4.2 The Laws of and Service

Law	Service Experience Collectives that are Realistically Gauged by Value	Service Experience
The first law	Inertia due to no changing effort $\vec{V} \propto \vec{E}$, or $\vec{V} = f(\vec{E})$; if $\Delta E = 0$, then $\Delta V = 0$	Inertia due to no changing effort ($\Delta E = 0$)
The second law	$A = E/R$, or $\vec{V} = f(\vec{E}), \vec{E} = \vec{R} \times \vec{A}$, that is, $\vec{V} \propto \vec{R} \times \vec{A}$, or $\vec{V} = f(\vec{R} \times \vec{A})$	$E = MA$
The third law	$E_2 = -E_1$, or $\lvert\vec{V}_1\rvert = \lvert\vec{V}_2\rvert$, $\vec{V}_1 = f(\vec{E}_1)$, $\vec{V}_2 = f(\vec{E}_2)$	$E_2 = -E_1$

direction. When the values of both service providers and customers are used, we have $|\vec{V}_1| = |\vec{V}_2|$, $\vec{V}_1 = f(\vec{E}_1)$, $\vec{V}_2 = f(\vec{E}_2)$.

Table 4.2 highlights the newly compiled laws of service. When we look into the difference we have made in Table 4.2, we know that we are surely heading into the right direction. The defined new relationships that describe how service values change with the changing efforts can be realistically and viably interpreted, defined, and managed in service organizations.

4.2 THE SERVICE ENCOUNTER SOCIOPHYSICS

As discussed earlier, a service is people-centric, truly cultural, and bilateral. From the observations of service rendering in practice and research outcomes from the service community, we know that the type and nature of a service dictates how the service is performed. The service system that offers the service accordingly defines how a series of service encounters could and should occur throughout its service lifecycle. The type, order, frequency, timing, time, efficiency, and effectiveness of the series of service encounters throughout the service lifecycles determine the quality of services perceived by customers who purchase and consume the services (Bitner, 1992; Chase and Dasu, 2008). To a given service system, it is well recognized that the perceived service value and quality by its customers substantially impact the satisfaction and loyalty of the customers in the long run.

The above-mentioned social and transactional interactions in service encounters can be direct or indirect in today's digital and global economy. Face-to-face interactions are direct; virtual interactions, which are mediated through technical applications (i.e., phone, online social media, or self-served device), are indirect (Bitner et al., 1990; Bitner et al., 2000; Svensson, 2004; Svensson, 2006; Meyer and Schwager, 2007; Schneider and Bowen, 2010). Hence, depending on the complexity and nature of an executed service, the social and psychological expectations and needs of a customer at the time that a service encounter occurs can be directly and/or indirectly, slightly or significantly influenced by the experiences of preceding service encounters perceived by the customer or other customers.

For instance, in a service encounter chain, a later encounter is inevitably influenced by the immediately preceding one; employees or customers could also be influenced by other previous encounters that they may have had before if those are somewhat functionally or sociopsychologically related. By relying on conventional wisdom, Ramdas et al. (2012) study four different dimensions in service encounters, including the structure of the interactions, the service boundary, the allocation of service tasks, and the delivery location. They suggest that service providers should define and deliver services by examining interactions among the four dimensions, aimed at creating more mutual values for both customers and providers. However, conventional wisdom is not necessarily true in today's digital and global economy, in which change is the only constant. In order to have a body of knowledge that can be scientifically applied in such a dynamic environment, we need Service Science that can help us foster the planning, design, engineering, delivery and operations of service encounters in a comprehensive and holistic manner.

More broadly, as the velocity of globalization accelerates, the changes and influences are more ambient, quick, and substantial, impacting us as providers or customers in dynamic and complex ways that have not seen before. Service encounters, as a matter of fact, form a time series network in service provision. The understanding of service encounter networks is essential for service providers to be able to design, offer, and manage services for competitive advantage.

The prior lack of means to monitor and capture people's dynamics throughout the service lifecycle has prohibited us from gaining insights into the service encounter chains or networks. Promisingly, the rapid development of digitization and networking technologies has made possible the needed means and methods to change this. Therefore, now is the time for us to rethink service encounters and explore those social and transactional interactions in a deeper and more sophisticated manner than ever before.

4.2.1 The Service Encounter Dynamics of a Service System

Once again, service is an application of relevant knowledge, skills, and experiences to benefit both service providers and customers. To a customer, the encounter of a service or "moment of truth" frequently is the service from the customer's perspective (Bitner et al., 1990). However, from the perspective of a system, service is a process of transformation of the customer's needs utilizing the operations' resources, in which dimensions of customer experience manifest themselves in the themes of a service encounter or service encounter chain. The dynamics of service encounter networks of a service system essentially describes the systemic behavior of the service system in the service engineering and management perspective.

As discussed earlier service encounters derived from human interactions in services are inherently dyadic and collaborative both socially and psychologically (Shostack, 1985; Solomon et al., 1985; Schneider and Bowen, 2010). Service

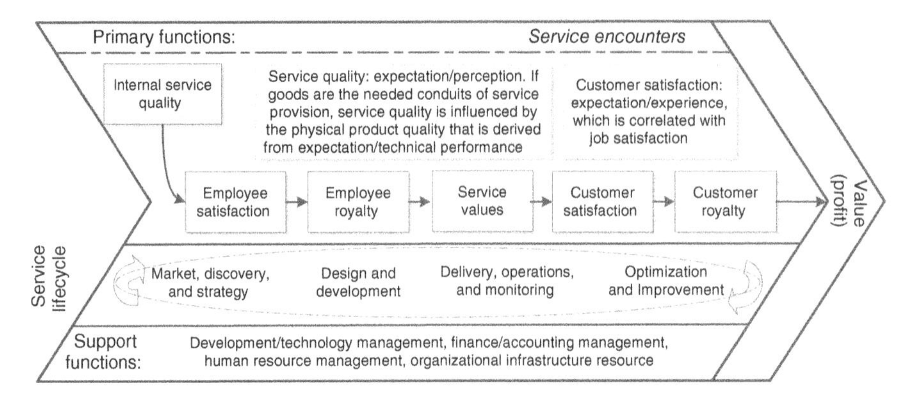

FIGURE 4.5 *The service organizational profit chain.*

encounters involve all interacting activities in the service delivery process, resulting in reciprocal influence between service providers and customers. For example, consumers are the customers in the retailing service sector; students are the customers in educational service systems; patients then are the customers in health care delivery systems. By simply observing the service cocreation processes in our daily life, we are sure that it is the service encounter that enables the necessary manifest function that engages the providers and the customers in order to show the "truth" of service.

Most service businesses understand that the service profit (or value) chain relies on the creation of loyal customers through their experience with excellent service encounters (Figure 4.5) (Qiu, 2013b). Significant research on the service profit chain has revealed the changing, complex, cascading, and highly correlated relationships among job satisfaction, customer satisfaction and loyalty, and business profitability in service firms (Taylor, 1977; Bolton and Drew, 1991; Heskett et al., 1994; Zeithaml, 2000; Lovelock and Wirtz, 2007; Gracia et al., 2010).

The majority of academics and practitioners in the service community uniformly agree that the service encounters between employees and customers are central to the delivery of services (Czepiel et al., 1985; Bitner, 1992; Heskett et al., 1994). Scholars in the service marketing field, in particular, have been interested in the dynamics of service encounters for decades (Solomon et al., 1985; Meyer and Schwager, 2007). At each point of customer interaction, the customer gains experience that essentially is the perceived effect of the fulfillments in both the utilitarian and sociopsychological dimensions (Larson, 1987; Chase and Dasu, 2008; Bradley et al., 2010). Utilitarian needs are met through the performance of the desired service utility functions that are typically defined in specifications (e.g., default general references, signed service agreements), frequently including both technical and functional units. However, meeting the social and psychological needs of customers and employees presents more challenges as the vast array of sociopsychological needs vary with time, duration, servicescape, and an individual's expectation and competency (Bitner, 1990; Bitner, 1992; Bitner et al., 1994; Keillor et al., 2004; Svensson, 2004; Bradley et al., 2010; Qiu, 2013b).

The quality of a firm's services is the overall perception that results from comparing the firm's actual performance with the customers' general expectations of how firms in that industry should perform. Empirical studies affirm the importance of the service encounters in evaluating the overall quality and satisfaction with a firm's service (Parasuraman et al., 1985, 1988; Bitner et al., 1990). The customers' satisfaction is mainly determined by their experience with the service provider; the users' experience, in turn, is their perception based on their experienced service encounters. Because customers are increasingly susceptible to the market changes, competitive service firms must engineer and execute service encounters that consistently yield superior user experience in order to win the allegiance of the time-pressed customers.

Service quality, customer experience, and job and customer satisfaction all are quite subjective. Recent studies have revealed that job satisfaction and customer experience are highly correlated; indeed, in many aspects, they mutually influence each other (Svensson, 2006; Bradley et al., 2010). However, in service encounters, many academics and practitioners over the years have recognized the interactions but have separated the customer and provider perspectives:

- From the customers' perceptive, scholars have developed a body of knowledge primarily in the service marketing arena. These contributions were mainly derived from empirical studies and are related to service quality, service encounters, service failure and recovery, and customer satisfaction and loyalty (Parasuraman et al., 1985; Surprenant and Solomon, 1987; Smith et al., 1999; Meyer and Schwager, 2007). The determinants of service quality surely vary with services (Bitner et al., 1990; Johnston, 1995).
- From the providers' perspective, scholars have also mainly relied on empirical research with a focus on the dynamics of the providers' service delivery systems. A body of knowledge relating to job satisfaction, provider performance, employee job stress, morale and commitment, and their wellbeing has been developed (Taylor, 1977; Bitner, 1990; Bitner, 1992; Bitner et al., 1994; Lloyd and Luk, 2009).

Czepiel (1990) advocates that both customer and provider perceptions should be the focus of any study in service encounters. Svensson (2006) also articulates that service encounters should be explored in a deeper and more sophisticated manner. By considering service encounter successes from both the employee and customer perspectives, Bradley et al. (2010) propose a dyadic psychosocial needs approach to exploring the practical insights into the effective resolution of common service delivery problems. They conclude that there are eight psychological needs in service encounters that require attention: cognition, competence, control, power, justice, trust, respect, and pleasing relations.

Indeed, most of the literature has looked upon service encounters mainly in the service delivery processes. In pursuit of further development of Service Science in this book, we study service encounters more broadly by exploring all stages of the service lifecycle, as service providers now demand the comprehensive understandings of

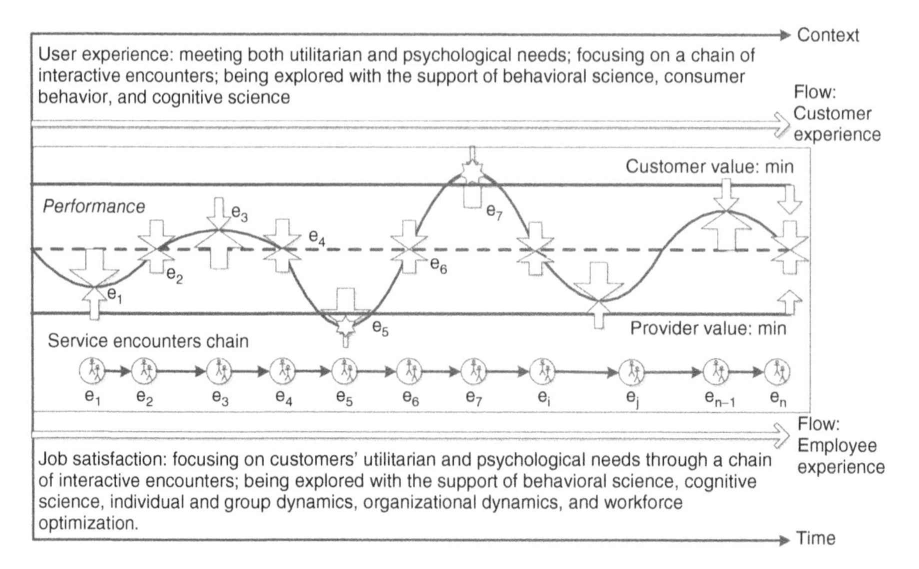

FIGURE 4.6 *The dynamics of service encounters with performance/time/context.*

social and transactional interactions between customers and employees in the areas of service market, design and development, delivery, and operations and management.

In summary, service interactions not only fulfill the manifest function of socioeconomic transactions in general, namely, satisfying the involved entities' material/utilitarian needs, but also their corresponding latent functions, namely, meeting the social and psychosocial needs of employees and customers (Bradley et al., 2010). Figure 4.6 illustrates the dynamics of service encounters in the dimensions of performance, time, and service context or servicescape (Qiu, 2013b). This book looks into the science of service by taking the lifecycle and systemic perspectives. It will be essential and promising for the compiled laws of service to be viably applicable for the governance and guidance of service encounters engineering and management throughout the service lifecycle.

4.2.2 The Laws of Service for Service Encounters

As mentioned earlier, Newton's three laws are useful approximations at the scales and speeds of daily life, in which physical objects move at the much slower speed than light. It is well understood that the first law of motion was used to establish a frame of reference for which the other laws are applicable. In other words, the other two laws of motion are not applicable for use in certain circumstances, such as one when objects travel at the speed of light or higher.

As indicated in Chapter 3, the worldwide service research community has conducted a variety of service studies in many focused areas for several decades, including service operations, marketing, and organizational structure and behavior, and economic transformation. Indeed, a lot of pioneer studies and fruitful research findings have helped the service industry tackle a variety of problems in practice over the years. However, as the service industries continue to evolve fast, service organizations are confronting more and more challenging and new issues in meeting the dynamic changes of services around the world. Emerging and under-researched areas in service have been explored and identified by Ostrom et al. (2010) through collecting viewpoints from over 200 academics across 15 relevant disciplines and from organizations in 32 countries.

More specifically, to spotlight service research priorities by capturing the state of the art and more importantly looking forward to understand and reflect the emerging future needs in academia and practice, Ostrom et al. (2010) spent 18 months collecting and interpreting the viewpoints of academic scholars and leading practitioners around the world. Their great efforts lead to identifying a set of global, interdisciplinary research priorities in the science of service. As the excellent result of their hard work, the following 10 overarching research priorities have been identified:

- *Fostering service infusion and growth*
- *Improving wellbeing through transformative service*
- *Creating and maintaining a service culture*
- *Stimulating service innovation*
- *Enhancing service design*
- *Optimizing service networks and value chains*
- *Effectively branding and selling services*
- *Enhancing the service experience through cocreation*
- *Measuring and optimizing the value of service*
- *Leveraging technology to advance service*

Note that their hard work confirms that the science of service has become exceedingly broad, interdisciplinary, and cross-functional. We understand that their intention was to call for swift action to study the emerging service research topics from academics and practitioners in the service community. Their findings are definitely subjective because a qualitative research rather than quantitative study method was radically applied during the compilation process. However, their interesting results are of great value, undoubtedly sparking discussion and spurring thinking about service research areas worldwide.

This book mainly takes quantitative approaches to address service issues in a quite unique way. Although it seems that this book covers several priorities in the above-mentioned list, we do understand that it is impossible that the above-compiled laws of service are universally applicable. Therefore, like Newton's laws of motion, we should establish a frame of reference in a similar way to make sure that the laws of service are applicable for the established reference frame. As this book mainly explore service encounters, we should further develop the laws of service and make sure they are truly applicable for the investigations that we will continue to explore and present in the following chapters.

We as the service provider know that customer interactions go beyond the service delivery processes and must include all influential in-person or virtual contacts during the design, development, and preparation of service encounters. In other words, we now fully understand that customers consume and perceive services through a list of service encounters that truly occur throughout the service lifecycle. Therefore, service research to gain a perspective of collective service and systemic behavior must focus on the dynamics of service systems across the service lifecycle. Although qualitative research certainly helps to identify certain managerial guidance in service engineering, operations, and management, quantitative research is certainly the best option in providing the insights into the dynamics of service with scientific rigor when service explorations must be carried out by taking the lifecycle and systemic perspective of service.

Please keep in mind, interactions between customers and providers can occur in different ways, face-to-face or virtually, directly or indirectly. As discussed earlier, personifying self-serving units thus becomes essential for a service provider to provide empathic and pleasing supports in service provision. As a result, the service provider can avoid creating apathy and negativity that people might feel when self-services are radically supplementary under some circumstances.

Furthermore, the value and service experience must be calculated in a collective, personifying, and systems manner. Theoretically and practically, the laws of service that are finally presented in this book should be applicable for describing the rules and guidelines for service interaction-oriented service networks (Figure 4.3b). Similar to econophysics with a root in statistical physics using probabilistic and statistical methods, sociophysics (Galam et al., 1982; Galam, 2008) deals with five main subjects of modeling, including democratic voting in bottom up hierarchical systems, decision making, fragmentation versus coalitions, and terrorism and opinion dynamics. In Service Science, we believe that the study of the dynamics of service systems shall develop service-focused sociophysics further, in which core variables must be definable and measurable quantities, directly or indirectly.

As we focus on the fundamental laws of sociophysics in service, we compile the laws of service for service interaction-oriented service networks as follows:

- *Service Encounter's First Law (A Frame of Reference—Systemic Inertia).* The human interactions that essentially create and drive the process of value cocreation in service characterize a service system. If there is no changing effort E (i.e., a systemic disruption force) on the collective service and systemic behavior of the service system, then the value V converted from the collective service experience perceived by customers and providers remains the same. V is proportional to or a function of E, that is, $\vec{V} \propto \vec{E}$ or $\vec{V} = f(\vec{E})$. If respective values from customers and providers are measured, we have $|\vec{V}_1|$ and $|\vec{V}_2|$, where $\vec{V}_1 = f(\vec{E}_1)$ and $\vec{V}_2 = f(\vec{E}_2)$. Therefore, we have $\vec{V} = \vec{V}_1 \cup \vec{V}_2 \propto \vec{E} = \vec{E}_1 \cup \vec{E}_2$, or $\vec{V} = f(\vec{E})$.
- *Service Encounter's Second Law (Systemic Disruption).* The service efficacy A of a service system is directly proportional to a systemic disruption force E, which is applied to the service system, is in the direction of the systemic disruption force, but is inversely proportional to the systemic resistance R of the

service system, that is, $A = E/R$. The systemic resistance essentially is the systemic inertia of a service system that can be approximately measured by its relative systemic complexity and/or inefficiency due to the bureaucracy embedded in the service system. In theory, the second law holds regardless of the constituent units. However, the quantity value of the systemic resistance of a service system is always measured collectively by including both customers and providers involved in the service system. In other words, as the dynamics of service networks must be described by the collective service and systemic behavior of the interactions between customers and providers, the variables in the second law of service for service encounters are always measured in a collective manner. Therefore, we have $\vec{V} = f(\vec{E})$, $\vec{E} = \vec{R} \times \vec{A}$, that is, $\vec{V} \propto \vec{R} \times \vec{A}$, or $\vec{V} = f(\vec{R} \times \vec{A})$ for the service system as a whole.

- *Service Encounter's Third Law (Value Cocreation or Mutuality).* When the customers apply an effort E_1 on the providers in a service system, the providers simultaneously apply an effort $E_2 = -E_1$ on the customers. This means that E_1 and E_2 are equal in magnitude and opposite in direction. When the values of both service providers and customers are used, we have $|\vec{V}_1| = |\vec{V}_2|$, $\vec{V}_1 = f(\vec{E}_1)$, $\vec{V}_2 = f(\vec{E}_2)$. Although the meanings and numbers in the measured values are surely different between the customers and providers, their socioeconomic values when converted must be comparatively equivalent.

Once again, we understand that the science of service relies on both quantitative and qualitative approaches. Many academics and practitioners have pioneered many managerial guidelines and methodologies to help the service industry to compete. However, the above-presented fundamental laws of service help us frame and investigate service issues in a unique way. As readers continue to read the remaining chapters, readers will find that the above-compiled laws of service for service interaction-oriented service networks lay a sound and solid foundation for this book in support of the further study of the needed quantitative and qualitative approaches to address the emerging and challenging service issues.

4.2.3 Service as a Value Cocreation Process: From SMART to SMARTER

In reality, the value of a service is the summation of socioeconomic benefits that providers and customers have accumulated along with the service process trajectory throughout the service lifecycle. Thus, it largely depends on the systemic behavior of the sociotechnical system that offers the service. The service-led economy is now facing the grand challenges of addressing green consumerism, pervasive and ubiquitous digitalization, and accelerated globalization. Note that the service-led economy surely conceptualizes a quantitative service value increase in the percentage of gross domestic product (GDP). Moreover, it indicates a substantive shift in which the services sector must become a main driving engine for future growth and innovation.

To stay competitive in the twenty-first century, service organizations have to rethink their business operations and organizational structures by focusing on *people* while leveraging the technology (e.g., implementing a novel approach to overcome geographic, social, and cultural barriers to cultivate and enhance the cultures and mechanisms of cocreation, collaboration, and innovations). This will ensure the prompt and cost-effective production and delivery of quality products and competitive and satisfactory services to customers (Qiu, 2012).

To stay competitive, service organizations understand that they must make their serving processes efficient and cost-effective. To improve their business competitiveness, they should explore a variety of flows that can highly reflect their service status quo in operations (Figure 4.7). These flows include customer experience, organizational behavior, physical support, and information assistance. A customer experience flow focuses on meeting both the utilitarian and psychological needs. An organizational behavior flow focuses on employees' job satisfaction by meeting the customers' utilitarian and psychological needs through enabling a chain of interactive and positive encounters. A physical flow deals with the conduits of service provision. An efficient and effective physical flow should provide employees and customers with the right tools, servicescape, and other necessary resource supports to facilitate the service encounters in meeting both utilitarian and psychological needs of customers while improving job satisfaction. The information assistance flow then aims to capture the right data/information in a timely manner and support the operational and managerial needs in a more intelligent way across the service lifecycle.

Lacking appropriate tools and methodologies, these highly correlated flows have been separately studied over the years. As discussed earlier, the fast development of digitization, computing, and networking technologies has made possible the needed

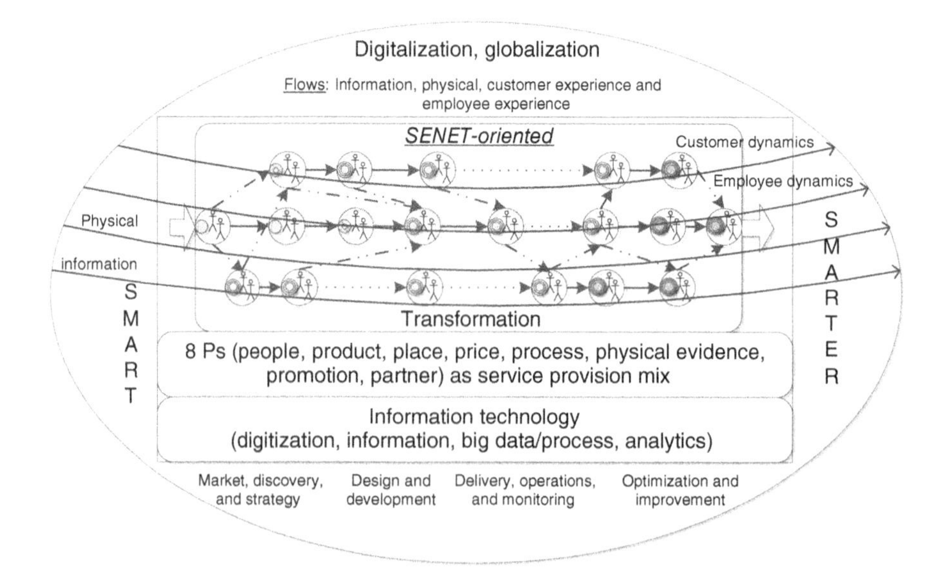

FIGURE 4.7 *SENET-oriented sociotechnical process-driven system.*

means and methods to change this. It is surely the time for us to rethink service encounters and explore those social and transactional interactions in a deeper and more sophisticated manner than ever before. As illustrated in Figure 4.7, only when these flows are fully explored in an integrated and collaborative manner, we can operate service systems in a competitive way, making service networks "SMART" and even "SMARTER" over time.

The mnemonic "SMART" is popularly used to guide people to set goals and objectives in practice (Doran, 1981). The common terms behind the letters of "SMARTERS" change from one user, either academic scholar or practitioner, to another, depending on the situation in which the mnemonic is actually applied. However, academics and practitioners have quite often focused on the common characteristics of "SMARTERS," referring to "specific, measurable, attainable, relevant, timely, evaluative, rewarding, and satisfactory." As compared to the mnemonic "SMART" that are popularly used to guide people to set goals and objectives in practice, we focus on investigating quantitative approaches by applying these identified characteristics in the models discussed in the remaining chapters of this book.

As indicated in Figure 4.7, virtual, indirect, and other kinds of service encounters should be fully explored and included in the service encounter research. Not just physical and information flows, customer experience and job satisfaction must also be included at the same time when service profit chains are analyzed, planned, and managed. Ultimately, we can efficiently and effectively monitor, capture, and analyze service encounter networks. Therefore, "SMART," "SMARTER," and "SMARTERS" are equivoques, which are truly expressions with two meanings. When we explore service systems by well adopting "specific, measurable, attainable, relevant, timely, evaluative, rewarding, and satisfactory" rules, real smart service operations can be fully engineered and delivered over time.

To have smart service operations, we have to make sure that each service encounter in service encounter networks is specific, measurable, attainable, relevant, and timely to both service providers and customers. Aligning service activities with corresponding service capability, competence, and necessary physical and information supports from both service providers and customers are thus essential for executing smart service operations. For example, Bradley et al. (2010) confirm that actions in a series of service encounters should be specifically planned, designed, and operated. In case that some of planned actions or events are not well performed, relevant remedy actions or changes should be pinpointed in real time to recover the unsatisfactory service to some extent or for the purpose of further improvement (Tax and Brown, 1998).

Indeed, service interactions must meet respective expectations from both service providers and customers although the expectations and then perceptions are truly different. Because of the existence of differences, the effectiveness of service encounters must be evaluated using a range of different metrics. For instance, we might use the following different metrics to understand their respective effectiveness of service encounters:

- *To the Customers*. Service time is the time that should be taken to complete a certain function or task. We could use the following measures to evaluate service time: mean service time, mean waiting time, etc.
- *To the Provider*. Mean waiting time, server utilization, throughput, and average number of customers waiting are frequently used to evaluate the efficiency of services. In addition, we might use mean time between failures (MTBFs) and mean time to recover (MTTR) to determine the reliability of services. MTBF is the mean time that a service might fail; MTTR is the mean time that should be taken to recover after a service failure.

To understand if service encounters meet the needs of both providers and customers, we have to evaluate their respective outcomes perceived from service encounters in service. Similarly, the satisfaction of service encounters must be evaluated using a range of different measures and metrics. For instance, we might use the following metrics to understand the satisfaction of service encounters in a customer perspective.

- *Utilitarian or Functional Requirements*. Functions or tasks should be performed based on signed contracts or default agreements and references. The contracts can be in the form of certain service-level agreements. To individual customers, default or implied service tasks might be performed, and general service references could be applied to manage the process of transformation and accordingly measure the service utilitarian outcomes.
- *Sociopsychological Needs*. The feelings that the customer has had during these service encounters. Pleasing and positive feeling perceived in service encounters frequently result in that the provided services excel in values that are simply evaluated based on utilitarian requirements.

Then, we could use the following metrics to understand the satisfaction of service encounters in a provider perspective.

- *Utilitarian or Functional Requirements*. This should be the same as one that is evaluated by the customers although the values generated from meeting these requirements are surely different.
- *Socioemotional Needs*. The feelings that the employee have had during these service encounters. As job satisfaction is highly correlated to customer satisfaction, we must ensure that the right tools, setting, and supports are in place to make employees appropriately empowered, capable, and cozy to serve the customers at point of service.

Surely a SENET-oriented sociotechnical process-driven system is effective and competitive only if all the necessary needs of customers can be fully met throughout the service lifecycle.

4.3 SERVICE SCIENCE: A PROMISING INTERDISCIPLINARY FIELD

In the service sector, what really matters in service offering comes to the happiness and satisfaction in service perceived by customers and providers throughout the service lifecycle. Surely, the determinants of service quality vary with services. For instance, the determinants in the banking industry, both satisfying and dissatisfying, primarily include attentiveness/helpfulness, responsiveness, care, availability, reliability, integrity, friendliness, courtesy, communication, competence, functionality, commitment, access, flexibility, aesthetics, cleanliness/tidiness, comfort, and security (Johnston, 1995). Those determinants can be radically categorized into two types. One is the instrumental determinants that define the performance of utilitarian or technical functions of the service; the other is the expressive determinants that then define the psychological performance of the service. Using an empirical study, Johnston identifies some determinants of quality predominate over others in the banking industry. The main sources of satisfaction are attentiveness, responsiveness, care, and friendliness, while the main sources of dissatisfaction are integrity, reliability, responsiveness, availability, and functionality.

We have discussed earlier that empirical examinations of service value drivers include a variety of factors throughout the service profit (or value) chain (Figure 4.5). We know that internal service quality, job satisfaction, customer perceived service quality, customer satisfaction and loyalty, and profitability in a service organization are highly correlated. In particular, employees' competence and their feeling of empowerment substantially influence the outcomes of service quality, accordingly impacting on customer satisfaction and loyalty (Heskett et al., 1994; Loveman, 1998; Durvasula et al., 2005). To a service organization, a virtuous circle is essentially what makes it competitive in the long run (Figure 4.8) (Heskett et al., 1990; Chesbrough, 2011b). As the primary function of service is to engineer, execute, and support service encounters, we must rethink service encounters and find the scientific ways to build and manage people-centric, information-enabled, cocreation-oriented, and innovative service organizations with a focus on the dynamics of service encounters throughout the lifecycle of service.

It becomes clear that both service providers and customers are interweaved in the process of service transformation, in which dimensions of customer experience manifest themselves in the themes of service encounters. Service encounters or service networks are thus dynamic, collective, and evolutionary in nature. Because their cocreation relationships become the general characteristics of the modern services, the consistent sensing, interaction, and creativity from customers' feedbacks, participations, or consumptions throughout the lifecycle of service play a pivotal role in service provision (Drogseth, 2012). The complexity in light of dynamics of systemic behavior in a service organization throughout the service profit chain (Figure 4.8) clearly shows that service encounter networks must be explored in an integrated manner with scientific rigor. It is the virtuous cocreation-oriented innovation circle that makes a service value chain competitive and profitable (Chesbrough, 2011a).

Quantitative research of service relies on methods and tools to capture and understand the dynamics of service networks in the setting described in Figure 4.8.

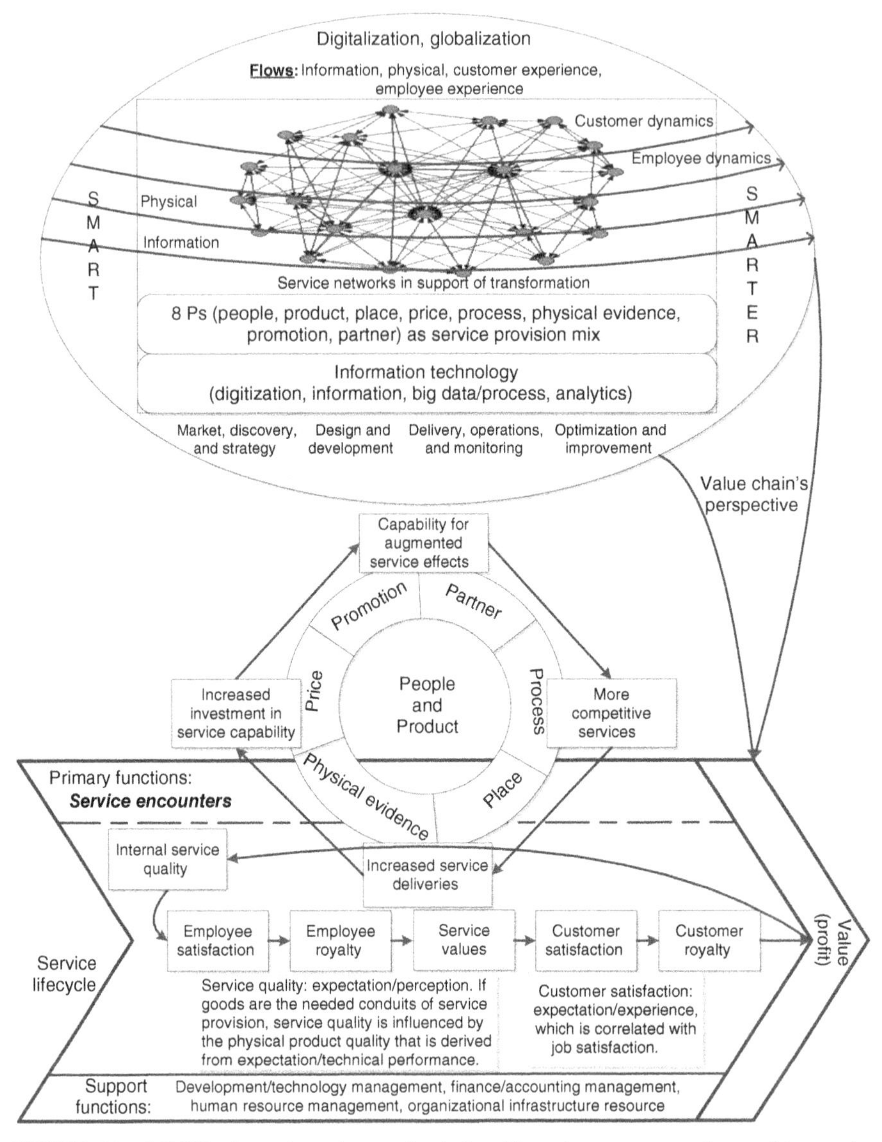

FIGURE 4.8 *SENET-oriented service profit chain with a virtuous cocreation and innovation circle.*

Chase and Dasu (2008) argue that behavioral science offers new insights into better service management. For instance, we should design and manage service encounters with a help of the understandings of how people experience social interactions, what biases they bring to affect their feelings, and how cognitive competency and experiences impact their storing memories. According to Chase and Dasu (2001, 2008), we have to consider the key points in both utilitarian and sociopsychological dimensions when service encounter networks are analyzed, so those exceedingly critical moments of truth can be well planned, designed, and managed.

We must make customer experience enjoyable and memorable. The reality is that perception is what really matters in customer experience. However, service quality, customer experience, and job and customer satisfaction all are quite subjective (Chase, 1978). Fortunately, over the years psychologists and cognitive scientists have poured tremendous efforts in interpreting human behaviors with fascinating outcomes that help us to understand the complex process of the formation of human's perceptions. For instance, we are surely concerned with the following perception perspectives in service (Chase and Dasu, 2001; Chase and Dasu, 2008)

- *Sequence Effects.* People typically conclude their assessments of feelings after a series of encounters based on the trend in the sequence of pleasure, a few significant moments, and the ending. People usually prefer to enjoy the steady and sequential improvement, fast response, and quick recovery to an unsatisfied event. The uptick and happy ending has been always the most intriguing experience in all encounters in service.
- *Duration Effects.* The perceptions of time's passage are surely subjective. For a service encounter, people largely appreciate the experienced pleasurable content and how the service encounter is arranged rather than how long it takes. It is typical that people will not notice how long it takes when they are pleasurably and mentally engaged in the service encounter. Of course, people's expectation of the length of processing time will impact their evaluation of the duration.
- *Rationalization Effects.* "The more empowered and engaged they feel, the less angry they are when something goes wrong." "[P]eople want explanations, and they'll make them up if they have to. The explanation will nearly always focus on something they can observe—soothing that is discrete and concrete enough to be changed in their if-only fantasies" (Chase and Dasu, 2001, p. 81).
- *Perceived Control.* The customer's comfort and pleasure in a service encounter are positively influenced by the responsibility the customer has during the service encounter.

We should optimally design and influence customers' reactions to the sequence of service encounters, properly arrange and control the segments of each service encounter, appropriately adopt the mechanisms to help educate, empower, comfort, and engage the customers, and turn over certain control to the customers to improve their engagements in a positive way whenever possible (Qiu, 2013b).

As discussed earlier, in order to help service organizations offer competitive services to customers, academics and practitioners worldwide have conducted service research in focused areas for many decades. Endeavors in certain focused fields have a long history, resulting in a lot of excellent work from many constituent parts of the study of services (Larson, 2011). Rigorous methodologies and frameworks allow us to conduct advanced descriptive and prescriptive research of a service spanning its lifecycle in an integral and quantitative manner. With the support of systems theory, marketing science, operations research, management science, advanced computing,

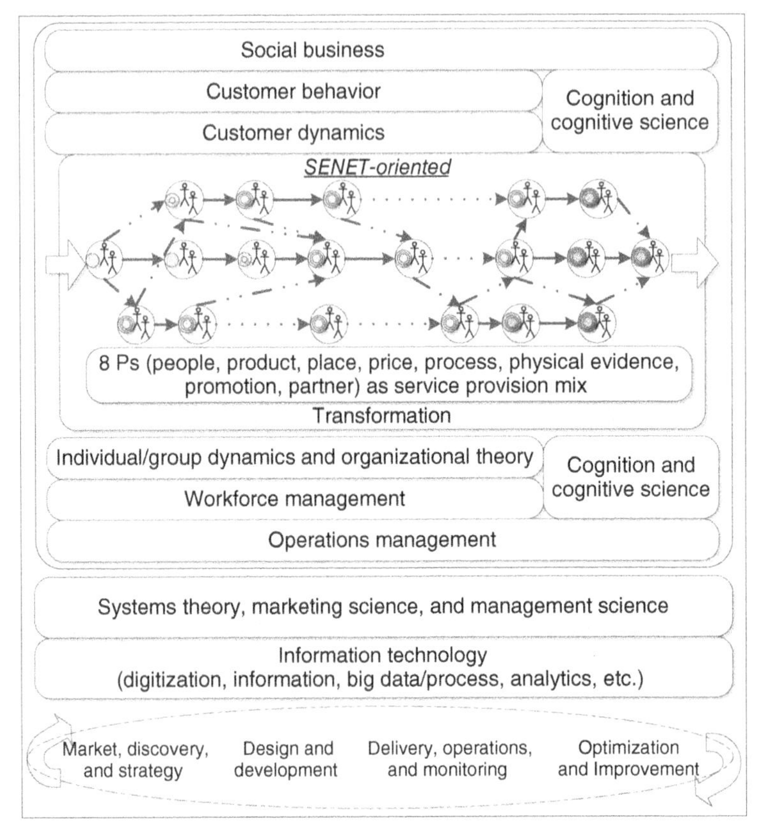

FIGURE 4.9 *Service Science as a promising interdisciplinary field.*

and communication technology, recent advancements in (online) social networks, informatics, and analytics, and others, the means do exist (Figure 4.9).

As discussed earlier, for any given phase in the service lifecycle, the literature has shown many outstanding works. For a given service, its technical requirements are most likely materialized in its service-bundled products with the support of the needed operational resources. Indeed, how service encounters substantially impact perceived service quality in a variety of dimensions in perspectives of the service providers, customers, or both has been well studied (Parasuraman et al., 1988; Bitner, 1990; Bitner, 1992; Bradley et al., 2010). However, these human interactions, in general, have not well explored by considering them throughout the service lifecycle in an integrative and collaborative manner. In other words, how these intensive and broad interactions at one phase impact the other phases of service provision have been largely ignored so far in service research. We must rethink service encounters by integrating these human interactions into a service encounter chain, from beginning to end across the service lifecycle, so that we can look into services defined in this book using a holistic, systems, and integrative approach (Qiu, 2013b).

Service Science is a promising interdisciplinary field. On the basis of the laws of service for service encounters, we can view the systems behavior of a service organization as the dynamics of cocreation-oriented service networks (Figure 4.10). The service encounter networks evolve over time, which must be evaluated, managed,

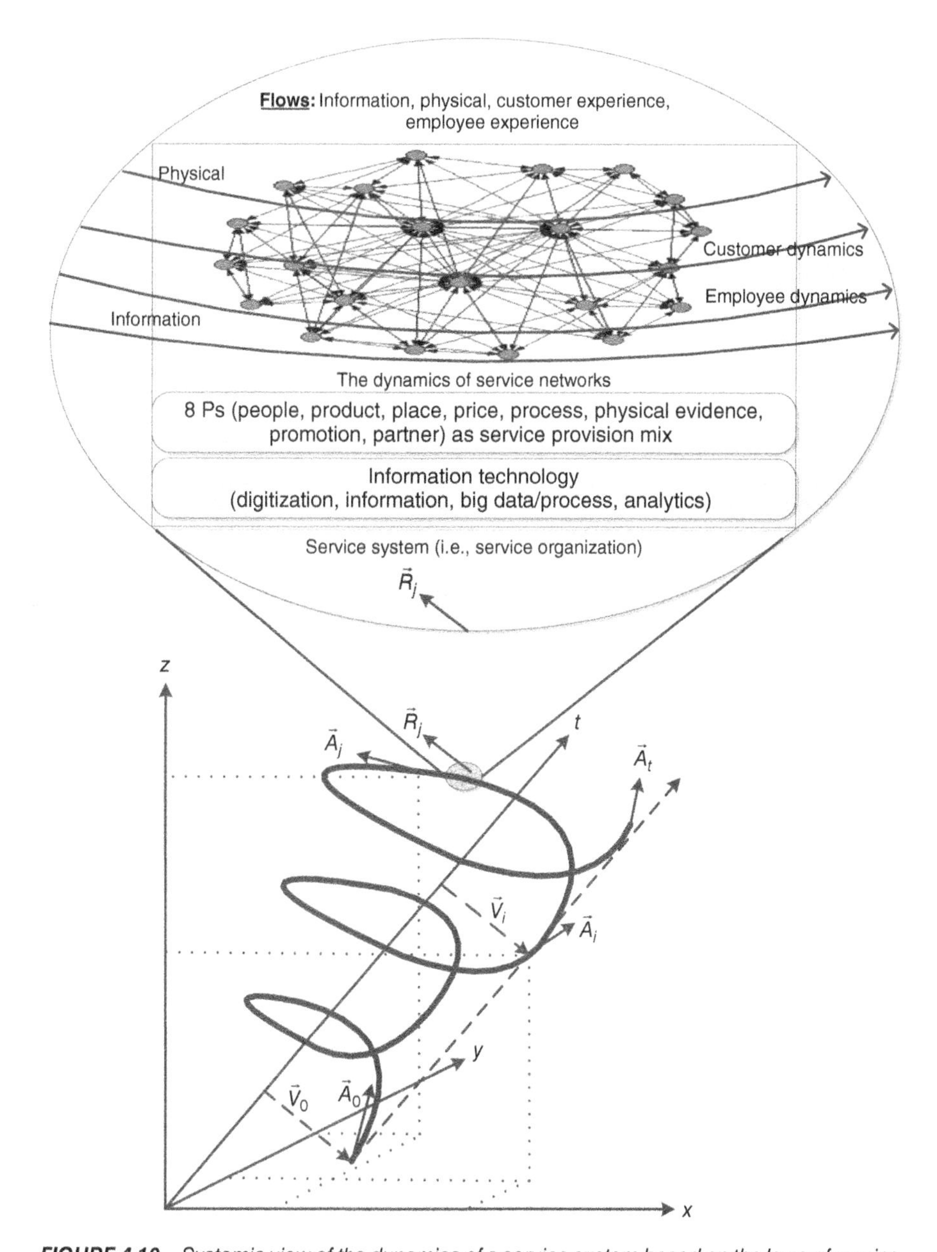

FIGURE 4.10 *Systemic view of the dynamics of a service system based on the laws of service.*

and controlled in a real-time and quantitative manner in order to stay competitive. To avoid platitudinous ponderosity, this book takes the following "SMART" points into consideration to develop the science of service.

- Service itself is a value cocreation process, being centered at service encounters. An interaction approach is truly (Web) 2.0 thinking—making everything truly collaborative because each service encounter surely engages both providers and customers, directly or indirectly, physically or virtually. The service "truth" is then revealed, ensuring the delivery of the value of the service.
- Holistic or systemic approach to exploring service systems and/or service encounter networks is essential for the laws of service to hold. Because of the systemic perspective adopted in study, a systemic resistance becomes potentially measurable and accordingly meaningful in a sociotechnical service world.
- By taking advantage of ubiquitous and pervasive computing, we can truly put people first, aimed at capturing and collecting real-time data on people's activities in service.
- Computational thinking plays a critical role in abstracting sociotechnical behavior of service encounter networks.
- Quantitative research is the key to interpret the systemic behavior of service encounter networks; qualitative research should also be taken into account when quantitative research is not feasible. Indeed, an integrated model of service encounter networks by fully leveraging both quantitative and qualitative researches is the appropriate approach that should be adopted in the science of service.
- Social capital has been long recognized as an important resource, which considerably complement to human capital in the modern society. Service is centered with people, varying with its social context and societal setting. As the globalization and digitalization continue to accelerate, understanding and leveraging social capital in the sociotechnical service world becomes more and more critical than ever before (Burt, 2000).
- Service encounter networks are indeed social and economic networks, rapidly evolving along with the evolution of technologies and societies (Jackson and Watts, 2002). To build customer loyalty, we must ensure that total quality service can be delivered to customers. Hence, we must examine the dynamic formation and stochastic evolution of service encounter networks, spanning the service lifecycle. The structures of service encounter networks significantly influence the dynamics of services that are engineered and managed by a service organization. Ultimately, a service organization can truly operate much smarter service businesses than its competitors by appropriately applying "SMARTER" to its all service encounter networks.

More specifically, we need to begin to rethink service encounters or service networks in general. In particular, there is an opportunity to explore in a comprehensive and holistic way the following four flows in the service profit chain (Figure 4.11):

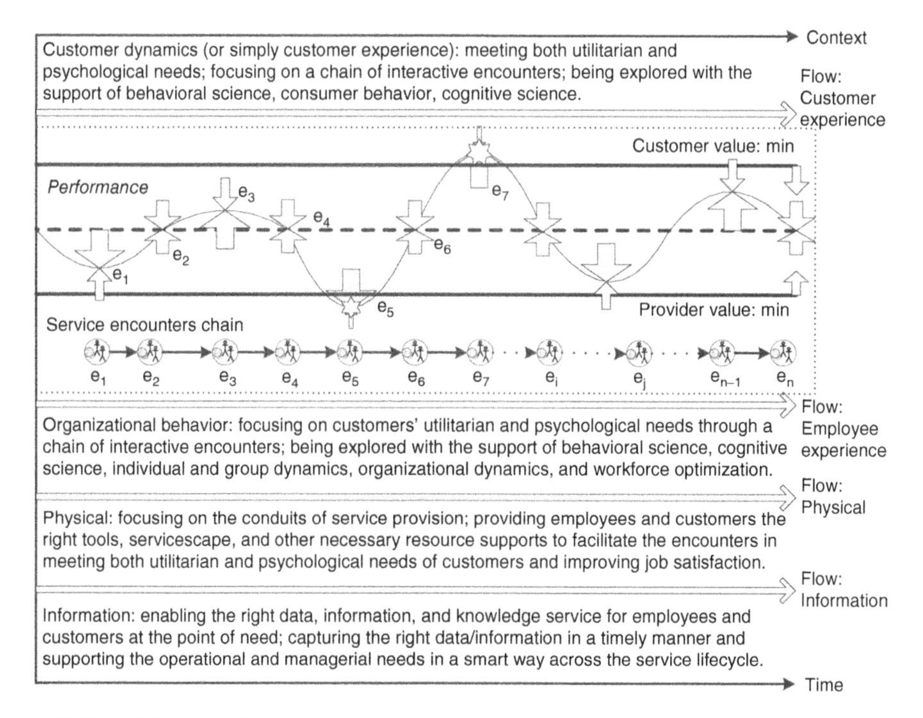

FIGURE 4.11 *Four interdependent and essential flows in support of service encounters.*

- *Customer Dynamics (Or Simply Customer Experience) Flow.* Meeting both the utilitarian and psychological needs of customers and focusing on a chain of interactive service encounters. Customer dynamics flow must be explored with the support of behavioral science, consumer behavior, and cognitive science to offer and truly create an excellent customer experience.
- *Organizational Behavior Flow.* Focusing on employees' job satisfaction by meeting the customers' utilitarian and psychological needs through enabling a chain of interactive and positive service encounters. Organizational behavior flow must be explored with the support of behavioral science, cognitive science, individual and group dynamics, organizational dynamics, operations management, and workforce optimization to improve job satisfaction and organizational behavior.
- *Physical Flow.* Focusing on the conduits of service provision. An efficient and effective physical flow should provide employees and customers with the right tools, servicescape, and other necessary resource supports to facilitate service encounters in meeting both utilitarian and psychological needs of customers while improving job satisfaction.
- *Information Flow.* Capturing right data/information in a timely manner and supporting the operational and managerial needs in a more intelligent way across the service lifecycle. An optimal information flow should enable the right data,

information, and knowledge service for employees and customers at the point of need.

Surely a well-defined and more advanced Service Science would better facilitate service engineering and management across service value-added networks, spanning all the areas in service provision from engineering and managing service marketing, design, creation, quality, compliance, operations, to innovation throughout the lifecycle of service. Capable and competitive service systems should be highly adaptable and sustainable to their service environment (when, where and who to deliver, and whom to be served, etc.). Hopefully, the articulated points in this book would help draw much more attention from scholars, managers, engineers, practitioners, and policy makers who are interested in service research, education, and practice around the world.

4.4 FINAL REMARKS

Note that people who are centered at the service production and consumption process in service offering and delivery have personal traits in the physiological and psychological perspectives, different cognitive abilities, and unique sociological constraints. It has been exceedingly challenging for the service research community to present the world of methods and tools that can be well applied to modeling and exploring people's behaviors in service. Hence, Service Science must welcome the contributions to the development of service theories and principles that can be applied in effectively managing and controlling systemic behavior, leveraging sociotechnical effects, and stimulating innovations throughout the service lifecycle (marketing, design and engineering, operations, delivery, benchmarking, and optimization for improvement).

In general, with the foundation of systems theory, operations research, management science, marketing science, advanced computing and communication technology, network theory, social computing, and analytics, Service Science is in need of descriptive and prescriptive research of a service spanning its lifecycle (i.e., market analysis, design, engineering, delivery, and sustaining) in an integral and quantitative manner. Service Science as a metascience of services should build on predecessors' excellent work from many of the above-mentioned disciplines. Like many constituent parts of the study of services, Service Science must follow the scientific method, and must be rigorous and scholarly (Larson, 2011; Qiu, 2012).

As discussed earlier, in general, advanced descriptive and prescriptive Service Science research surely relies on the continual development of systems theory, operations research, management science, marketing science, advanced computing and communication technology, network theory, social computing, and analytics. However, this book takes an innovative and unique approach to contribute to the development of Service Science (Figure 4.12). We take a holistic view of the service lifecycle and explore the real-time dynamics of service systems and networks. By defining service as a cocreation transformation process, we holistically analyze the performance of service systems that enable and execute complex and heterogeneous service processes.

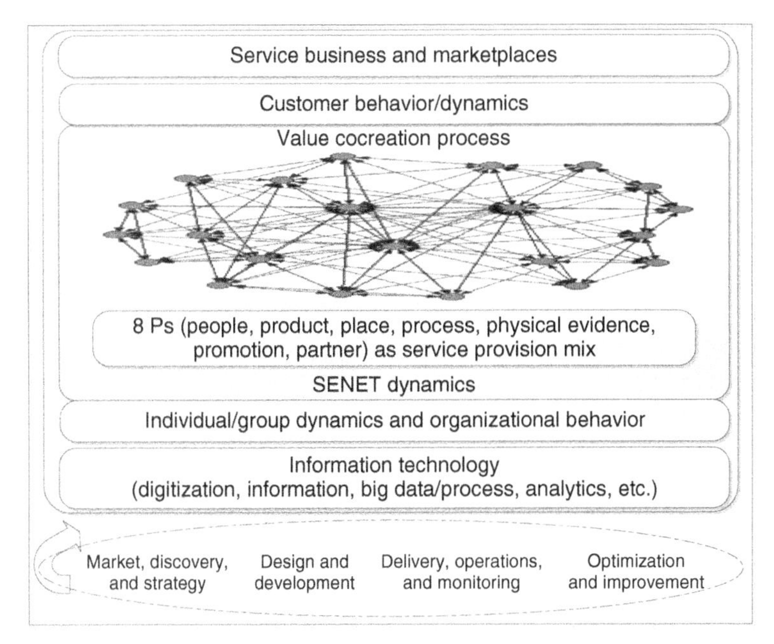

FIGURE 4.12 *A holistic and integrated approach to contribute to the development of Service Science.*

By leveraging the advances in computing and network technologies, social science, management science, and other relevant fields, we present the concept and principles of putting people first in service. We demonstrate that service networks in light of service encounters can be comprehensively explored in a closed-loop and real-time manner with the help of the advanced computing methods and tools, aimed at helping service organizations understand and capture market trends, design and engineer service products and delivery networks, operate service operations, and control and manage the service lifecycles for competitive advantage.

In the later chapters of this book, we will use case studies to demonstrate how different models of service systems and networks can be applied well and with scientific rigor to engineering and managing services in service organizations in an effective and efficient manner.

REFERENCES

Andriessen, J. E., & Verburg, R. M. (2004). A Model for the Analysis of Virtual Teams. Chapter XV in *Virtual and Collaborative Teams: Process, Technologies and Practice*, ed. by S. Godar and S. Ferris. Hershey, PA: Idea Group Publishing.

Bitner, M. J. (1990). Evaluating service encounters: the effects of physical surroundings and employee responses. *Journal of Marketing*, 54(2), 69–82.

Bitner, M. J. (1992). Servicescapes: the impact of physical surroundings on customers and employees. *Journal of Marketing*, 56(2), 57–71.

Bitner, M. J., Booms, B. H., & Mohr, L. A. (1994). Critical service encounters: the employee's viewpoint. *The Journal of Marketing*, 58, 95–106.

Bitner, M. J., Booms, B. H., & Tetreault, M. S. (1990). The service encounter: diagnosing favorable and unfavorable incidents. *The Journal of Marketing*, 54, 71–84.

Bitner, M. J., Brown, S. W., & Meuter, M. L. (2000). Technology infusion in service encounters. *Journal of the Academy of marketing Science*, 28(1), 138–149.

Bolton, R. N., & Drew, J. H. (1991). A multistage model of customers' assessments of service quality and value. *Journal of Consumer Research*, 17(4), 375–384.

Bradley, G. L., McColl-Kennedy, J. R., Sparks, B. A., Jimmieson, N. L., & Zapf, D. (2010). Service encounter needs theory: a dyadic, psychosocial approach to understanding service encounters. *Research on Emotion in Organizations*, 6, 221–258.

Burt, R. S. (2000). The network structure of social capital. *Research in Organizational Behavior*, 22, 345–423.

Chase, R. B. (1978). Where does the customer fit in a service operation? *Harvard Business Review*, 56(6), 137–142.

Chase, R. B., & Dasu, S. (2001). Want to perfect your company's service? Use behavioral science. *Harvard Business Review*, 79(6), 78–85.

Chase, R. B., & Dasu, S. (2008). Psychology of the Experience: The Missing Link in Service Science in *Service Science, Management and Engineering Education for the 21st Century*, 35–40, eds. by B. Hefley and W. Murphy. US: Springer.

Chesbrough, H. W. (2011a). *Open Services Innovation: Rethinking Your Business to Grow and Compete in a New Era*. San Francisco, CA: Jossey-Bass.

Chesbrough, H. W. (2011b). Bringing open innovation to services. *MIT Sloan Management Review*, 52(2), 85–90.

Czepiel, J. A. (1990). Service encounters and service relationships: implications for research. *Journal of Business Research*, 20(1), 13–21.

Czepiel, J. A., Solomon, M. R., & Surprenant, C. F. (1985). *The Service Encounter: Managing Employee/Customer Interaction in Service Business*. Lexington, MA: Lexington Books.

Dietrich, B., & Harrison, T. (2006). Serving the services industry. *OR/MS Today*, 33(3), 42–49.

Doran, G. T. (1981). There's a SMART way to write management's goals and objectives. *Management Review*, 70(11), 35–36.

Drogseth, D. (2012). User experience management and business impact—a cornerstone for IT transformation. *Enterprise Management Associates Research Report*. Retrieved Oct. 10, 2012 from http://www.enterprisemanagement.com/research/asset.php/2311/.

Durvasula, S., Lysonski, S., & Mehta, S. C. (2005). Service encounters: the missing link between service quality perceptions and satisfaction. *Journal of Applied Business Research*, 21(3), 15–25.

Galam, S. (2008). Sociophysics: a review of Galam models. *International Journal of Modern Physics C*, 19(3), 409–440.

Galam, S., Gefen, Y., & Shapir, Y. (1982). Sociophysics: a new approach of sociological collective behaviour. I. mean-behaviour description of a strike. *Journal of Mathematical Sociology*, 9(1), 1–13.

Gracia, E., Cifre, E., & Grau, R. (2010). Service quality: the key role of service climate and service behavior of boundary employee units. *Group & Organization Management*, 35(3), 276–298.

Heskett, J. L., Jones, T. O., Loveman, G. W., Sasser, W. E., & Schlesinger, L. A. (1994). Putting the service-profit chain to work. *Harvard Business Review*, 72(2), 164–174.

Heskett, J. L., Sasser, W. E., & Hart, C. W. (1990). *Service Breakthroughs: Changing the Rules of the Game*. New York, NY: The Free Press.

Jackson, M. O., & Watts, A. (2002). The evolution of social and economic networks. *Journal of Economic Theory*, 106(2), 265–295.

Johnston, R. (1995). The determinants of service quality: satisfiers and dissatisfiers. *International Journal of Service Industry Management*, 6(5), 53–71.

Keillor, B. D., Hult, G. T. M., & Kandemir, D. (2004). A study of the service encounter in eight countries. *Journal of International Marketing*, 12(1), 9–35.

Larson, R. C. (1987). OR forum—perspectives on queues: social justice and the psychology of queueing. *Operations Research*, 35(6), 895–905.

Larson, R. (2011) Foreword in *Service Systems Implementation*, ed. by H. Demirkan, J. Spohrer, and V. Krishna. Springer.

Lloyd, A. E., & Luk, S. T. (2009). Interaction behaviors leading to comfort in the service encounter. *Journal of Services Marketing*, 25(3), 176–189.

Lovelock, C. H., & Wirtz, J. (2007). *Service Marketing: People, Technology, Strategy*, 6th ed. Upper Saddle River, NJ: Prentice Hall.

Loveman, G. W. (1998). Employee satisfaction, customer loyalty, and financial performance—an empirical examination of the service profit chain in retail banking. *Journal of Service Research*, 1(1), 18–31.

Meany, W. (2012). *The Seven Laws of Service*. Retrieved Dec. 26, 2012 from http://www.asset.net/news/7laws.asp.

Meyer, C., & Schwager, A. (2007). Understanding customer experience. *Harvard Business Review*, 85(2), 116–126.

NewtonWiki. (2013). *Newton's Laws of Motion*. Retrieved Jan. 10, 2013 from http://en.wikipedia.org/wiki/Newton's_laws_of_motion.

Ostrom, A. L., Bitner, M. J., Brown, S. W., Burkhard, K. A., Goul, M., Smith-Daniels, V., Demirkan, H., & Rabinovich, E. (2010). Moving forward and making a difference: research priorities for the science of service. *Journal of Service Research*, 13(1), 4–36.

Oxford. (2001). Law of Nature in *Oxford English Dictionary*, 3rd ed. Oxford University Press.

Parasuraman, A., Zeithaml, V. A., & Berry, L. L. (1985). A conceptual model of service quality and its implications for future research. *The Journal of Marketing*, 49(Fall), 41–50.

Parasuraman, A., Zeithaml, V. A., & Berry, L. L. (1988). SERQUAL: a multi-item scale for measuring customer perceptions of service quality. *Journal of Retailing*, 64(1), 12–40.

Qiu, R. G. (2009). Computational thinking of service systems: dynamics and adaptiveness modeling. *Service Science*, 1(1), 42–55.

Qiu, R. G. (2012). Editorial column—launching service science. *Service Science*, 4(1), 1–3.

Qiu, R. G. (2013a). *Business-Oriented Enterprise Integration for Organizational Agility*. Hershey, PA: IGI Global.

Qiu, R. G. (2013b). We must rethink service encounters. *Service Science*, 5(1), 1–3.

QRWiki. (2012). *Quantitative Research*. Wikipedia. Retrieved Oct. 10, 2012 from http://en.wikipedia.org/wiki/Systems_theory.

Ramdas, K., Teisberg, E., & Tucker, A. L. (2012). Four ways to reinvent service delivery. *Harvard Business Review*, 90(12), 98–106.

Schneider, B., & Bowen, D. (2010). Winning the Service Game in *Handbook of Service Science*, 31–59, eds. by P. Maglio, C. Kieliszewski, and J. Spohrer. US: Springer.

Shostack, G. L. (1985). Planning the Service Encounter in *The Service Encounter: Managing Employee/Customer Interaction in Service Businesses*, 243–254, ed. by J. Czepiel, M. Solomon, and C. Suprenant. Lexington, MA: Lexington.

Smith, A. K., Bolton, R. N., & Wagner, J. (1999). A model of customer satisfaction with service encounters involving failure and recovery. *Journal of Marketing Research*, 36(3), 356–372.

Solomon, M. R., Surprenant, C., Czepiel, J. A., & Gutman, E. G. (1985). A role theory perspective on dyadic interactions: the service encounter. *The Journal of Marketing*, 49(1), 99–111.

Surprenant, C., & Solomon, M. (1987). Predictability and personalization in the service encounter. *Journal of Marketing*, 51(2), 86–96.

Svensson, G. (2004). A customized construct of sequential service quality in service encounter chains: time, context, and performance threshold. *Managing Service Quality*, 14(6), 468–475.

Svensson, G. (2006). New aspects of research into service encounters and service quality. *International Journal of Service Industry Management*, 17(3), 245–257.

Taylor, J. C. (1977). Job satisfaction and quality of working life: a reassessment. *Journal of Occupational Psychology*, 50(4), 243–252.

Tax, S., & Brown, S. (1998). Recovering and learning from service failure. *Sloan Management Review*, 40(1), 75–88.

Wyckoff, D., & Maister, D. (2005). *The Laws of Service Business*. Retrieved on Dec. 10, 2012 from http://www.davidmaister.com.

Zeithaml, V. A. (2000). Service quality, profitability, and the economic worth of customers: what we know and what we need to learn. *Journal of the Academy of Marketing Science*, 28(1), 67–85.

5

Organizational and IT Perspectives of Service Systems and Networks

Without question, a competitive service organization is a system that is always well developed, cost-effectively controlled, and efficiently managed. We fully understand that a service organization must be positioned in offering service products to its prospective customers. However, the real value of service is not created until the offered service products are utilized by customers who acquired the services. As discussed in the preceding chapters, to a service provider in today's service-led economy, it is the people (customers and employees) rather than physical goods that must be put first and are at the center of its organizational structures and operations. To the service customers, service satisfaction in light of meeting their meets is what truly matters although the eventual satisfaction is subjective and varies with a variety of factors. Regardless of the size and nature of a service business, the daily operations of a service organization are commonly run and managed through some fundamental while relevant business processes (Figure 5.1).

Because we must focus on engineering and delivering services using all available means to meet the technical functional and socioeconomic needs and hence realize respective values for both providers and consumers, competitive service organizations essentially are social-technical service systems that must optimally leverage both the strengths of people and technologies under given serving circumstances. As discussed in Chapter 3, the fast advancement in distributed computing and interconnected network has significantly increased the role and power of IT and communications, transforming the ways how the service industry operates

Service Science: The Foundations of Service Engineering and Management, First Edition. Robin G. Qiu.

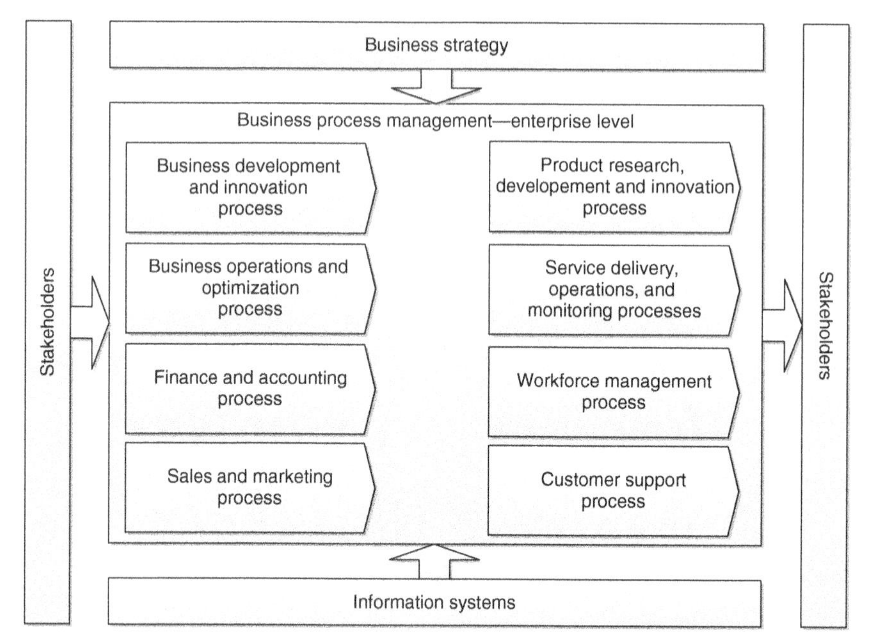

FIGURE 5.1 *Core business processes in support of business operations in service organizations.*

(Berman, 2012). By using service-dominant thinking while leveraging the increased flexibility, responsiveness, and capability of service business operations and management, service organizations can be realistically operated as effective sociotechnical service systems, improving business operational productivities and delivering new high levels of job and customer satisfaction (Qiu, 2013a).

Before we delve into the insights of system business structure and operations in a service organization or system, let us briefly review some core service science concepts we discussed in the earlier chapters. A service system essentially consists of service providers, customers, service products, and processes. As compared to a producing-goods system, a service system must be people-centric. The value of a service that is created along with the process trajectory throughout the lifecycle of the service largely depends on the sociotechnical dynamic behaviors of the service system that offers, engineers, and delivers the service. When we trace a service trajectory, we find that the service trajectory is nothing but a service encounter chain.

Although technical functions in service are fundamental, service is largely measured using a list of subjective measures, varying with customers and the corresponding service context. On a service encounter chain, a later encounter is inevitably influenced by the immediately preceding one; employees or customers could also be influenced by other previous encounters that they may have had before if those are somewhat functionally or sociopsychologically related. Therefore, a service provider should effectively account for the relevant and correlated Ps from all the 8 Ps at the point of interaction. It is critical for the service provider to ensure that each service encounter can be carried out in a satisfactory manner.

For example, Ramdas et al. (2012) suggest that four different dimensions in service encounters, including the structure of the interactions, the service boundary, the allocation of service tasks, and the delivery location, should be well identified, aligned, and executed in service provision. Mutual values for both customers and providers must be met at the same time. However, conventional wisdom is not necessarily effective and efficient as each interaction at its point of service is sociopsychologically unique. To ensure that each interaction in a service system can become effective and efficient, we have to have the service system developed, controlled, and managed well and with scientific rigor.

To ensure readers to get a comprehensive understanding of service provision in the systems perspective, in this chapter we fully discuss the following areas:

- Service, first and foremost, as an offering of a service system that can fulfill the customer's need through a process of transformation
- Putting people first in a service system: a focus on service interactions and social capitals
- Service system dynamics with a focus on smarter operations: the empowerment of systems and technologies
- Competitiveness, sustainability, and innovation: themes in the social and technical dimensions

When all the above-mentioned areas are well considered in building, operating, and sustaining a service system, each service provision from beginning to end within the service system can thus be performed in a satisfactory manner.

5.1 SERVICE AS AN OFFERING OF A SERVICE SYSTEM

First, let us briefly look at the list of some essential daily life services discussed in Chapter 1:

- *Restaurant Food Services*. Catering service is surely driven by the quality of foods and customer's perceived pleasure and service satisfaction. Frequently word of mouth plays a key role in choosing a restaurant when we plan a dinner for a party. Although many factors would influence the perception of service perceived by customers, there is no doubt that, it is the time and experience the customers enjoy during their stay in the restaurant that matters the most.
- *Car Services*. For a regular maintenance, we call a car service shop and schedule the needed service. On the scheduled day, we take the car scheduled for service to the shop. After we confirm with a receptionist on the needed maintenance, we drop the car there and leave for work. We will be informed of the completion of the service. We pick up the car after we pay the due. Getting the

requested maintenance done well surely is the highest priority for the shop. However, appropriately managing a customer's each interaction throughout the designated maintenance service process plays a critical role in ensuring that the market competitive level of service quality and satisfaction is positively perceived by the customer.

- *Residential Gas or Electricity Services.* We call a local office of a gas or an electricity service provider we choose and inform the service provider of the date when we move in. When we move out, we simply do the same. We pay the bill based on the monthly usage of gas or electricity. We interact with the service provider only if there would be a problem with power lines, gas pipes, or a discrepancy in a monthly bill statement.
- *Resident Education.* We register a course and then go to school to attend instructor-led lectures or laboratory sessions as scheduled. We surely work on assignments and take exams or finish projects in due course. It is well recognized that student–instructor and student–student interactions throughout the course are academic and pedagogical engagements that are considerably appreciated by the students.
- *Online Training.* We register a training course. No matter where we are, we can log on whenever we have time and an Internet connection. We read lecture notes and watch or listen to recorded lectures via a variety of online social media. Without question, online training is quite different from resident instruction-based education. Student–instructor and student–student interactions throughout an online course are often conducted asynchronously over the telecommunication means or networks. The training quality of online training perceived by the trainees is also considerably influenced by the involved virtual interactions.
- *Federal Bureaus or State Agencies.* Again, we can use a driver license renewal service as a typical example of utilizing state-level governmental services. We fill in renewal forms online. Letters from Department of Transportation of the state we live will arrive in a few days, which informs us of the time and location to have our driver licenses renewed. We show up at the designated office on the date indicated in the appointment letter. A staff at the office will assist us to finish the whole renewal process. Photographs will be taken, our signatures will be required, and accordingly new driver licenses will be issued. Service encounters essentially manifest as a variety of social and transactional interactions.
- *Global Project Development.* A software project development virtual team has six small groups of people, populating in six different geographic regions. Each group has certain unique skill sets of from 5 to 15 talent employees. A top-level management group manages the entire virtual project team. A project draft specification might be brainstormed when the top-level management group meets with a group of customer representatives. The project specification might be revised and enriched as time goes. Customer representatives could be directly or indirectly contacted by group members if necessary. A series of interactions

and coordination, physically and/or virtually, were necessary. Surely each interactive activity in the process of transformation adds a value into the solution in an integral and cocreative manner.

- *Health Care Service Networks.* The example of an outpatient in Chapter 1 shows how a typical US health care service is performed. Regardless of how many facilities or doctors and specialists the outpatient has to visit, each interaction plays a key role in the patient's recovery process.

Evidently, service, first and foremost, is an offering of a service system that truly has the potential of fulfilling the customer's needs through a customer processing operations (CPO) process with the support of the necessary operational resources. Its value is cocreated through service provision in a systemic and collaborative manner. Offering competitive service has truly become a business goal manifesto in any service organization, while in reality customers deserve better service than most service organizations are prepared to provide them (Fisk, 2009). Offerings that well meet the customer needs require systematic and effective executions of service strategy, marketing, design, and operations in service organizations. Figure 5.2 shows the operational perspective of service, highlighting the systems operations that must be well addressed and supported by smart service systems.

For a service system to be competitive, the real challenge is how its service practices and operations can be always and appropriately aligned with its offerings throughout the service lifecycle. Although we had quite detailed discussions of service system dynamics at a high and abstract level in Chapter 3, in practice we must drill down all operations and interactions in the service system (Figure 5.2),

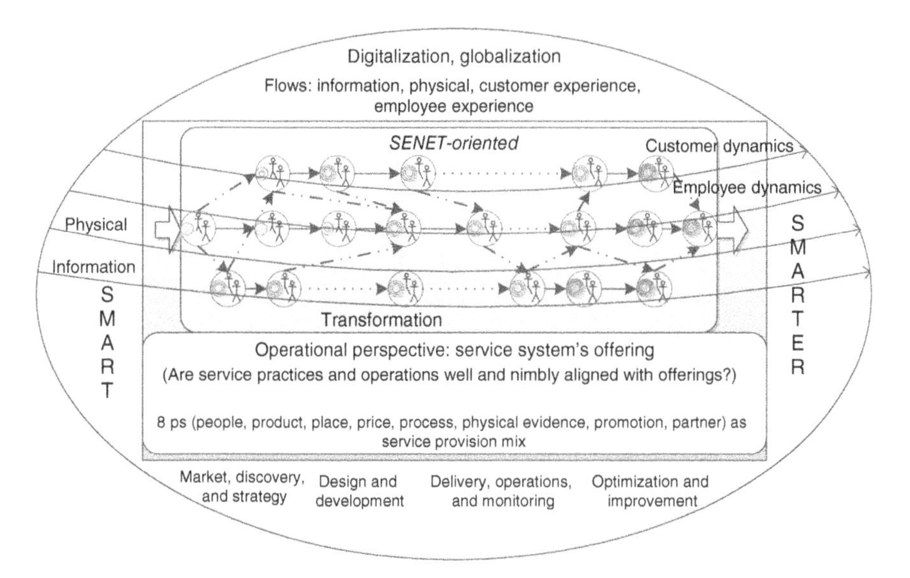

FIGURE 5.2 *The operational perspective of service.*

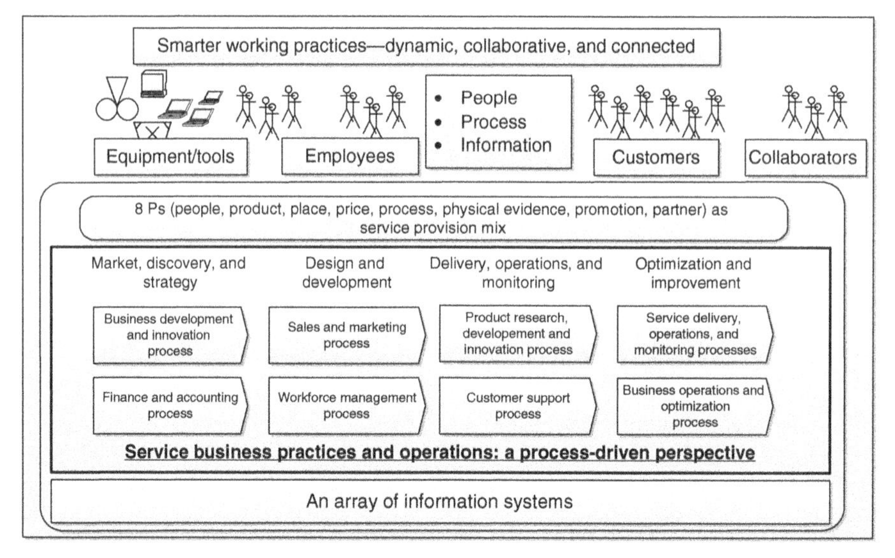

FIGURE 5.3 *Service business practices and operations: an end-to-end process-driven perspective.*

internally and externally, to an operational level at which all actions must be implementable, executable, and controllable.

Figure 5.3 shows that service business practices and operations can be operated and managed by adopting an end-to-end process-driven approach (Qiu, 2013a). As analyzed earlier, an offering of a service system truly has the potential of fulfilling the customer's needs. However, by the end of a day, it is a CPO process with the support of the necessary operational resources that governs the value cocreation of service during service provision in a systemic and collaborative manner. In other words, all operational actions in phases throughout the lifecycle of service should be well aligned with all operational business processes in a service system. It becomes necessary to explore effective approaches to help us understand and manage service planning, marketing, design, and operations in service organizations. Note that we focus on service offering and essential corresponding supports in a service system in the remaining part of this chapter; we will explore service provision in light of service encounter networks in a great detail starting from Chapter 6.

5.1.1 Service Business Strategy and Planning

As an organization is diving into building a highly profitable service-oriented business by taking advantage of its own unique engineering expertise and service knowledge, aimed at shifting gears toward creating superior outcomes to best meet customer needs, an adequate business service strategy will be vital for the organization's growth in the long run. As discussed earlier, it is the mainstream for service organizations to collaborate with their worldwide partners to deliver best-of-breed services to their customers, in particular, in their international marketplaces.

Despite the recognition of the importance of service engineering and management research, the shift to focus on services in the information era has created a research gap because of the overwhelming complexity of interdisciplinary issues across service business modeling, design, engineering, operations and management, information technology (IT), and workforce management. Filling the gap is essential. "We can move the field forward not only by understanding and serving the customer but by designing efficient systems of service delivery; training and motivating service providers; using new service technologies; and understanding how service affects the marketplace, the economy, and government policy" (Rust, 2004). The development of a business strategy meeting the long-term growth of a service enterprise should ensure that the defined business roadmap organically integrates corporate strategy and culture with organizational structure and functional strategy, and allows managing the interface of strategy, organization, resource, and technology in a flexible and cost-effective manner as illustrated in Figure 5.3.

In general, the development of business strategy for enterprises adaptable to a current service business environment requires extensive understandings of incorporation of solutions to address at least the following challenges in the service-led economy (Cherbakov et al., 2005; Wright et al., 2005; IBM, 2011; Bharadwaj et al., 2013; Korsten et al., 2013):

- *Maximizing the Total Value Across the Service Value Chain.* The outcome of the value chain nowadays is clearly manifested through customer satisfactions, which are mainly dependent on the capability of providing on demand, customizable and innovative services across the service value chain.
- *The International Transferability to Stay Competitive.* Enterprises reconstruct themselves by taking advantages of globalization in improving their profit margins, resulting in that subcontracting and specialization prevails. Radically relying on efficient and cost-effective collaborations, a service provider essentially becomes a global ecosystem in which the international transferability plays a critical role. The international transferability could cover a variety of aspects from human capital, worldwide trade and finance, social structures, and natural resources, to cultures, and customs.
- *Organizational Learning as a Competitive Advantage.* The globalization of the service workforces creates new and complex issues because of the differences in cultures, time, and skills. Leveraging all aspects of resources in the five capitals model, regionally and/or internationally, plays a critical role in keeping service business competitive.
- *Coping with the Complexities, Uncertainties, and Changes.* Change is the only certain thing today and tomorrow. As the complexities, uncertainties, and changes are reconfiguring the business world, an enterprise should be able to quickly adapt to the change.
- *Aligning Business Goals and Technologies to Execute World-Class Best Practices.* Business componentization cultivates value chains embracing for best-of-breed components throughout collaborative partnerships. The value

chains essentially are social-technical systems and operate in a network characterized by more dynamic interactions, real-time information flows, and integrated IT systems. Apparently, aligning business goals and IT is indispensable to the success execution of applied world-class best practices in services enterprises.

- *Connecting people in real time so as to have customers, partners, and employees engaged in an effective and positive manner.* By leveraging the advances in communication and network technologies, service organizations can get better understanding of the needs of people who are involved in the service profit chain across the service delivery network. Putting people first should truly be a part of business strategies that are technically and practically executable.

5.1.2 Service Marketing

Rust (2004) remarks "[Today's] business reality is that goods are commodities; the service sells the product." It is not a secret; that the quality of services essentially leads to a high level of customer satisfaction. It is satisfaction characterized as a superior outcome that further drives customer purchase decisions. In other words, the service-led total solutions that are measured by performances of delivering customers' final benefit rather than the functionality of physical goods lead to winning the competition in the global service-led marketplaces.

There are many new business opportunities in many newly expanded areas under the new concept of service, for example, e-commerce, e-service, auctions, and IT consulting (Menor et al., 2002). Although these emerging services have gained much popularity in consumers, a variety of new issues solicit further explorations for better understandings of service marketing to ensure that business goals can be met in the long run. Rust and Lemon (2001) discuss that the Internet-based e-service can better serve consumers and exceed their expectations through real-time interactive, customizable, and personalized services. To a service provider, effective e-service strategy and marketing play a significant role in growing the overall value of its service profit chain. A set of research questions in many customer-centric areas is proposed, aimed at leading to a stronger understanding of e-service and consumer behavior. Cao et al. (2003) model the relationships between e-retailer pricing, price satisfaction, and customer satisfaction, aimed at helping service organizations to operate more competitive businesses than ever before.

According to Rangaswamy and Pal (2005), service marketing as a fundamental service value driver is much less understood compared to product marketing. Typically, a service outcome is freshly "manufactured" or "remanufactured" at the customer's site at the time when it is delivered; its quality heavily depends on a well-defined and consistent process applied by trained personnel time after time. Hence, it is hardly an easy transition from product marketing to service marketing. In fact, we should adopt service-dominant logic thinking in service marketing. Conducting effective service marketing highly relies on effective mechanisms of connecting customers. Indeed, leveraging real-time communications and a sea of data in today's information era proves to be effective in leading effective customer analytics, which helps deliver future innovations in service marketing for service organizations over time (IBM, 2011; Korsten et al., 2013).

5.1.3 Service Design and Engineering

There have been many publications in the literature illustrating a variety of approaches to services design and engineering across industries. Although some of them present their scientific methodologies in focused areas to realize the targeted goals specified by customers, the majority of them simply show applying empirical and heuristic methods to execute their service design and engineering processes. However, there is a great need for generic methodologies for the design and engineering of high quality and sustainable services in order to meet the defined business strategy of a service provider in the long run.

Zhang and Prybutok (2005) study the design and engineering factors impacting the service quality in the e-commerce service sector. Introducing new products/services indeed would certainly help new revenue generation. However, to retain customers' high level of satisfaction and to allure them for purchasing further products/services are highly dependent on other numerous critical factors, for instance, system reliability, ease of use, localization and cultural affinity, personalization, and security. As the level of price satisfaction might not be increased simply by lowering prices, competing on price hence is not a viable long-term strategy for online retailers. Cao et al. (2003) model the relationships between e-retailer pricing, price satisfaction, and customer satisfaction through analyzing the whole services process. They find that the design and engineering of a satisfactory ordering process generates higher overall ratings for fulfillment satisfaction, which better retains loyal customers and accordingly helps a service provider stay competitive over time.

As discussed earlier, service sectors cover from traditional services (e.g., commercial transportation, logistics and distribution, health care delivery, retailing, hospitality and entertainment, issuance, and product after-sale services) to contemporary services (e.g., supply chain, knowledge transformation and delivery, financial engineering, e-commerce, and consulting). The competitiveness of today's services substantially depends on the efficient and effective operations of service delivery networks that are constructed using talents and comprehensive knowledge; service systems or delivery networks across the profit chain must be well integrated and aligned with people, management, and technology. The service products and corresponding transformation processes should be flexibly designed, engineered, and unified through effectively bridging the science of modeling/algorithms on one hand, and business processes, people skills, and diversified cultures on the other hand.

5.1.4 Service Delivery, Operations, and Management

Operations research and management with a focus on business internal efficiency has made significant progress and developed a huge body of knowledge during the past 65 years or so. The relevant research and algorithm development has been mainly conducted in the areas of optimization, statistics, stochastic processes, and queuing theory. Current applications cover areas from vehicle routing and staffing, supply chain modeling and optimization, transportation modeling, revenue management, risk management, services industry resource planning and scheduling, to airline optimization, and forecasting. In general, operations research has unceasingly improved living standards as it has been widespread applied in practice for the improvement of production management and applications productivity.

Operations research and management originated from practices and has been growing as a more quantitative, mathematical, and technical things. Larson (1989) argues that practice makes perfect operations research. As new problems are identified, when framed, formulated, and solved by applying operations research approaches, tremendous impact will be provided and accordingly a new theory might be created. Sociotechnical service systems show more practical natures and are extremely complex, which are typically modeled and formulated using qualitative approaches. Understanding of such a complex problem involves a deep and thoughtful discussion and analysis using common sense, basic principles, and modeling. As new initiatives, the operations research body of knowledge can be perfectly applied to these practical problems. Service operations and management is essentially operations research and management applied to service settings.

As discussed earlier, on one hand, the research and development of IT is a service. On the other hand, when IT helps enterprises streamline their business processes to deliver quality and competitive goods and services, it essentially functions as a knowledge service. However, for efficient IT service delivery to meet the needs of adaptive enterprises requires talents and comprehensive knowledge with combination of business, management, and IT. Therefore, the need for service-based operations research and management is on demand as it matches the emerging realization of the importance of the customer and a more customer-oriented view of operations. Service operations and management fits well with the growing globalization economic trend, which requires operations research in services practice.

According to Bell (2005), operations research applied to services has much to offer, which could improve the lives of everyone. He presents seven really useful operations research frameworks that can be effectively used in addressing practical and complex problems like the ones in service delivery networks. Moreover, services operations are closely synchronized with business operations of other collaborative partners as well as customers aimed at cocreating the value for customers in a satisfactory manner while meeting the business objectives across the value chains. Given the fact of the industrialization of services and the economy of globalization, reorganizing, realigning, redesigning, and restructuring of enterprises' strategies, processes, IT systems, and people for the challenges ahead are essential for ensuring services providers to be agile and adaptive and stay competitive (Karmarkar, 2004).

In summary, given the increasing complexity of building sociotechnical services systems for improving living standards by applying operation research and management science in practice, services operations and management should cover more initiatives of the rooted practical aspects of research, linking operational performance to business drivers, performance measurement and operations improvement, service design, service technology, human capitals, the design of internal networks and managing service capacity (Johnston, 2005). The study should also take into consideration high performance, distributed computing, humans and systems behavioral and cognitive aspects (which emerges as a new look of the interface to systems engineering), and highly collaborative interaction natures.

5.2 PUTTING PEOPLE FIRST

No matter what services are offered and how they are delivered, in reality, high living standard with better quality of life is what we as human beings are pursuing. When the communities in which we are living are deeply studied, we understand that our communities are truly IT-driven service-oriented in today's information era. Here are a few daily noticeable, inescapable, and more contemporary service examples that could be on demand at any time and place (Dong and Qiu, 2004; Qiu, 2005):

- A passenger traveling in a rural and unfamiliar area suddenly has to go to a hospital because of his sickness, so local hospital information is immediately needed at the point of need. He and his companions wish to get the local hospital information through their cellular phones. Generally speaking, when travelers are in an unfamiliar region for tourism or business, handy and accurate information services on routes and traffic, weather, restaurants, hotels, hospitals, and attractions and entertainments in the destination region become very captivating and helpful.
- A truck fully loaded with hazardous chemical materials is overturned on a city suburban highway. As the chemicals could be "pretty poisonous," people on site need appropriate knowledge (i.e., intelligent assistance) to quickly perform life saving and other critical tasks after one calls 911 (in the United States). However, people on site most likely cannot perform the tasks effectively because of the limited knowledge and resources. Situations could be worse if the task is not done appropriately, which could lead to an irreversible and horrible result. Intelligent assistant services are necessary at the point of need. Obviously, the situation demands a quick response from the governmental IT-driven emergency responsive systems.
- Transportation plays a critical role in warranting the quality service and effectiveness of a supply chain. When a truck is fully loaded with certain goods, certain attentions might be required from the driver from time to time, for instance, air, temperature, and/or humidity requirements. Only when the requirements are fully met on the road, can the goods in-transit be maintained with good quality. Otherwise, the provided transportation service could be unsatisfactory. Owing to the existence of a variety of goods, it is impossible for the drivers to master all the knowledge on how the goods can be best monitored and accordingly protected on the highway as many uncertain events might occur during the transportation of goods. On-demand services to assure the warranty are the key for an enterprise to lead the competitors. As manufacturing and services become global, more challenges are added into this traditional service.

- The growing elderly population draws much attention throughout the world, resulting in issues on the shortage of labors and more importantly the lack of effective health care delivery methods. Study shows that elder patients (65 or older) are twice likely to be harmed by a medication error because they often receive complex drug regimens and suffer from more serious ailments that make them particularly vulnerable to harmful drug mistakes. Outpatient's prescription drug-related injuries are common in elder patients, but many could be prevented. For instance, about 58% adverse drug events could be prevented if the continuity of care record plan and related health care information systems are adopted for providing prompt assistant services; over 20% drug-related injuries could be prevented if the given medication instructions are provided at the point of need so that the instructions are adhered by the patients.

Apparently, the real-time flow of information and quick delivery of relevant information and knowledge at the point of need from an information service provider/system is essential for providing quality services to meet the on-demand needs described in all the above-mentioned scenarios. Putting people first surely is the norm. Noah (2010) argues that "well-established research on the service-profit chain—the link between companies with a reputation for excellent service and profit—married with the assertion that happy staff equals happy customers, is disputed by few."

Obviously, putting people first in a service system is the natural response to the shift from manufacturing to service. The philosophy of putting people first essentially promotes the people-centric concept. Indeed, we have been cultivating the concept of people-centric service systems from Chapter 1. Hence, focusing on service interactions and leveraging social capitals are fundamental tasks in building, operating, and managing today's service systems.

According to Pfeffer and Veiga (1999), research, experience, and common sense all increasingly point to a direct relationship between an organization's success and its commitment of treating people as the most important in business operations. However, lacking effective mechanisms of capturing people's behavioral dynamics in real time made organizations difficult in directly connecting people in organizational management practices before. Promisingly, the significant advances in networks, telecommunication, and computing technologies have made it possible for us to have effective mechanisms of capturing people's behavioral dynamics in real time.

Because of the pervasiveness of networking and information technologies, people, organizations, systems, and heterogeneous information sources now can be linked together more efficiently and cost-effectively than ever before. We all have witnessed that the quick advances of IT has significantly transformed not only science and engineering research but also people's expectation on how to live, learn, and work since the turn of this new millennium. Surely life at home, work, and leisure gets easier, better, and enjoyable (Qiu, 2007).

As a variety of devices, hardware, and software become network aware, almost everything is capable of being handled over the network. Thus, many tasks can be done onsite or remotely, in the same manner so are a variety of services provided, or even self-performed over the Internet. In the business world, because of the enabled rich information linkages, the right data and information in context can be delivered to the right user (e.g., people, machine, device, software component, etc.) in the right place, at the right time, resulting in the substantial increase of the degree of business process automation, continual increment of production productivity and services quality, reduction of services lead time, and improvement of end users satisfaction. At the end of a day, end users or consumers do not care about how and where the product was made, by whom, and how it was delivered; what the end users or consumers essentially care about is that their needs are met in a satisfactory manner (Qiu, 2007).

5.2.1 The Digitalization Approach to Capture People's Behavioral Dynamics

As a service organization puts people rather than physical resources and goods in the center of its organizational structure and operations, scientific exploration on a service system should go beyond conventional technical and physical thinking, which has been used to deal with technical-oriented systems over centuries. To act accurately and effectively on the need for scientific exploration of sociotechnical systems, in addition to the full use of ubiquitous digitalized business process data and information, we know that we need a novel approach to model systems dynamics by substantially harnessing people-centric sensing for capturing and collecting human behavioral data.

In manufacturing, the deployment of integrated information systems is accelerating (Qiu, 2004). A typical IT-driven manufacturing business can be created by deploying enterprise-wide information systems managing the lifecycle of both "e" and "business," that is, an order is taken over the Internet, and the products are made and delivered as promised. For instance, customers submit their orders via Internet browsers directly through a sales force automation center, which automatically triggers the generation of the appropriate material releases and production requirements. It also informs all the other relevant planning systems, such as advance production schedule, finance, supply chain, logistics, and customer relationship management of the new order entry. The scheduler then assigns or configures an onsite or remote production line through the production control in the most efficient way possible, taking into account raw material, procurement, and production capacity. A shop floor production execution schedule is then generated, where problems are anticipated and appropriate adjustments are made accordingly in a corresponding manufacturing execution system. In the designated facility, the scheduled work is accomplished automatically through a computer-controlled production line in an efficient and cost-effective manner. As soon as the work is completed, the ordered product gets automatically warehoused and/or distributed. Ultimately, the customers should be provided the least cost and best quality goods, as well as the most satisfactory services (Qiu et al., 2003).

As compared to manufacturing, service in this new millennium is fully recognized to be people-centric, truly cultural, and bilateral. From the preceding chapters, we come to know that the type and nature of a service dictate how the service is performed, while a service system that offers the service accordingly defines how a series of service encounters could and should occur throughout its service lifecycle. In general, the type, order, frequency, timing, time, efficiency, and effectiveness of the series of service encounters throughout the service lifecycle determine the quality of service perceived by customers who purchase and consume the service (Bitner, 1992; Chase and Dasu, 2008). To ensure that quality services can be performed as promised, we have to monitor and control the processes of transformation in a real-time and effective manner, wherein promptly capturing and managing peoples' dynamics throughout the service lifecycle plays a key role in operating a today's service business. Therefore, for a service system, the implementation of appropriate approaches to collect quality data that capture people's digital footprinting in real time throughout the service lifecycle surely becomes essential for service provision.

Before the emergence of this information era, the lack of means to monitor and capture people's dynamics throughout the service lifecycle had prohibited us from gaining insights into the service lifecycle. Since the dawn of world wide web and pervasive mobile computing, the rapid development of digitization and networking technologies has made possible the needed means and methods to overturn the technical, application, and social barriers. According to Girardin et al. (2008), "Along with the growing ubiquity of mobile technologies, the logs produced have helped researchers create and define new methods of observing, recording, and analyzing a city and its human dynamics. In effect, these personal devices create a vast, geographically aware sensor web that accumulates tracks to reveal both individual and social behaviors with unprecedented detail. The low cost and high availability of these digital footprints will challenge the social sciences, which have never before had access to the volumes of data used in the natural sciences, but the benefits to fields that require an in-depth understanding of large group behavior could be equally great."

Truly a variety of novel methods and tools have been developed to capture and explore the significance of pervasive while overwhelming people-related spatiotemporal, socioeconomic, and sociopsychological data in addition to technical and functional systems data in service. With the help of digitalization and globalization, it is indeed possible to gather every service activity and behavior of every user throughout the service lifecycle regardless of where and when the user interacts with the service. Therefore, now is the time for us to bear in mind that the philosophy of putting people first in service is not just a fantasy; it can be substantively and effectively adopted and implemented through methods and tools. As a matter of fact, to make service systems and networks competitive, we can and indeed we must rethink the whole service lifecycle and explore those interwoven social and transactional interactions in a deeper and more sophisticated manner than ever before.

5.2.2 Supplementary Approaches to Capture People's Behavioral Dynamics

Although mobile computing and sensory technologies are ubiquitous and pervasive, approaches to collect people-related data that can be used for people-centric services explorations vary with service systems. As a result, the collected data might be incomplete under the changing circumstances. We understand that questionnaire-based surveys have been used in the fields of marketing research, psychology, health professionals, and sociology for over a century. Although the forms and means to conduct surveys have considerably changed over the years with the development of technologies, markets, and societies, the purpose of understanding peoples' opinions on the studied subjects does not change.

An effective survey relies on systematic panning for data from a variety of sources, including questionnaires, interviews, observation, existing records, and electronic devices. In service, surveys can be undertaken with a focus on making statistical inferences about the served customers and the employees who interact with the served customers. Practical examples of quantitative research based on contemporary survey methodologies to understand the studied question of a focused population include polls about public opinions, public health surveys, and market research surveys. In both academia and practice, cross-sectional and longitudinal surveys are frequently applied to study services. Cross-sectional surveys essentially use a single questionnaire or interview to get a corresponding response from a participant, while longitudinal surveys repeatedly collect information from the corresponding participants over time. In general, longitudinal surveys have significant analytical advantages, although they are more challenging to carry on than cross-sectional surveys.

Surely the methods and means to conduct surveys have been considerably improved over the years with the emergence of new technologies in sensing, computation, and communications. For instance, ubiquitous and pervasive mobile phones are turned into global mobile sensing devices. According to Campbell et al. (2008), people-centric sensing with the support of technological advances enables "a different way to sense, learn, visualize, and share information about ourselves, friends, communities, the way we live, and the world we live in." The rise of people-centric sensing enables an array of new applications, including personal, public, and social sensing, which might replace the traditional ways to conduct surveys. As people are the key architectural and system component in service systems, appropriate and successful implementations of people-centric sensing would entail collecting responses necessary and real time from people who are involved in the processes of service transformation.

5.2.3 Putting People First

In general, putting people first is a philosophy for service management, engineering, and operations. To have this philosophy well applied to a service business, we thus should pay attention to the following areas:

- *Management*. Meeting people's needs should be the goal of a service business.
- *Operations*. Making work environment pleasant and enjoyable should be well considered before and during service operations.

- *Technology*. Assuring the support tools ready, handy, and effective throughout the service lifecycle significantly impact job and service satisfaction and customer royalty.
- *Data Collection*. Enabling ways and mechanisms to capture and collect data relevant to people's behavior is indispensable for us to understand and accordingly serve customers well from beginning to end. In addition to technical data, physiological and psychological as well as personal data are also necessary for executing services. Rich data has the potential of helping service organizations to assist their employees and customers to do better service, and thus they can truly enjoy the work and personal life.

Psychometrics has been well applied to the understanding of individual differences. "Psychometrics is the field of study concerned with the theory and technique of psychological measurement, which includes the measurement of knowledge, abilities, attitudes, personality traits, and educational measurement. The field is primarily concerned with the construction and validation of measurement instruments such as questionnaires, tests, and personality assessments" (WikiPsychometircs, 2013). By referring to psychometrics with a focus on people's dynamics while considering service technical and functional attributes in service, we should fully develop service metrics. Service metrics will help us understand the competitiveness of service systems and networks.

On the whole, it is the data produced through people's interactions while capturing human, social, or environmental states (Campbell et al., 2008; Girardin et al., 2008; Qiu, 2009; Guo et al., 2012; Nunes et al., 2012) that allows us to understand and accordingly manage and control a service system in an efficient and effective manner. Therefore, the performed services are able to overcome sociological and cultural barriers, resulting in such a way that the cultures of cocreation, collaboration, and innovation can be cultivated and fostered.

5.3 CONTROLLABLE AND TRACTABLE SERVICE SYSTEMS IN PURSUIT OF SMARTER OPERATIONS

Enterprises are eagerly embracing for building highly profitable service-oriented businesses through properly aligning business and technology and cost-effectively collaborating with their worldwide partners so that best-of-breed services can be generated to meet the ever-changing needs of customers. To be competitive in the long run, it is critical for enterprises to be adaptive, given the extreme dynamics and complexity of conducting businesses in today's global economy. In an adaptive enterprise, people, processes, and technology shall be organically integrated across the enterprise in an agile, flexible, and responsive manner. As such, the enterprise can quickly turn changes and challenges into new opportunities in this on-demand business environment.

IT as a service is a high value service area, which also plays a pivotal role in support of business operations in a service-oriented enterprise. The delivery of IT service for a service business requires knowledge workers who have sound and deep understandings of IT, organizational structures, and behavior, as well as human behavior and cognition science in general (IBM, 2004). For IT systems to better serve a service enterprise, service-oriented business components based on business domain functions are necessary (Cherbakov et al., 2005). However, the real question to us is what a systematic approach and adequate computing technologies will ensure that engineered IT-enabled service systems can lead to the success of building an adaptive and people-centric enterprise, resulting in that competitive service engineering, operations, and management across the business can be realized.

The remaining parts of this section briefly discuss how enterprise service computing is evolving (Qiu, 2007). The quick advances of enterprise service computing make possible enable IT to control, manage, and empower service systems in a cost-effective and adaptive manner. Hence, IT-enabled service systems can be truly people-centric while computational, resulting in the realization of smarter service business engineering, management, and practices.

5.3.1 Overview of Enterprise Service Computing

Computing technologies (e.g., software development) unceasingly increase their complexities and dependencies in order to capture and help maneuver the increased complexity of business operations within and across enterprises because of the accelerated globalization economy. For instance, aiming to find a better approach to manage complexities and dependencies within an IT-enabled system, the practice of software development has gone through several methods (e.g., conventional structural programming, object-oriented method, interface-based model, and component-based constructs). The emergence of developing coarse-grained granularity constructs as a computing service allows components to be defined at a more abstract and business semantic level. Technically, a group of lower level and finer grained object functions, information, and implementations within software objects/components can be choreographically composed as a coarse-grained computing component or service. As a result, deployed computing services can support and be well aligned with daily operational activities conducted by employees or customers.

The componentization of a business is the key to the construction of best-of-breed components for delivering superior services to customers. Successful operations of a componentized business require seamless enterprise integration. Technically, a service-oriented IT-enabled system makes more sense in support of people-centric service businesses as it is able to deal with more types of people-centered interactions among heterogeneous while interconnected computing components. As a result, a service-oriented IT-enabled service system is more flexible and adaptive than an IT-enabled service system based on other approaches.

As computing technologies evolve over time, adaptive and semantic computing services that represent and align business functions and activities meet the needs of developing service-oriented IT-enabled service systems. Indeed, when computing components manifest themselves as operational and support services at the business

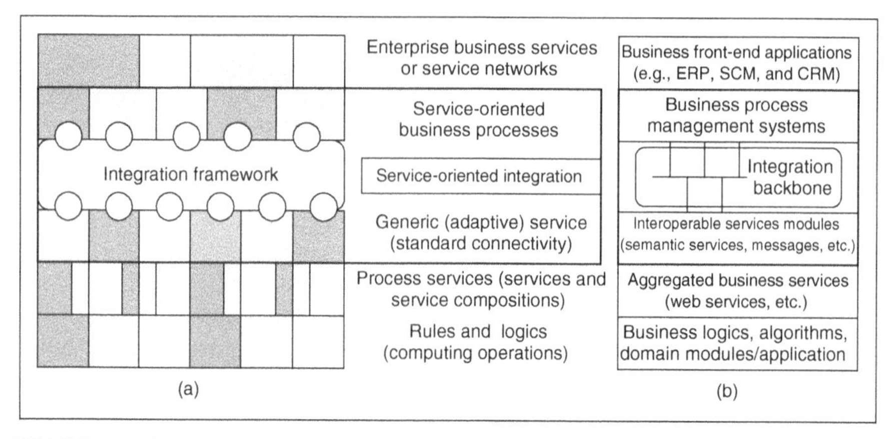

FIGURE 5.4 Service-oriented and business component-based network architectural model. (a) The enterprise service computing architectural model and (b) an implementation.

level, an IT-enabled service system becomes a component-based network, fundamentally illustrating a logic assembly of interconnected service computing components. "The need for flexibility across the value net requires that the component network be flexible; that is, the enterprise can 'in-source' an outsourced component and vice versa; replace, on demand, a current partner with a different partner; change the terms of the contract between the two components, and so on" (Cherbakov et al., 2005).

A generic service-oriented IT computing architecture in support of the development of a component-based service network (i.e., enterprise business-level services) is illustrated in Figure 5.4. The top two layers represent service operations from the business process perspective while the bottom three layers show the value-added service processes from the computing perspective. Apparently, how to optimally align enterprise-level business strategies with value-added operations/activities is the key to the success of the deployment of an agile enterprise service-oriented IT-enabled service system (Qiu, 2007).

However, the exploitation, establishment, control, and management of dynamic and inter- and cross-enterprise resources that are highly related and significantly contribute to the realization of the agility of a service-oriented IT-enabled service system require new methodologies, technologies, and tools. The remaining discussion hereafter focuses on the following three evolving synergic IT research and development areas, aimed at providing some fundamental understandings of the emerging methodologies, technologies, and tools in support of the deployment of IT-enabled services. Indeed, the following methodologies, technologies, and tools are essentially consisting of and delivering the adaptive enterprise service computing discussed in this section.

- *Service-Oriented Architecture (SOA).* SOA is considered as the design principle and mechanism for defining business services and computing models and thus effectively aligning business and IT.

- *Component Process Model (CPM).* Component business process model facilitates the construction of the business of an enterprise as an organized collection of business components (Cherbakov et al., 2005).
- *Business Process Management (BPM).* BPM essentially provides mechanisms to transform the behaviors of disparate and heterogeneous systems into standard and interoperable business processes, aimed at effectively facilitating the conduct of IT-enabled system integration at the business semantic level (Smith and Fingar, 2003).

5.3.2 Service-Oriented Architecture

According to Datz (2004), "SOA is higher level of [computing] application development (also referred to as coarse granularity) that, by focusing on business processes and using standard interfaces, helps mask the underlying complexity of the IT environment." Simply put, SOA is considered as the design principle and mechanism for defining business services and computing models and thus effectively aligning business and IT (Figure 5.5) (Newcomer and Lomow, 2005).

On the basis of the concept of SOA, a deployed service-oriented IT system can establish a standard framework for cost-effectively and efficiently managing and executing distributed heterogeneous services including human tasks within an enterprise and across service networks. To properly implement service-oriented IT systems that are well complied with SOA principles, three major levels of abstraction (as shown in Figure 5.4) throughout collaborated IT systems are necessary (Zimmermann et al., 2004):

- *Business Processes.* A business process typically consists of a set of business actions or activities that are aligned with specifically defined short- and long-term business goals. A business process thus requires a variety of computing services. Service invocations frequently involve business components

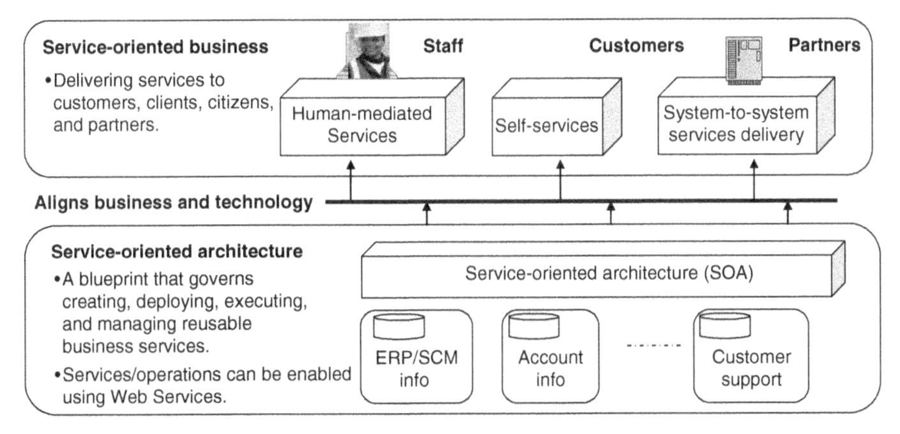

FIGURE 5.5 Aligning business and information technology to empower a people-centered service system.

across the formed service network. Examples of daily business processes could be: initiate new market campaign, sell products or services, coordinate projects, and order fulfillment management.

- *(Computing) Services.* A service represents a logical group of computing operations at a designated organizational level. For example, if customer profiling is defined as a service, then, lookup customer from data sources by telephone number, list customers by name and postal code on the web, and update data for new service requests could represent the associated operations.
- *(Computing) Operations.* A computing operation represents a single logical unit of computation. In general, the execution of an operation will cause one or more data sets to be read, written, or logically processed. In a well-defined SOA implementation, operations have a specific, structured interface, and return structured responses. An SOA operation can also be composed of other SOA operations in support of high level structures and enhanced maintainability.

Technically, SOA as a design principle essentially is concerned with designing and developing integrated systems using heterogeneous network addressable and standard interface-based computing services. Over the last decade or so, SOA and service computing technologies have gained tremendous momentum with the introduction of Web services (a series of standard languages and tools for the implementation, registration, and invocation of services) (Thomas, 2005). Enterprise-wide integrated IT systems based on SOA ensure the interconnections among integrated applications in a loosely coupled, asynchronous, and interoperable manner. It is believed that BPM (as transformative technologies) and SOA enable the best platform that can fully leverage existing information assets while facilitating enterprises in positing their IT systems capable of adapting to the future investments in an amenable way (Bieberstein et al., 2005).

5.3.3 Component Process Model

Given the increasing complexity and uncertain dynamics of the globalized business environment, the success of a business highly relies on its underlying IT systems to support the evolving best practices in an organization. In adaptive enterprise service computing, the appropriate design of IT-driven business operations mainly depends on well-defined constructs of business processes, computing services, and operations. Hence, to make this promising SOA-based component network architectural model implementable, it is essential to have a well-defined process-driven analytical and computing model that can help analysts and engineers understand and optimally construct the operational model of an enterprise with a consideration of its implementation of needed IT supports.

A business process typically consists of a series of computing services. As a business process acts in response to business events, the process should be dynamically supported by a group of services that are invoked in a logic sequence. To ascertain the dynamic and optimal behavior of a process, the group of underlying computing services should be selected, sequenced, and executed in a choreographed rather than predefined manner according to a set of dynamics business rules. A computing service is made of an ordered sequence of computing operations. Therefore, in support of adaptive enterprise service computing across an enterprise, CPM serves as a design and analytical method and platform to ensure that well-designed operation, service, and process abstractions can be characterized and constructed systematically (Cherbakov et al., 2005; Kano et al., 2005; Zimmermann et al., 2004).

More technically, we articulate that CPM provides a framework for organizing and grouping business functions as a collection of business components in a well-structured manner, so the components based on business processes can be modeled as logical business service building blocks that can appropriately represent corresponding business functions as desired. Figure 5.6 schematically illustrates a simplified components process model for a service provider (Cherbakov et al., 2005). Just like many business analysis diagrams, CPM can also be refined in hierarchy. In other words, a process can be composed of a number of refined processes in a recursive manner.

As CPM can accurately model the business operations using well-defined computing services in SOA terms, CPM helps to analyze a business and develop its componentized view of the business. Furthermore, the developed model for the business will define components that clearly describe the interfaces and service-level agreements between coordinated and collaborated services. Promisingly, each business component will be fully supported by a set of IT-enabled services, while meeting the requirements of the deployment of adaptive enterprise service computing (Cherbakov et al., 2005).

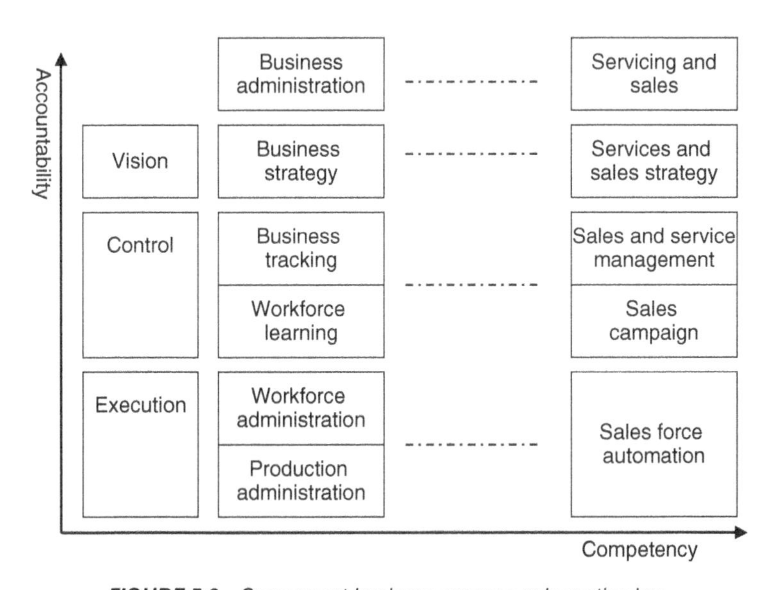

FIGURE 5.6 Component business process schematic view.

5.3.4 Business Process Management

BPM emerges as a promising guiding principle and technology for integrating existing assets and future deployments. BPM is new in the sense that it can orchestrate among existing disparate and heterogeneous systems by promoting business process services to increase business agility when it is applied to conducting IT system integration; it differs from other approaches that simply focus on integrating those systems using EAI (enterprise application integration), API (application programming interface), Web services coordination, etc. By providing mechanisms to transform the behaviors of disparate and heterogeneous systems into standard and interoperable business processes, BPM essentially aims at enabling a platform effectively facilitating the conduct of IT system integration at the semantics level (Smith and Fingar, 2003). As an SOA computing service at the system level essentially is the business function provided by a group of components that are network addressable and interoperable, and might be dynamically discovered and used, BPM and SOA computing services can be organically while flexibly and choreographically integrated, which is schematically illustrated in Figure 5.7 (Newcomer and Lomow, 2005).

In essence, BPM takes a holistic approach to promote and support enterprise service computing from the business process execution perspective, substantially leveraging the power of standardization, virtualization, and management. BPM initiatives include a suite of protocols and specifications, business process definition metamodel (BPDM), business process modeling notation (BPMN), and business process execution language (BPEL). By treating business process executions as real-time data flows, BPM provides the capability of addressing a range of choreographic business challenges, encompassing people-centered services, and supporting business operations execution in real time.

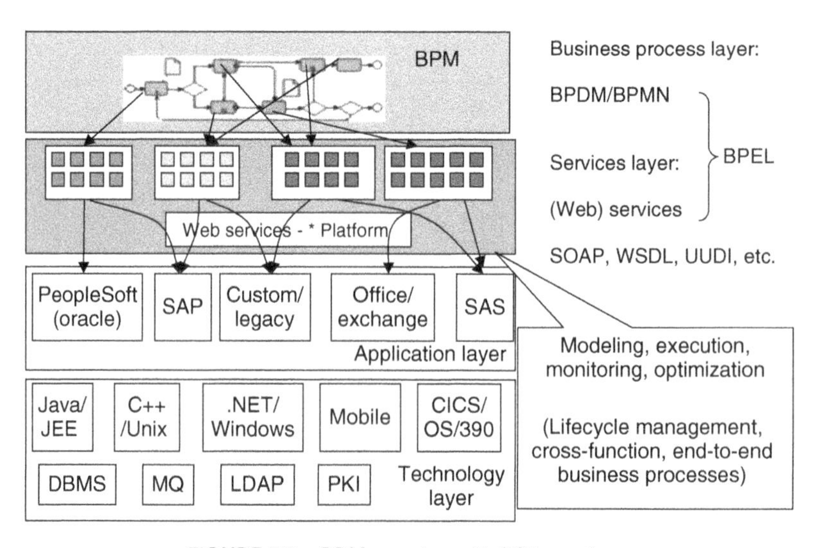

FIGURE 5.7 *BPM merging with SOA services.*

BPDM is defined for modeling complex business processes. Using the BPDM specification to describe the business process metamodel of an enterprise provides the abstract model of the enterprise. The abstracted model is programmatically structured and represented using extensible markup language (XML) syntax to express the defined executable business processes and supporting entities for the enterprise. Relying on business process diagrams, BPMN provides the capability of defining and understanding internal and external business operations for an enterprise. Through visualization it gives the enterprise the ability to communicate these modeling and development procedures in a standard manner. BPEL for Web services then defines a standard way of representing executable flow models, which essentially extends the reach of business process models from analysis to implementation through leveraging the power of Web service technologies.

The emergence and evolution of BPM enable an innovative platform for conducting IT system integration. BPM enables service-oriented IT systems over the interconnected networks to be able to dynamically and promptly coordinate the behaviors of disparate and heterogeneous computing services across enterprises. It is the BPM that business agility is retained while the return of IT investment gets maximized. Most importantly, BPM with the supports of SOA and CPM can surely make IT-enabled service organizations truly people-centric and computational. Hence, service science can be well applied in the realization of smarter service business engineering, management, and practices in the service organizations.

5.4 COMPETITIVENESS, SUSTAINABILITY, AND INNOVATION: SYSTEMS APPROACHES TO EXPLORE THE SOCIOTECHNICAL NATURES OF SERVICE SYSTEMS AND NETWORKS

Dynamic changes and uncertainties are prevalent in the business world. Whether a service organization (or simply called a service system) can fast adapt to the changes and uncertainties relies on approaches that can be applied effectively to address competitiveness, sustainability, and innovation issues in its corresponding business practice and operations. According to Becker et al. (2001), General Electric (GE) had a team to identify such an approach by looking into hundreds of team-based problem-solving and employee empowerment programs (i.e., project development services) within GE in the late 1980s and early 1990s. The team found that a high quality technical solution was insufficient to guarantee success. As a matter of fact, an extremely high percentage of failed programs had excellent technical strengths. The team further identified that paying little attention to the challenges in the sociopsychological dimension actually derailed the programs.

GE's exploration resulted in the so-called change effectiveness equation, $Q \times A = E$, which is widely used as a model to describe a solution to address change acceleration phenomena. Essentially, the effectiveness (E) of a solution to meet the changing needs of end users is equal to the quality (Q) of the technical attributes of the solution and the acceptance (A) of that solution by the end users. Our service value diagram absolutely matches GE's change effectiveness model (Figure 5.8).

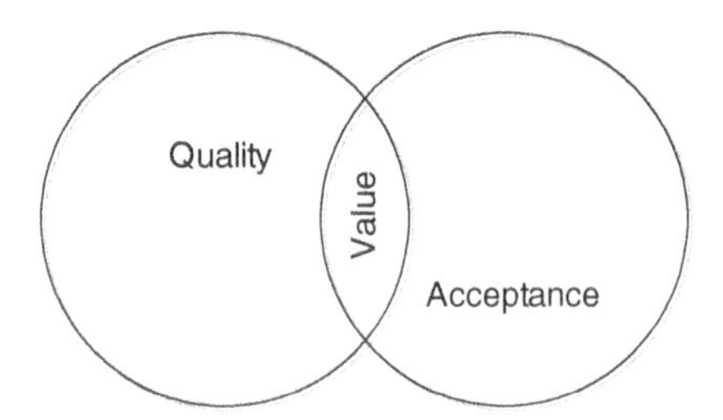

FIGURE 5.8 *Service value diagram corresponding to GE's change effectiveness model.*

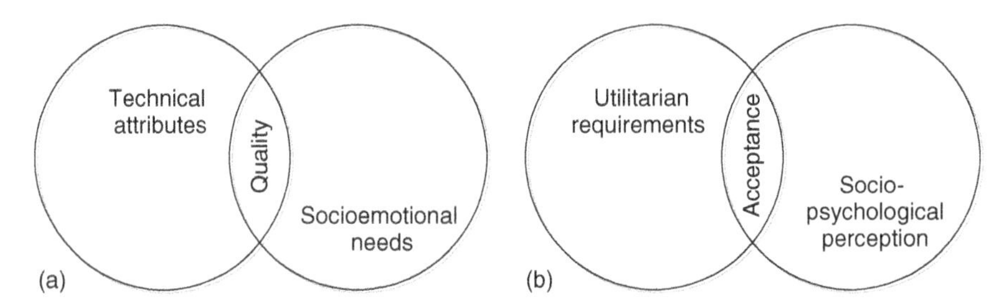

FIGURE 5.9 *Service value diagram with drill-down viewpoints in multiple dimensions. (a) Quality driven by the provider and (b) acceptance driven by the customer.*

The service value equation, $V = Q \times A$, means that paying attention to the people side of the equation (i.e., acceptance by end users) is as important to success as the technical side (i.e., quality of technical attributes).

If the service value equation ($V = Q \times A$) is further drilled down from the viewpoints of both providers and customers (Figure 5.9), we will have the following refined equations:

- *For Service Providers.* $Q = T \times S_p$, T is a set of technical attributes that defines the service, and S_p is a set of socioemotional needs that contribute to job satisfaction.
- *For Service Customers.* $A = U \times S_c$, U is a set of utilitarian requirements that satisfy the agreements and specifications understood by the customers, and S_c is a set of sociopsychological perceptions that significantly contribute to customer satisfaction.

Therefore, in both organizational and operational perspectives, service organizations must allocate right resources and operate their businesses in such a way that the service value equation reaches an optimal value by maximizing the overlapping areas of Q and A defined in Figures 5.8 and 5.9.

In practice, competitive and innovative service delivery models are essentially derived by working closely with customers to cocreate creative and unique solutions best meeting customer inevitably changing needs. According to Rangaswamy and

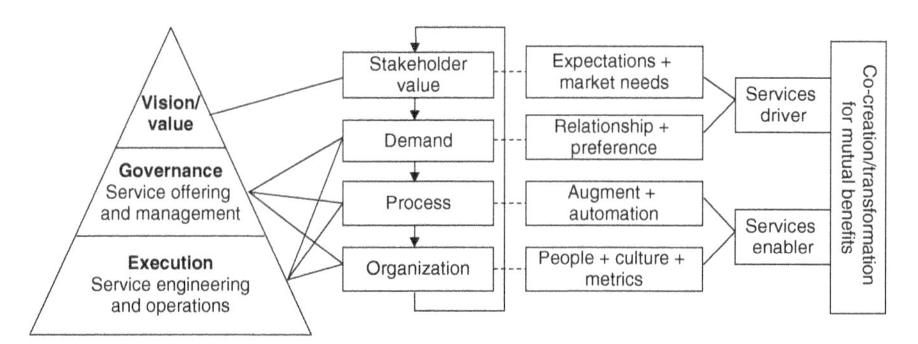

FIGURE 5.10 *Service innovation framework and modeling.*

Pal, a competitive service business model for an enterprise should be clearly described using a service innovation framework in organizational perspective (Figure 5.10). "The framework can guide the creation of customer value and demand, and the processes and organizations that deliver services successfully—all of it catalyzed by emerging technologies" (Rangaswamy and Pal, 2005).

In Chapter 2, we discussed the complex relationships between employee satisfaction, customer retention, and profitability and emphasized that we must rethink service encounters and find scientific ways to build and manage people-centric, information-enabled, coceration-oriented, and innovative service organizations in the service-led economy (Heskett et al., 1994; Lovelock and Wirtz, 2007; Chesbrough, 2011; Qiu, 2013b). Given the increasing complexity and variation from service to service, vertical service domain knowledge of modeling and frameworks should be first investigated. Only when a better understanding of a variety of services domains is accomplished, can an integrated and comprehensive methodology to address the services model and innovation framework across industries be explored and thereafter acquired. Vertically or horizontally, there is a need for a systematic approach to address how such a service innovation framework and modeling shown in Figure 5.10 can be optimally applied in practice, which is what we mainly discuss in this section.

Systems approaches to manage and control service systems essentially are methodologies for assisting us to make optimal decisions of transforming service business practices while reducing corresponding transformation risks. Competitive edges can be created for a service system only if we can deliver closed-loop and real-time controls of service business engineering, management, and practices. Figure 5.11 illustrates such a desirable systems approach, which is derived from Figure 4.4 with a focus on operational specifics in practice, technically and managerially. The following closed-loop steps highlight how we can manage and control a service system in real time by leveraging the technological advances as of today.

Loop Step 1. The dynamic, collaborative, and connected working practices with the support of computing and networking technologies are monitored and captured. As a result, data relevant or irrelevant to service provision become available.

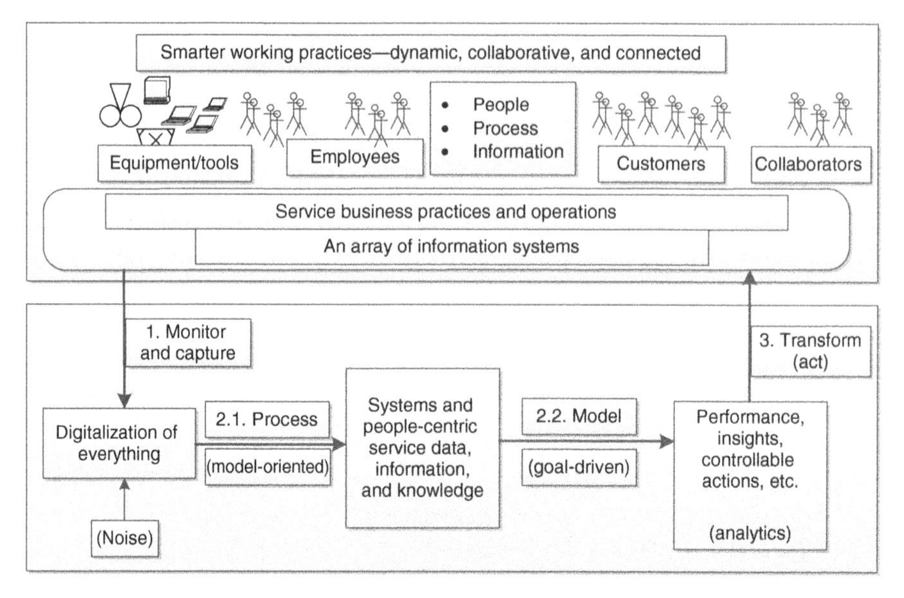

FIGURE 5.11 *Closed-loop systems approach to manage and control a service system.*

Loop Step 2. Data are overwhelming. Hence, the collected big data must be processed in order to feed into the models that are mainly driven by the decision-making and business analytics across the operated service system.

Loop Step 3. Both service rendering actions and organizational transformation procedures for improving systems' competitiveness and sustainability and service innovation are recommended.

As soon as the recommended service rendering actions and organizational transformation procedures are executed, a new loop with innovations starts over.

In summary, service organizations must offer and deliver innovative services in order to stay competitive and sustainable. As service attributes reveal across sociopsychological and technical dimensions, we must explore service systems in both social and technical dimensions. A well-defined services model and innovation framework will effectively guide and enable service organizations to design, develop, and execute not only their daily business operations but also their well-defined long-term growth strategic plans. Simply put, by executing right service management and engineering practices throughout the service lifecycle, they can ensure that they perform SMARTER service operations than their competitors.

5.5 FINAL REMARKS

In complying with the developed concept and principles of service science in the preceding chapters, in this chapter we presented organizational and IT perspectives of service systems and networks. First, we articulated that service is as an offering of a service system in fulfilling the customer's need through a process of transformation. In order to foster people's interactions and leverage social capitals in service, we then emphasized that we must put people first in any given service system. We further discussed that the empowerment of systems and technologies can facilitate desirable service system dynamics with a focus on smarter management, engineering, and operations. Finally, we concluded that the themes of competitiveness, sustainability, and innovation in service systems must be fully described and well addressed in both the social and technical dimensions.

Historically, we know that rules of thumb have been quite effective when applied to a variety of business situations. In particular, at a time when situation information cannot be monitored and captured in real time, empirical studies surely make sense and prove extremely effective. For instance, the Pareto principle is one of the rules of thumb, which has been widespread applied in many causal analysis and decision-making circumstances. The Pareto principle (a.k.a., the 80–20 rule, the law of the vital few, and the principle of factor sparsity) essentially articulates that, for many situations, about 80% of the effects come from 20% of the causes (Koch, 2011).

Koch (2011) provides many good evidences and further argues that the stated empirical distribution has been roughly correct in a descriptive manner under many circumstances, ranging different aspects from business management, product engineering, to business operations. Some convincing examples in a service business organization could be as follows:

- About 80% of profits earned by the organization come from about 20% of the customers the organization served.
- About 80% of complaints come from 20% of the customers.
- About 80% of profits come from 20% of the time the organization spent.
- About 80% of sales come from 20% of products provided by the organization.
- About 80% of sales are generated by 20% of sales staff in the organization.

Truly many businesses have applied this rule of thumb to improve their profitability. The outcomes were dramatic when the known effective areas were considerably scrutinized for improvements. Indeed, the available resources could be fully leveraged when the rest was appropriately eliminated, ignored, or retrained. However, the rapid change in today's business world gradually shrinks the role of rules of thumb as we are confronted with significantly more sophisticated competitions than ever before.

As discussed earlier, service organizations offer service products to their prospective customers. The total value of a service is created only if the offered service is completely consumed by its designated customer. The service value equation described as $V = Q \times A = (T \times S_p) \times (U \times S_c)$ is then applied in evaluating the outcome of the completed service. It is clear that the service value depends on service products' technical attributes, that is, performed service functions that meet the customer's functional needs, and also the systemic behavior of its transformation process that satisfies service provision participants' socioemotional and sociopsychological needs.

In today's service sector, it is the people (customers and employees) rather than physical goods that must be put first and are in the center of the organizational structure and operations. As the value of a service lying along the process trajectory throughout the lifecycle of the service largely depends on the sociotechnical dynamics of the service system, we must track and trace the service trajectory, which is nothing but a service encounter chain as discussed in the preceding chapters. The interconnected service encounter chains created within a given service system essentially forms a corresponding service network.

However, a descriptive approach in a qualitative way is not sufficient to meet the ever-changing challenges. In other words, a better understanding of the systemic dynamics of a service network is necessary. Indeed, as discussed in Chapter 4, the holistic or systemic viewpoint is necessary, which allows us to focus on the "big picture," which is interacting relationships, and long-range view of service network dynamics. On the basis of Figure 4.4 that shows our envisioned qualitative and quantitative approach (Andriessen and Verburg, 2004; Qiu, 2009), in this book we advocate that an innovative, integrative, and interactive approach that is illustrated in Figure 5.12 can transform a service network throughout its lifecycle for competitive advantage.

A prescriptive approach in an integrated qualitative and quantitative manner is adopted to look at both the systemic dynamics of service systems and networks. As shown in Figure 5.12, a service system is considered as a sociotechnical system, which can be modeled as the number of daily business operations. A suite of mathematical models in the form of integrated structured equation model (SEM) and social network analysis (SNA) will then be applied to explain the dynamics of occurred service transformation processes, which ultimately helps to identify

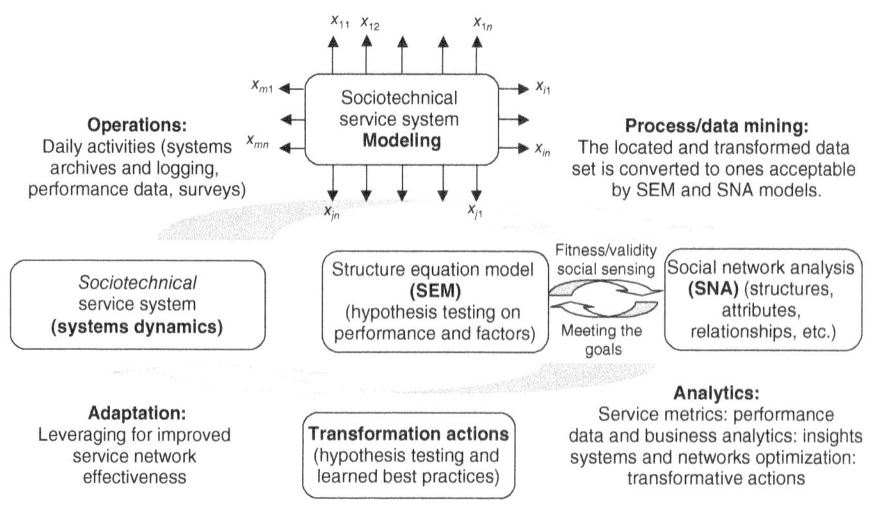

FIGURE 5.12 *Integrated people-centric approach to transform daily service business practices for competitive advantages.*

optimal transformative actions for service improvements and accordingly prescribe the service networks to be formed from the actions.

As shown in Figure 5.12, a two-step exploratory approach is essentially adopted in this book. A systems approach to gain the fundamental understanding of how a service system as a whole behaves will be first investigated. Specifically, SEMs can be applied to describe the system's performance and/or conduct necessary hypotheses testing and/or confirmatory factor analyses. Service network approaches such as SNAs to explore the interactions and insights of people-centered service networks will be then applied, aimed at understanding how service networks have been formed and behaved and how the service networks could evolve over time. The combined systems and network approach focuses on identifying the areas for service improvements across all service system constituents in a holistic and comprehensive manner.

Computational thinking-based theoretical discussions and applied examples are discussed in great detail in the next few chapters.

ACKNOWLEDGMENT

A significant portion of this chapter is derived from the chapter entitled "Information Technology as a Service" in *Enterprise Service Computing: From Concept to Deployment*—a book that was published by Idea Group Publishing (Hershey, PA) in 2007, Copyright 2012, IGI Global, www.igi-global.com. Reprinted by permission of the publisher.

REFERENCES

Andriessen, J. E., & Verburg, R. M. (2004). A Model for the Analysis of Virtual Teams. Chapter XV in *Virtual and Collaborative Teams: Process, Technologies and Practice*, ed. by S. Godar and S. Ferris. Hershey, PA: Idea Group Publishing.

Bharadwaj, A., El Sawy, O. A., Pavlou, P. A., & Venkatraman, N. (2013). Digital business strategy: toward a next generation of insights. *MIS Quarterly*, 37(2), 471–482.

Becker, B., Huselid, M., & Ulrich, D. (2001). *The HR Scorecard; Linking People, Strategy, and Performance*. Boston, MA: Harvard Business School Press.

Bell, P. (2005). Operations research for everyone (including poets). *OR/MS Today*, 32(4), 22–27.

Berman, S. (2012). Digital transformation: opportunities to create new business models. *Strategy and Leadership*, 40(2), 16–24.

Bieberstein, N., Bose, S., Walker, L., & Lynch, A. (2005). Impact of service-oriented architecture on enterprise systems, organizational structures, and individuals. *IBM systems journal*, 44(4), 691–708.

Bitner, M. J. (1992). Servicescapes: the impact of physical surroundings on customers and employees. *Journal of Marketing*, 56(2), 57–71.

Campbell, A., Lane, N, Miluzzo, E., Peterson, R., Lu, H., Zheng, X., Musolesi, M., Fodor, K., Eisenman, S., & Ahn, G. (2008). The rise of people-centric sensing. *IEEE Internet Computing*, July–August, 12–21.

Cao, Y., Gruca, T. S., & Klemz, B. R. (2003). Internet pricing, price satisfaction, and customer satisfaction. *International Journal of Electronic Commerce*, 8(2), 31–50.

Chase, R. B., & Dasu, S. (2008). Psychology of the Experience: The Missing Link in Service Science. in *Service Science, Management and Engineering Education for the 21st Century*, 35–40, eds. by B. Hefley and W. Murphy. US: Springer.

Cherbakov, L., Galambos, G., Harishankar, R., Kalyana, S., & Rackham, G. (2005). Impact of service orientation at the business level. *IBM Systems Journal*, 44(4), 653–668.

Chesbrough, H. W. (2011). Bringing open innovation to services. *MIT Sloan Management Review*, 52(2), 85–90.

Datz, T. (2004). What you need to know about service-oriented architecture. *CIO Magazine*, 78–85.

Dong, M., & Qiu, R. G. (2004). An approach to the design and development of an intelligent highway point-of-need information service system. *2004 Proceedings of the 7th International IEEE Conference on Intelligent Transportation Systems*, Washington, D.C., 673–678.

Fisk, R. (2009). A customer liberation manifesto. *Service Science*, 1(3), 135–141.

Girardin, F., Blat, J. Calabrese, F., Fiore, F., & Ratti, C. (2008). Digital footprinting: uncovering tourists with user-generated content. *Pervasive Computing*, October–December, 36–43.

Guo, B., Yu, Z., Zhou, X., & Zhang, D. (2012). Opportunistic IoT: exploring the social side of the internet of things. *2012 IEEE 16th International Conference on Computer Supported Cooperative Work in Design (CSCWD)*, Wuhan, China, 925–929.

Heskett, J. L., Jones, T. O., Loveman, G. W., Sasser, W. E., & Schlesinger, L. A. (1994). Putting the service-profit chain to work. *Harvard Business Review*, 72(2), 164–174.

IBM. (2004). Service Science: A New Academic Discipline? Retrieved on Feb. 4, 2006 from http://www.research.ibm.com/ssme.

IBM. (2011). The Social Business Advent of a New Age. IBM Software Group. Retrieved on Dec. 6, 2012 from http://www.ibm.com/smarterplanet/global/files/us_en_us_social business_epw14008usen.pdf.

Johnston, R. (2005). Service operations management: return to roots. *International Journal of Operations & Production Management*, 25(12), 1278–1297.

Kano, M., Koide, A., Liu, T. K., & Ramachandran, B. (2005). Analysis and simulation of business solutions in a service-oriented architecture. *IBM Systems Journal*, 44(4), 669–690.

Karmarkar, U. (2004). Will you survive the services revolution? *Harvard Business Review*, 82(6), 100–107.

Koch, R. (2011). *The 80/20 Principle: The Secret to Achieving More With Less*. Crown Business.

Korsten, P. J., Lesser, E., & Cortada, J. W. (2013). Social business: an opportunity to transform work and create value. *Strategy & Leadership*, 41(3), 20–28.

Larson, R. (1989). OR/MS and the services industries. *OR/MS Today*, April, 12–18.

Lovelock, C. H., & Wirtz, J. (2007). *Service Marketing: People, Technology, Strategy*, 6th ed. Upper Saddle River, NJ: Prentice Hall.

Menor, L. J., Tatikonda, M. V., & Sampson, S. E. (2002). New service development: areas for exploitation and exploration. *Journal of Operations Management*, 20(2), 135–157.

Newcomer, E., & Lomow, G. (2004). Understanding SOA with Web Services Addison-Wesley Professional.

Noah, C. (2010). Putting people first. *Rail Professional Magazine*, June 2010. Retrieved on Dec. 6, 2012 from http://www.railpro.co.uk/magazine/?idArticles=241.

Nunes, D., Tran, T. D., Raposo, D., Pinto, A., Gomes, A., & Silva, J. S. (2012). A web service-based framework model for people-centric sensing applications applied to social networking. *Sensors*, 12(2), 1688–1701.

Pfeffer, J., & Veiga, J. F. (1999). Putting people first for organizational success. *The Academy of Management Executive*, 13(2), 37–48.

Qiu, R. G. (2004). Manufacturing grid: a next generation manufacturing model. *2004 IEEE International Conference on Systems, Man and Cybernetics*, 5, 4667–4672.

Qiu, R. G. (2005). An Internet computing model for ensuring continuity of healthcare. *2005 IEEE International Conference on Systems, Man and Cybernetics*, 3, 2813–2818.

Qiu, R. G. (2007). Information Technology as a Service. Chapter 1 in *Enterprise Service Computing: From Concept to Deployment*. Hershey, PA: Idea Group Publishing.

Qiu, R. G. (2009). Computational thinking of service systems: dynamics and adaptiveness modeling. *Service Science*, 1(1), 42–55.

Qiu, R. G. (2013a). *Business-Oriented Enterprise Integration for Organizational Agility*. Hershey, PA: IGI Global.

Qiu, R. G. (2013b). Rethinking service encounters. *Submitted to Harvard Business Review*.

Qiu, R. G., Wysk, R., & Xu, Q. (2003). Extended structured adaptive supervisory control model of shop floor controls for an e-manufacturing system. *International Journal of Production Research*, 41(8), 1605–1620.

Ramdas, K., Teisberg, E., & Tucker, A. L. (2012). Four ways to reinvent service delivery. *Harvard Business Review*, 90(12), 98–106.

Rangaswamy, A., & Pal, N. (2005). Service innovation and new service business models: harnessing e-technology for value co-creation. *An eBRC White Paper*, 2005 Workshop on Service Innovation and New Service Business Models, Penn State.

Rust, R. (2004). A call for a wider range of service research. *Journal of Service Research*, 6, 211.

Rust, R., & Lemon, K. (2001). E-service and the consumer. *International Journal of Electronic Commerce*, 5(3), 85–101.

Smith, H., & Fingar, P. (2003). *Business Process Management: The Third Wave*. Tampa, FL: Meghan-Kiffer Press.

Thomas, E. (2005). *Service-Oriented Architecture: Concepts, Technology, and Design*. Prentice Hall.

WikiPsychometircs. (2013). Psychometircs. *Wikipedia*. Retrieved on June 6, 2013 from http://en.wikipedia.org/wiki/Psychometrics.

Wright, M., Filatotchev, I., Hoskisson, R., & Peng, M. (2005). Strategy research in emerging economies: challenging the conventional wisdom. *Journal of Management Studies*, 42(1), 1–33.

Zhang, X., & Prybutok, V. R. (2005). A consumer perspective of e-service quality. *Engineering Management, IEEE Transactions on*, 52(4), 461–477.

Zimmermann, O., Krogdahl, O., & Gee, C. (2004). Elements of service-oriented analysis and design. Retrieved Apr. 4, 2012 from http://www-128.ibm.com/developerworks/library/ws-soad1/.

6

Computational Thinking of Service Systems and Networks

Regardless of categories of services that a service system plans to offer, a service must be designed, developed, and delivered. Whether the service can stay competitive in its marketplace truly depends on whether it meets the needs of customers whom the service system intends to serve. As discussed in preceding chapters, the level of satisfaction that is received by the served customers depends on the efficient, effective, and smart operations of the service system. The people-centric dynamics of a service system in business operations essentially manifests itself in the theme of the service interactions between customers and providers. In both theory and practice, a service network is then used to describe the collective service and systemic behaviors of the involved service interactions.

We now understand that a competitive service system must put people (customers and employees) rather than physical goods in the center of its organizational structure and operations. The people-centric concept is explicitly reflected in the service value equation, $V = Q \times A = (T \times S_{\mathrm{p}}) \times (U \times S_{\mathrm{c}})$, which holds in any service industry. Indeed, the equation expresses that we must focus on service design, development, and delivery using all available means to realize respective values for both service providers and service customers, technically and sociopsychologically.

Depending on the exploratory goal under study, there are a variety of approaches to explore service systems. This book, as concluded in Chapter 5, is interested in a prescriptive approach to explore service systems in an integrated qualitative and quantitative manner, aimed at helping service organizations look at their systemic

Service Science: The Foundations of Service Engineering and Management, First Edition. Robin G. Qiu.

dynamics of service systems and networks. As we like to have a suite of mathematical models in the form of integrated structural equation modeling (SEM) and social network analysis (SNA) to be applied to capture and explain the dynamics of ongoing service transformation processes, computational thinking of service systems and networks must be fully adopted. Ultimately, we can identify optimal transformative actions for continuous service improvements, resulting in that prescribed and managed service systems and networks can serve both customers and service providers in a competitive manner.

In this chapter, we first discuss the technical supports of real-time monitoring and capturing people-centric dynamics in a service system. We then articulate the need for computational thinking in developing appropriate approaches to model service systems and networks. Using computational thinking, in Section 6.3, a service system is conceivably constructed based on the business process management (BPM) concept while being formularized using a structured workflow language and π-calculus. The new model called a (computational and configurable service system) C^2S^2 model mainly focuses on the future enablement of system configurability by taking into account human interactions and consequences. In Section 6.4, a brief discussion of metrics and methods to determine whether a given service system is operating on track (e.g., satisfaction level) will be provided, aimed at enabling quantitative, predictive, and social-technical analytics at the point of need throughout the lifecycle of services. In Section 6.5, we discuss an example of C^2S^2 modeling using the PDGroup case study we had earlier. In Section 6.6, a brief conclusion is given, and the further study of transformation mechanisms for reconfiguration, continuous and real-time optimization of a service system is also presented.

6.1 MONITORING AND CAPTURING PEOPLE-CENTRIC SERVICE NETWORK DYNAMICS IN REAL TIME

Change is inevitable. A highly vibrant, value-driven postrecession economy in the United States is surely on the rise. To get comprehensive and timely understanding of the postrecession consumers, Gerzema and D'Antonio (2010) travel across the States to examine the value shifts sweeping the nation. Through in-depth observation and interviews with experts, and also by conducting comprehensive analytics using voluminous historical brand data, they study the changes of consumer expectations, explore the shifting values and consumer behaviors, and explain what the shift means to businesses and leaders.

"This value-led consumerism is not a small, isolated target market. Over half the U.S. population is now embracing these value shifts. They are seeking better instead of more, virtue instead of hype, and experience over promises. The postcrisis consumer, already highly marketing-savvy and armed with the leveling powers of social connection and critique, is now an even more potent and unpredictable force in the marketplace. People are looking for value and values" (Kotler, 2010).

Exploring the people components of service systems, we find that two constituents (i.e., S_p and S_c) in the corresponding service equation $V = Q \times A =$

$(T \times S_p) \times (U \times S_c)$ are unceasingly and unpredictably changing as time goes. In addition to the change in how people value what they are buying, consuming, and retaining, changes occurs rapidly in how organizations should be managed and operated. For instance, by relying on the database of consumers, Gerzema and D'Antonio (2010) investigate how people felt about the government, the economy, and the (mostly male) leaders making decisions at work. From their study, they find that substantial majorities were not well satisfied with their organizations and pessimistic about their quality of life. They are particularly intrigued by the observations about gender at work. "Two-thirds said the world would be a better place if men thought more like women. Gerzema [and D'Antonio] also asked consumers to characterize 125 traits as male, female, or neutral and to indicate those most desirable in modern leaders. Topping the list of most desirable traits includes patience, expressiveness, intuition, flexibility, empathy, and many other traits identified by respondents as feminine" (Buchanan, 2013).

With a focus on the change of leadership and leader's styles in organizations, we have surely witnessed the progression of leadership from "command-and-control (roughly through the 1980s) to empower-and-track (the 1990s to mid-2000s) to connect-and-nurture (today)" (Buchanan, 2013). Power has more influence rather than control in many circumstances, which indeed is the trend of leadership transformation in this new millennium. The following seven identified traits are popularly perceived as what a leader needs today (Buchanan, 2013):

- *Empathy.* Being sensitive to the thoughts and feelings of others.
- *Vulnerability.* Owning up to one's limitations and asking for help.
- *Humility.* Seeking to serve others and to share credit.
- *Inclusiveness.* Soliciting and listening to many voices.
- *Generosity.* Being liberal with time, contacts, advice, and support.
- *Balance.* Giving life, as well as work, its due.
- *Patience.* Taking a long-term view.

> Feminine traits and values are a new form of innovation. They are an untapped form of competitive advantage (Gerzema and D'Antonio, 2010).

Note that S_p and S_c in the service value equation are highly correlated in the service sector. In addition to heavily investing on leadership development in service organizations, we must pay much attention to employees' job satisfaction. According to Buchanan (2013), "The Holy Grail in business today is engagement: employees' energy, enthusiasm, and commitment to their companies. Engagement has a powerful effect not only on productivity but also on profitability and customer metrics, numerous studies show." Engagement is not something we can buy as it must be nurtured and nourished over time through promoting and fostering employees' physical, emotional, and social well-being.

In summary, changes happen fast, everywhere, and unpredictably. Without exception, this is particularly true about service systems and networks we mainly study in this book. Unless we can capture the changes of service systems in real time, we cannot get the insights of the dynamics of the service networks and transform service encounters adaptively. Promisingly, when we put people first and build and operate IT-empowered service systems, monitoring and capturing relevant service encounters become possible. Indeed, because of the fast advances of computing, networks, and big data technologies, computational thinking of service systems and networks has its practical meaning. Generally speaking, IT makes possible for us to conduct real-time explorations of sociotechnical service systems with scientific rigor.

6.1.1 Computational Thinking of Service Systems and Networks: A Necessity in Service Science

Let us briefly recap what we discussed in Chapter 3. We understand that service-dominant thinking is essential. We know we must leverage all the necessary means to explore services and accordingly service systems and networks:

- Our approach must be process-driven and people-centric. Once again, service is a process of transformation of the customer's needs utilizing the operations' resources, in which dimensions of customer experience manifest themselves in the themes of a service encounter or service encounter chain. As compared to manufacturing, service is people-centric, which must be cocreated by customers and providers.
- Our approach must be holistic. The holistic or systemic viewpoint focuses on the "big picture" and the long-range view of systems dynamics, looking at a service organization as a collection of domain systems that constitute a whole. The systematic view allows us to see how each and every service activity is operated. We analyze the efficiency and effectiveness of each activity and accordingly control and manage them in a decisive manner. By paying attention to the whole, a holistic perspective thus allows us to understand and orchestrate service encounters among business domains across the service lifecycles.

As we discussed earlier, the prior lack of means to monitor and capture people's dynamics throughout the service lifecycle has prohibited us from gaining insights into the service encounter chains or networks. However, the rapid development of digitization and networking technologies has made possible the needed means and methods for us to change this.

Service networks essentially are the networks view of the behavioral dynamics of a service system, describing how interwoven service encounters in service meet the customers' needs and how the service system might evolve over time. As recognized, computational thinking that fully leverages today's ubiquitous digitalized information, computing capability, and computational power has evolved as an optimal way of solving problems, designing systems, and understanding human behavior. Computational thinking promotes qualitative and quantitative thinking in terms of abstractions, modeling, algorithms, explorations, and understanding the consequences of scale and adaptation, not only for reasons of efficiency and effectiveness but also for economic and social reasons. Therefore, computational thinking of service systems and networks becomes a necessity in developing service science.

6.1.2 Big Data in Support of Computational Thinking of Service Systems and Networks

Today, the term of big data is pervasive. Truly, big data still engenders confusion as different people might have different interpretations. In fact, data have been overwhelming over the years. Many organizations have had varieties of challenges in storing, processing, and making use of the massive data they collected intentionally or unintentionally. Because mobile computing devices, social media, and Web 2.0 applications help to collect people-centric data and information, organizations become interested in big data and start to explore and invest on new solutions to process and analyze this vast array of data and information. For instance, through capturing customers consuming dynamics in real time, organizations want to identify the market trend from the archived big data and promptly transform by aligning their business practices with the changes (IBM Global Business Services, 2012).

According to Schroeck (2013), "it's not until recently that three important trends are converging to usher in a new era of big data—one that will fundamentally transform how businesses operate and how they engage with customers, suppliers, partners, and employees." The three trends identified by Schroeck (2013) include the emerging mass digitization of things, the growing popularity of social media, and the significant technological advances in the areas of mobile computing, data storage, and networking and telecommunications. Schroeck argues that the convergence of these three trends is enabling an organization to effectively utilize these new streams of big data in assisting the management and employees in making informed decisions throughout the organization so that the desired values can be delivered to providers and customers, respectively, and optimally.

To better understand how big data are currently utilized to benefit businesses in the industry, in 2012 the IBM Institute for Business Value worked together with the Saïd Business School at the University of Oxford in surveying 1144 business and IT professionals around the world, and interviewing more than two dozen academics, subject matter experts, and business executives. According to IBM Global Business Services (2012), "To compete in a globally-integrated economy, today's organizations need a comprehensive understanding of markets, customers, products, regulations, competitors, suppliers, employees and more. This understanding demands the effective use of information and analytics. Next to their employees, many companies consider information to be their most valuable and differentiated asset."

The Internet of things is gradually becoming reality. Evidently, the use of advanced sensor-based devices and instruments is pervasive; organizations and individuals can monitor in real time everything from the status of a freight fleet across the countryside to hourly power usage in facilities and even individual appliances at their homes or irrigation devices in remote farms. Moreover, by further leveraging the advances in mobile computing and social networks, service providers and customers around the world can now easily and in real time communicate and interact with each other.

Because of the availability of streams of massive data, an array of advanced analytics methods and techniques come into being, enabling organizations to extract insights from big data with previously unachievable levels of sophistication, accuracy, efficiency, and effectiveness (Schroeck, 2013).

Indeed, today organizations are strongly interested in leveraging big data to realize customer-centric business objectives across industries (IBM Global Business Services, 2012; Partnership for Public Service, 2011, 2012). Over time, they will surely apply the newly emerging analytics methods and techniques to address other business objectives, including satisfying employees' socioemotional needs and aligning physical and information flows in support of satisfactory service encounters at the point of need. Simply put, because of big data technologies, comprehensive and overwhelming data on people-centric dynamics can be well and real time collected, transformed, and stored. As a result, service organizations can convert big data and analytics insights into results and opportunities with the support of newly developed sciences, tools, and methodologies, including service science.

6.2 COMPUTATIONAL THINKING OF SERVICE SYSTEMS AND NETWORKS

As the world becomes more complex and uncertain socially and economically, computational thinking that fully leverages today's ubiquitous digitalized information and the availability of massive data, computing capability, and computational power has evolved as an optimal way of solving problems, designing systems, and understanding human behavior. As discussed earlier, computational thinking essentially is qualitative and quantitative thinking in terms of abstractions, modeling, algorithms, explorations, and understanding the consequences of scale and adaptation, not only for reasons of efficiency and effectiveness but also for economic and social reasons (CMU, 2009).

In general, the adaptive capability or sustainable competitiveness of a service system largely depends on right principles and methods, and appropriate tools employed for conducting quantitative, predictive, and social-technical analytics at the point of need throughout the lifecycle of services. Hence, we must fully leverage computational thinking to develop necessary approaches and tools, which can be put into use by service organizations. As a result, with the help of the developed approaches and tools, service organizations can explore, model, capture, and manage systemic behaviors, interactions, connections, complex relations, and interdependencies of their service systems.

In this book, given a service system with known service products, resources, and operations, we mainly explore a computational model of the operational dynamics and system adaptiveness of the service system by looking into its systemic operations, behaviors, and interactions. As a breakthrough in delivering a resolution to the people-centric enablement in a service system, end users (service consumers) and employees (service enablers) must be simultaneously taken into consideration throughout the lifecycle of services. An adopted computational approach should be able to model people's physiological and psychological issues, cognitive capability, and sociological constraints (to a certain extent at the very beginning). People-centric sensing is the fundamental enablement, providing all potentials of collecting human activities data throughout the lifecycle of services. More specifically, appropriate mechanisms should be developed, aimed at helping service organizations

1. timely capture end users' requirements, changes, expectation, and satisfaction in a variety of technical, social, and cultural aspects;
2. efficiently and cost-effectively provide employees right means and assistances to engineer services while promptly responding the changes; and
3. allow involved people consciously infuse as much intelligence as possible into all levels and aspects of decision making to assure necessary system adaptiveness from time to time.

6.3 MODELING OF A CONFIGURABLE AND COMPETITIVE SERVICE SYSTEM

BPM mainly focuses on managing changes to improve business processes. By embracing core principles of striving for collaboration, agility, innovation, and integration with state-of-the-art technology (Weske, 2007, Qiu et al., 2008), BPM is considered as a holistic management and business process operation approach toward cutting-edge business competitiveness (Figure 6.1). BPM activities can largely be grouped into five categories from design, modeling, execution, monitoring, to optimization, aimed at ensuring the continuous process improvement on operation effectiveness and efficiency in order to stay competitive.

As the twenty-first century becomes an information- and knowledge-based service-led economy, countless new products and services have been spawned, creating new opportunities that often change the very nature of businesses and organizations. Many world-class business organizations have been transforming themselves by taking the BPM approach for competitive advantages. However, there lacks a novel science that can govern and guide the transformation of a service organization (i.e., service system) to ensure that the organization will be (i) people-centric, information-driven, e-oriented, and satisfaction-focused, and (ii) able to cultivate people to collaborate and innovate.

Note that a service system usually integrates different types of resources (capital, labor, technology, and innovation), realizing different scales of revenues and profits, and most importantly different competitiveness during a competition (Figure 6.2). Although it is common to use a profit equation to measure the competitiveness of a given service system at a given time, it might make more sense to measure its viability as the sustainable competitiveness of the service system. For instance, today's globally integrated economy is highlighted with dynamics and uncertainty; thus, system viability might be calculated using a suite of performance factors and business environmental indicators collected during operations in a comprehensive and scientific manner.

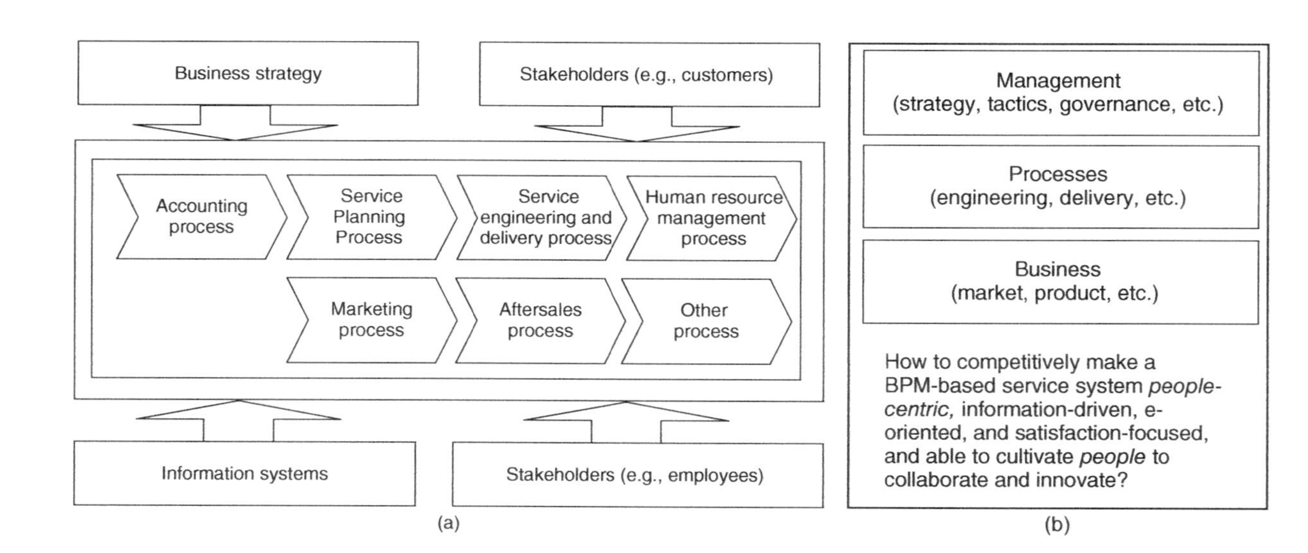

FIGURE 6.1 *Business process management — a process-driven approach. (a) Organizational level model and (b) operational view.*

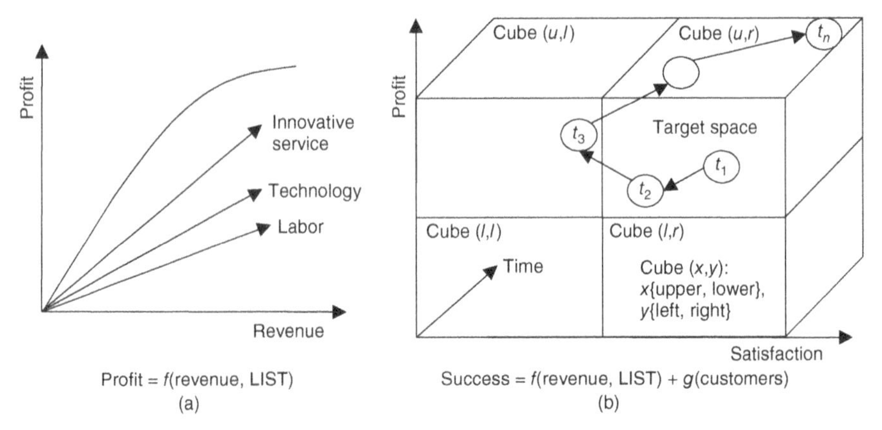

FIGURE 6.2 *Service system's competitiveness. (a) Profit model and (b) system trajectory model.*

More specifically, at a given time and market for a given service, the competitiveness of a service system might mainly rely on a combination of both profit and user satisfaction as shown in Figure 6.2(b). To retain its competitive advantage, it is obvious that the service system's trajectory should be well controlled as time goes. In other words, the system must navigate in its defined business target space during operations in order to outperform its competitors. Under different circumstances, the success might be measured using different or more dimensions of measures, such as an equation of profit, satisfaction, and productivity. Owing to the existence of a variety of uncertainties, the business goal at a given period might also require dynamic adjustment by navigating the system across different designated target spaces.

In the remaining section, a service system is conceivably constructed based on the BPM process-driven concept while being computationally formularized using an automaton-based structured workflow language and π-calculus. The newly proposed C^2S^2 model mainly focuses on the enablement of system configurability by taking into account service system's characteristics (e.g., people-sensing, cocreation values, human interactions, and consequences).

6.3.1 The Systemic View of a Service System

The reality is that many aspects in the market are correlated in today's integrated global economy, so are the service systems. To better understand how a service system performs, it becomes clear that the levels and details of analyses should be broad enough and comprehensive enough to reveal all the necessary interactions, interdependencies, and relationships within the service system. The process-driven BPM approach appears as an appropriate choice as drilling down into specific processes and their nested subprocesses is a necessity to reveal the details in a process-driven system.

However, as mentioned earlier, no matter where, why, when, who, what, and how, by the end of the day the real value of a delivered service most likely lies in its ability to satisfy customer's need from a business competitiveness perspective. Extreme customer service helps businesses survive; BusinessWeek (2009) publishes a list showing that 25 companies get it right in a tough year. Thus, the understanding of how process activities (or tasks) performed as individuals and a whole during the lifecycle of service affects customer satisfaction becomes essential. The understanding is also a necessity for systemic decision making on how a *service system* should be transformed for improved customer satisfaction or a competitive advantage in a technically capable, financially available and justifiable, and socially amiable and adaptable manner. In other words, the systemic view of a service system capturing the issues that are mainly related to operations, integration, human behaviors, and globalization will play a key role in computationally modeling the service system.

Yet Another Workflow Language (YAWL) has been developed to directly support all control flow patterns required in a workflow system (van der Aalst and ter Hofstede, 2005). "Corporations are notorious for introducing technology without considering the human consequences" (Kanellos, 2004). This is quite true for the existing workflow models and languages. In this chapter, the set of symbols in YAWL is revised and expanded to better support BPM-based process flows (Weske, 2007), incorporating human tasks and human interactions to meet the needs of our modeling approach. Figure 6.3 shows the revised set of symbols that will be used in the presented C^2S^2 model.

Figure 6.4 shows a typical example of a virtual project development system (VPDS) using the revised set of symbols (van der Aalst and ter Hofstede, 2005; Weske, 2007). This VPDS is composed of multiple teams working across geographic, political, cultural, and enterprise organizational boundaries, responsible for conducting research and development projects in a global high tech bellwether service organization. One team takes leadership, managing project development overall issues, such as customer contacts and requirement solicitation, service product design and architecture, work breakdown structure design, progress supervision,

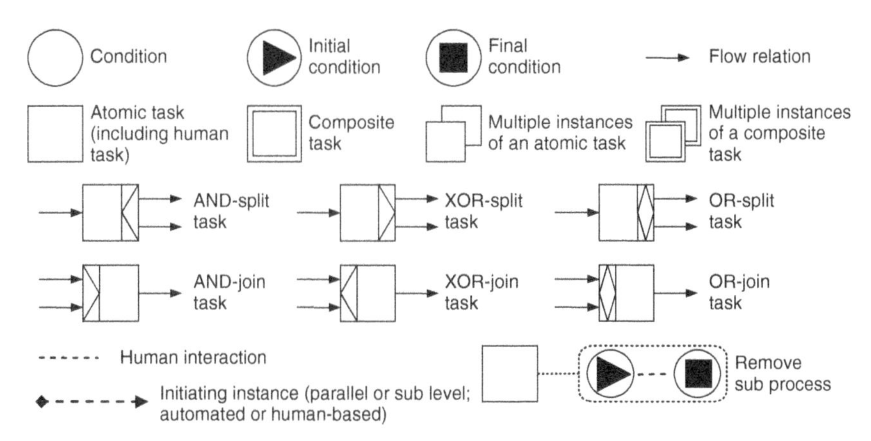

FIGURE 6.3 *Symbols used in C^2S^2 models. (Source: Revised and adapted from YAWL; van der Aalst and ter Hofstede, 2005).*

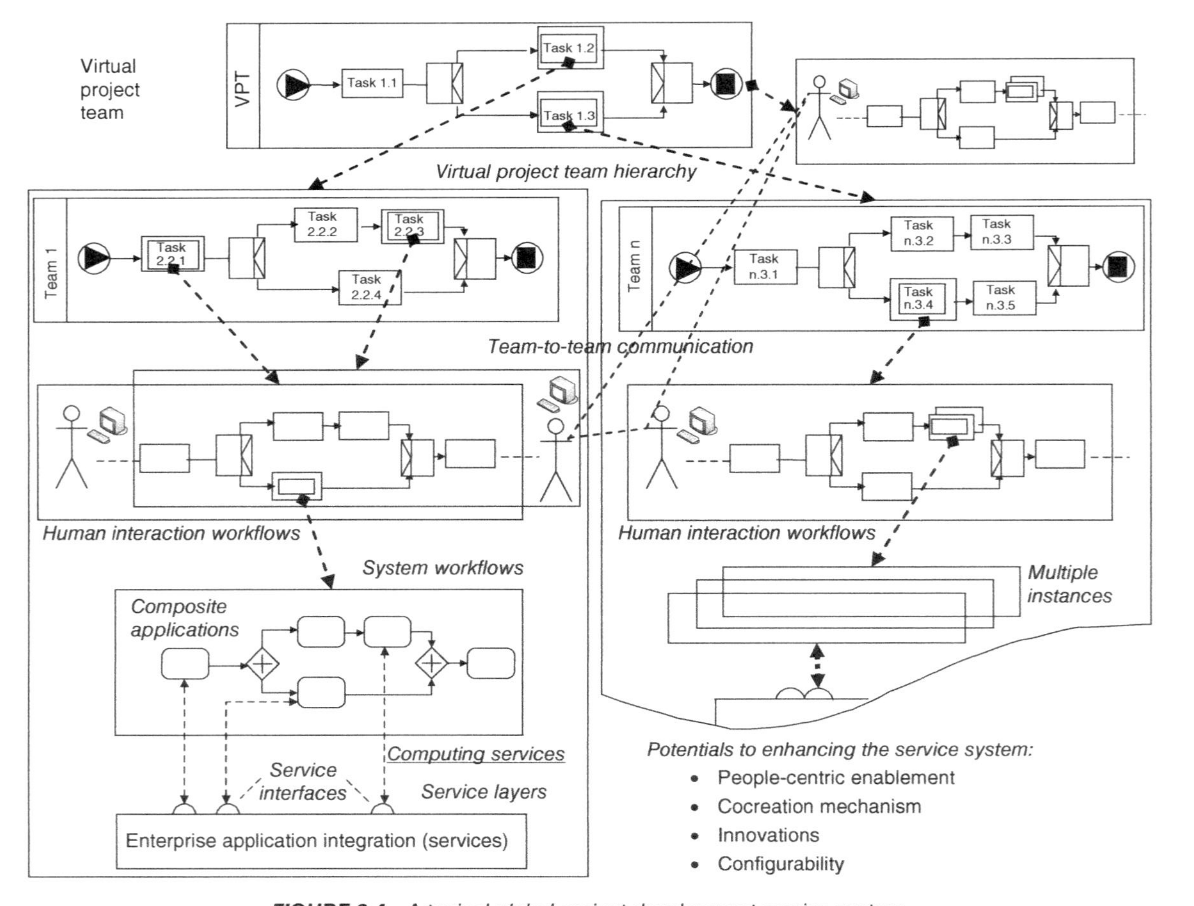

FIGURE 6.4 *A typical global project development service system.*

coordination among teams, and other managerial needs. Other teams are typically located in different regions or countries and focus on developing specific task components (i.e., part of projects or subprojects) based on their respective skill sets. By spanning institutional, geographical, and cultural boundaries, this VPDS as a typical sociotechnical service system aims at leveraging the best-of-breed talents in an integrated and collaborative way for a competitive advantage in the integrated global marketplace.

For simplicity, while without losing the generality in modeling service systems, this VPDS essentially is a scaled-down version of the global project development group (PDGroup) discussed in Chapter 2. Discussions based on the PDGroup will be provided in later sections of this chapter.

In general, the success of the VPDS operations largely depends on how people who are involved in the processes perform collectively, how team members individually follow the identified best practices, how they collaborate with each other by leveraging the best-of-breed talents, and how the VPDS navigates with uncertainties that surely exist from time to time. The following possible measurements collectively reflect how the VPDS is doing at the point of assessment across the VPDS:

- Conflicts (indicated by culture issues, managerial styles, personalities, etc.)
- Communication effectiveness (indicated by language barrier, customs, infrastructures, meeting setting, etc.)
- Project matter (indicated by size, complexity, geographical locations, the number of teams and members, team competency, etc.)
- Project management (indicated by project management method, tools, cost, commitment, risk matrices, control, etc.)
- Project goal (indicated by targeted marketplace, timeframe, etc.); and service satisfaction (indicated by customer feedbacks, loyalty, etc.).

Given the complexity and uncertainties, the VPDS' presumable superiority over centralized project development systems is not warranted if we do not have a scientific method and a suite of tools to efficiently and cost-effectively manage its end-to-end operations on a regular basis. A computational model employed for capturing the operational dynamics and trajectory of the VPDS becomes the key to explore its systemic operations, behaviors, and interactions. Being able to navigate in its target space at a given time, the VPDS would yield a more predictable, controllable, and sustainable service business.

6.3.2 The Dynamics of Processes in a Service System

Processes are the building blocks in a BPM service system. A process in a service system is a collection of related, ordered, and structured activities or tasks, which is typically organized for producing a designated service to meet a particular business operational need. When the service requires a divide-and-conquer approach by leveraging system resources (e.g., best-of-breed talents), the process can be recursively decomposed into subprocesses as shown in Figure 6.4. No matter how many process levels a service system has, each process should have its boundaries, inputs and outputs, dependencies, and communication channels clearly defined in the system hierarchy.

Dependent on circumstances, levels of processes could take a hierarchical structure following the defined work breakdown structures so as to ensure that service work dependencies can be well controlled and managed. However, beyond communications facilitated by the top-level process, personal communications among processes at all levels should be encouraged to leverage the diversity and culture of best-of-breed approach and ultimately cultivate people to innovate.

As a VPDS example seen in Figure 6.4, no matter how many projects or subprojects are under research and development in a team, the team follows a given business (project development) process flow. The flow dynamics of a team in a VPDS can be schematically illustrated using a systemic process diagram (Figure 6.5). Each flow is essentially a logic sequence of different task operations and obedient to its designated control patterns. For a task, it could be just a variety of activities performed at a given discrete time, by a given group of people. Using the Unified Modeling Language, Figure 6.6 conceptually shows the relationships and dependencies among the entities in a given process.

In reference to the extended workflow definition using YAWL by Weske (2007), a *service system* process in the proposed C^2S^2 model can be formally defined as a seven-tuple workflow net

$$S = (C, i, o, T, F, A, \pi) \tag{6.1}$$

where

- C is a set of conditions;
- $i \in C$ is the initial condition;

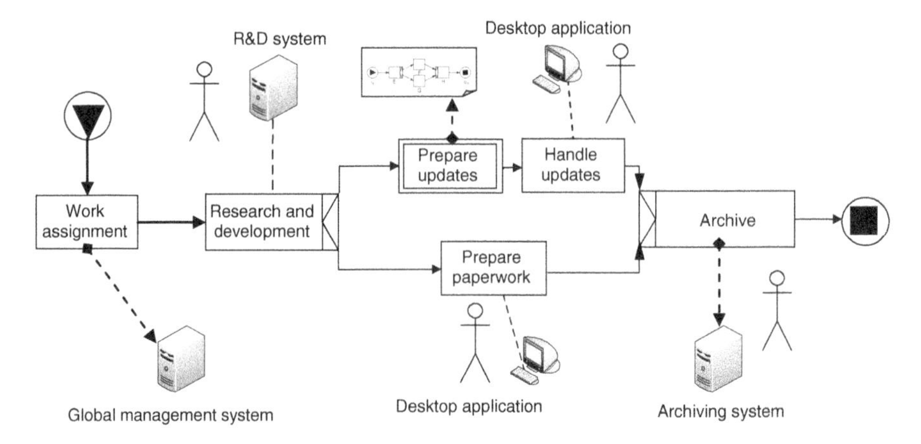

FIGURE 6.5 *A typical VPDS process flow. (Source: Revised based on Weske 2007.)*

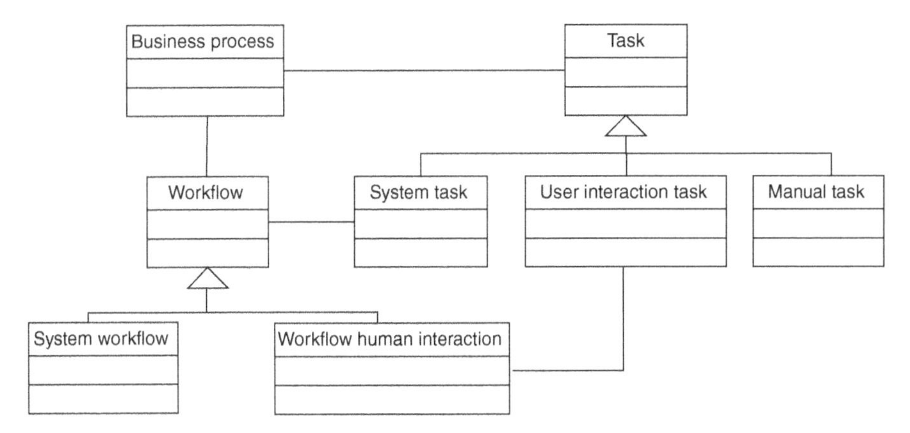

FIGURE 6.6 *Process conceptual entity model. (Source: Revised based on Weske 2007).*

- $o \in C$ is the final condition;
- T is a set of tasks, such that $C \cap T = 0$;
- $F \subseteq (C - \{o\} \times T) \cup (T \times C - \{i\}) \cup (T \times T)$ is a flow function, such that every node in the defined graph $(C \cup T, F)$ is on a directed path from i to o;
- A is a family of finite sets of task-oriented attributes $\{A(q)\}_{q \in C \cup T}$, where $A(q) = \{\text{split}, \text{join}, \text{rem}, \text{nofi}, \Phi(q)\}$:
  - *Split.* $t^{\text{split}} \in T \rightarrow \{\text{And, Xor, Or, Null}\}$, which specifies the split behavior of a task;
  - *Join.* ${}^{\text{join}}t \in T \rightarrow \{\text{And, Xor, Or, Null}\}$, which specifies the join behavior of a task;
  - *Rem.* $t^{\text{rem}} \in T \nrightarrow \wp$ for $\wp \in 2$, $S = (C, i, o, T, F, A, \pi)$, which specifies the subset of the net that should be removed when the task is executed;
  - *Nofi.* $t^{\text{nofi}} \in T \nrightarrow N \times N^{\text{inf}} \times N^{\text{inf}} \times \{\text{dynamic}, \text{static}\}$ specifies the number of instances of each task (min, max, threshold for continuation) and its dynamic/static creation of instances, where N^{inf} indicates that it is a set including an infinite in addition to the natural numbers;
  - $\Phi(q)$ for $q \in C \cup T$ and $\Phi(S) = \cup_{q \in C \cup T} \Phi(q)$ is the task hierarchy (e.g., a work breakdown structure) map function given at node q, where $\Phi(S)$ maps out all the tasks defined in the hierarchy.
- π is a set of collaborative communications defined using π-calculus. $\pi = \cup_{q \in C \cup T} \pi_q^{\Phi(q)}$, where $\pi_q^{\Phi(q)}$ for $q \in C \cup T$ is a collaborative communication with other concurrent processes by receiving and/or sending activity-related data through automated or manual channels: s, where $s = \cup_{q \in C \cup T}(s_q + \bar{s}_q)$. More specially,
  - $\pi_i^{\Phi(i)} = s_i(i^{\Phi(i)}).S$ indicates that process S gets instantiated and initiated after receiving a service task;
  - $\pi_o^{\Phi(o)} = \bar{s}_o\langle o^{\Phi(o)} \rangle.0$ indicates that process S gets terminated and removed after sending out the outcomes of the completed service task;

- $\pi_q^{\Phi(q)} = \overline{s_q} < a^{\Phi(q)} > .S + s_q(a^{\Phi(q)}).S$ for $q \in (C - i - o) \cup T$ indicates receiving and/or sending activity-related data at node q through automated or manual channels during the operations of this instantiated process:
  - If $\overline{s_q}$ is a newly established channel at q, $\overline{s_q}$ is defined as $(\gamma \overline{s_q})$; if no channel is needed, then $\overline{s_q}$ is not defined, that is, Λ (no sending channel).
  - If s_q is a newly established channel at q, s_q is defined as (γs_q); if no channel is needed, then s_q is not defined, that is, Λ (no receiving channel).

Generally speaking, a process instance is instantiated from its predefined process model whenever there is a new BPM-enabled service; the new service is operationally created as a new input (Figure 6.7). This instantiated BPM process instance will be terminated or removed as soon as the service is completed or has an exception requiring a process removal. Data related to all the activities (tasks, communications, system behaviors, and so on and so forth; automated or manual) during the process operations should be timestamp logged. This should be enforced for all levels of processes for a given *service system* although in reality it is well understood that only part of the needed data and information might be saved and collected.

6.3.3 The Dynamics of a Service System

A formal computational model of a *service system* is an unambiguous description of the system dynamics in light of control, communications, and interactions across all

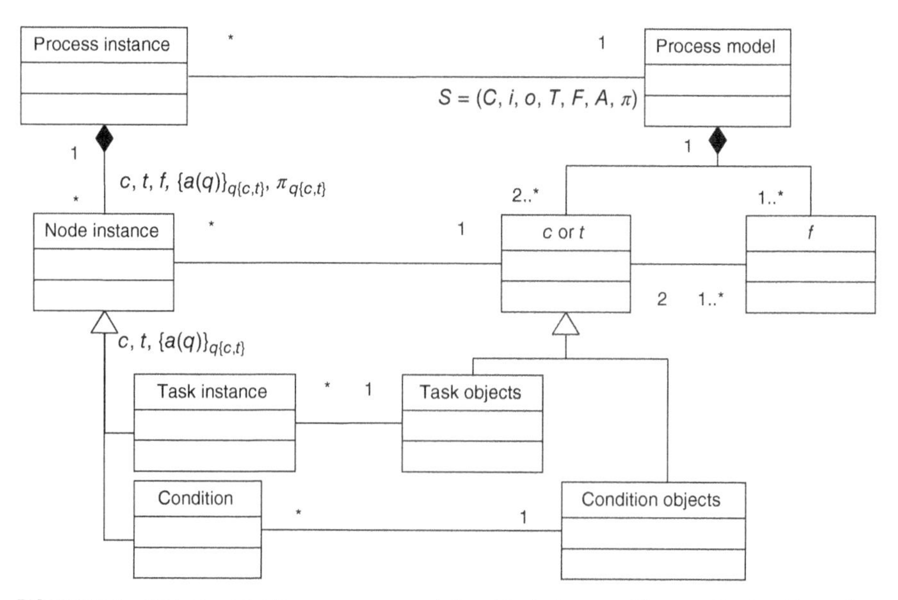

FIGURE 6.7 *Object model for a process model and its instances. (Source: Revised based on Weske 2007.)*

the involved processes. It is an abstract view of the system, which specifies the functionality and behavior of the system without being constrained by implementation details.

Formally, a *service system* can be defined as a five-tuple computational machine

$$M = (Q, \text{top}, T^{\diamond}, \text{map}, \Pi) \tag{6.2}$$

where

- $Q = \cup S$ is a set of *service system* processes;
- $\text{top} \in Q$ is a set of top-level processes;
- $T^{\diamond} = \cup_{S \in Q} T_S$ is the set of all tasks, such that $(\forall i)(\forall j)(S_i, S_j \in Q), i \neq j, S_i \neq S_j \Rightarrow (C_{Si} \cup T_{Si}) \cap (C_{Sj} \cup T_{Sj}) = 0$;
- *Map*. $T^{\diamond} \nrightarrow \Phi(Q\text{-}\{\text{top}\})$ is a mapping function that specifies the task hierarchies for each composite task in the *service system*;
- $\Pi = \cup_{S \in Q} \pi_S$ is the set of all collaborative communications in the *service system*.

6.3.3.1 *The Top Level of a Service System* When multiple services are engineered and managed throughout a *service system* simultaneously, there will be multiple processes at the top level. Each unique type of service engineered and delivered by the *service system* could require a unique process model as it might require a unique collection of related, ordered, and structured tasks. However, the same type of services should be cocreated by following the same process model. In other words, multiple instantiated process instances from the same process model can be executed to deal with the same types of services at the same time. Figure 6.8 schematically illustrates the top level of a *service system* in which three projects are developed at the same time. As shown in Figure 6.8, two out of three belong to the same type of services.

According to the above-proposed *service system* formal model M, its parallel computational process models executed at its top level can be further defined by

$$\text{top} = \left(S_1^1 \,||\, S_2^1 \,||\, \ldots \,||\, S_i^1 \,||\, \ldots \,||\, S_m^1\right) \text{ or}$$

$$\text{top} = \left(\bigcup_{j=1}^{n_1} S_{1j}^1 \,||\, \bigcup_{j=1}^{n_2} S_{2j}^1 \,||\, \ldots \,||\, \bigcup_{j=1}^{n_i} S_{ij}^1 \,||\, \ldots \,||\, \bigcup_{j=1}^{n_m} S_{mj}^1\right) \tag{6.3}$$

where superscript 1 indicates level 1 (i.e., the top level); $i = 1, \ldots, m$, m is the number of unique processes running simultaneously at the top level. For each process model, if multiple instances are instantiated, then the process can be further defined by

$$S_i^1 = \bigcup !S_{ij}^1 = \bigcup_{j=1}^{n_i} S_{ij}^1 \tag{6.4}$$

where $!S_{ij}^1$ is a process instance (i.e., process replication) for the ith type of service, $j = 1, \ldots, n_i$, n_i is the number of instances for the same type of service.

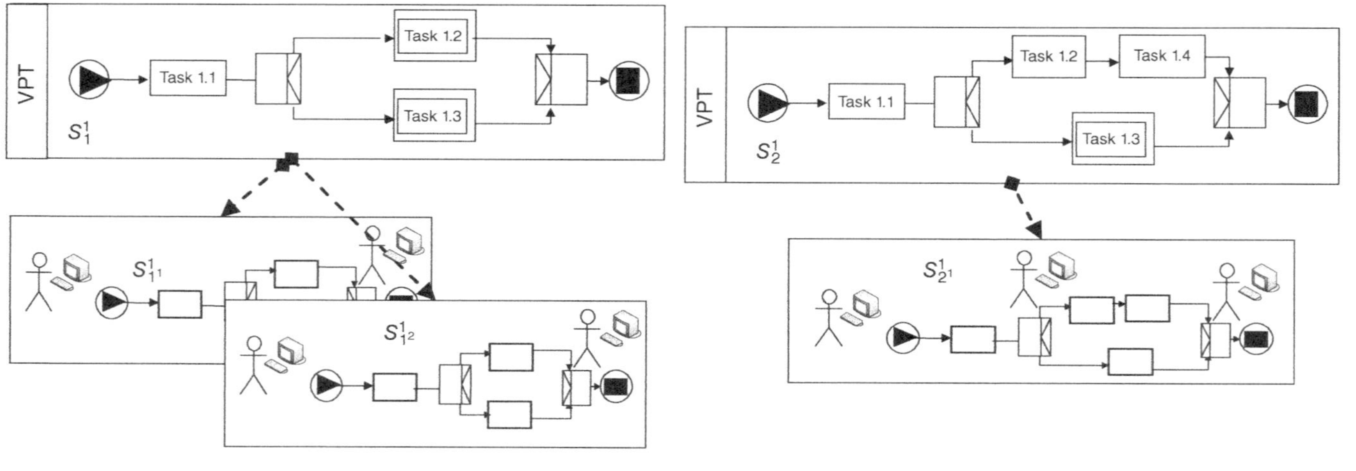

FIGURE 6.8 *Parallel process instances at the top level.*

The top level of the example shown in Figure 6.8 can be then defined as

$$\text{top} = \left(S_1^1 \,||\, S_2^1\right) \; or \; \text{top} = \left(\bigcup_{j=1}^{2} S_{1j}^1 \,||\, \bigcup_{j=1}^{1} S_{2j}^1\right) \tag{6.5}$$

6.3.3.2 *The Hierarchy of a Service System* As discussed earlier, it is extremely typical for a service system to have a hierarchical structure, aimed at effectively taking advantage of divide-and-conquer approach to deal with complex and/or global services. Each lower process is typically created by focusing on developing specific task components, so right skill sets and resources can be identified and allocated for efficient and cost-effective use. For instance, the above-discussed VPDS that spans institutional, geographical, and cultural boundaries is such a sociotechnical service system. It is established for gaining competitive advantage through leveraging best-of-breed talents across the continents.

On the basis of the definition of a generic process, a process model at the lth level ($l \geq 2$, i.e., lower level) can be then defined by

$$S^l = \left(C^l, i^l, o^l, T^l, F^l, A^l, \pi^l\right) \text{ iff } \exists \left((t^{\cdot\text{rem}})^{l-1}\right) \text{ and } (t^{\cdot\text{rem}})^{l-1} \in T^{l-1} \rightarrow S^l \tag{6.6}$$

where S^l is essentially a subprocess of process S^{l-1}. S^{l-1} might have several subprocesses, depending on the number of specific subtasks defined in S^{l-1}. Of course, lower level processes can be further defined if a composite task requires descending numerous levels from the top.

Assume without loss of generality that there is only one business process in the leading team and two processes executed by two other support teams in the presented VPDS example. As each of them is operated in a different country, any project will thus be divided and formed as a composite task as shown in Figure 6.9. Accordingly, a hierarchy of the service system can be established. For this simple illustration (Figure 6.10), no matter whichever level it is, only one process instance is instantiated from its corresponding model.

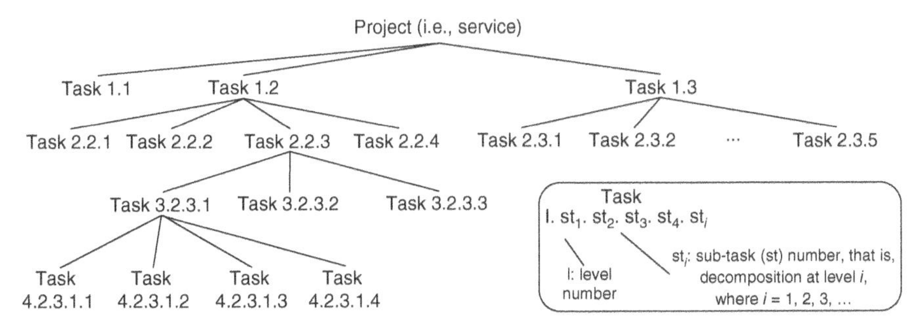

FIGURE 6.9 *Task breakdown hierarchical structure and tree mapping scheme.*

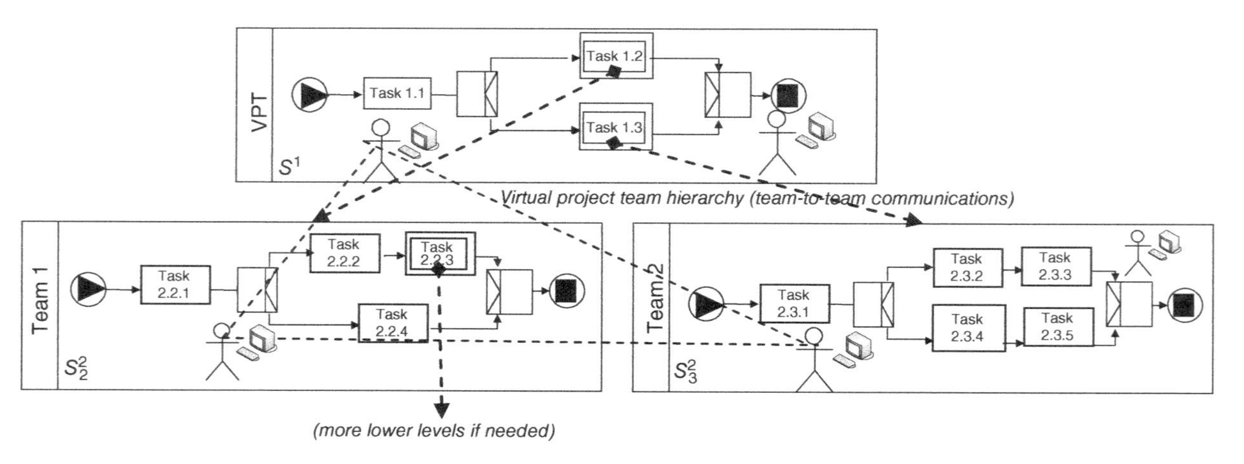

FIGURE 6.10 *Process hierarchy model of the VPDS.*

6.3.3.3 Ad hoc Processes Typically, service activities are performed in conformance with team's existing best practices that might be a suite of defined processes and/or guidelines for *ad hoc* operations. In other words, although it is more effective and sustainable for a VPDS to have its best practices defined as IT-enabled or IT-facilitated processes, in reality many business activities are still conducted at participants' discretion at run time. Business activities occur during service operations without strictly bound to defined process control flow patterns essentially constitute these *ad hoc* processes, which is more difficult to manage and control compared to IT-enabled or IT-facilitated processes (Dustdar, 2004; Dustdar et al., 2005). In Figure 6.9, one task might be executed following a defined process; another might be carried out at members' discretion at run time.

Generally speaking, operating *ad hoc* processes would generate inconsistency in business operations and management, creating issues related to service quality, job satisfaction, and customer satisfaction. It is extremely challenging to address the inconsistency issues throughout the lifecycle of service as people-centered data related to their activities in services could not be well collected, processed, and understood. In addition, we understand that the collected data and information from a variety of legacy systems would typically be incomplete, so significant data shall be collected through conducting surveys to ensure that the needed data for analytics are scientifically sufficient.

This discussed C^2S^2 model investigates an innovative approach to enable people-centric sensing by fully leveraging the advancement of cyber-infrastructures, focusing on the collection of structured, semistructured, and nonstructured data throughout the lifecycle of services (Dustdar, 2004; Dustdar et al., 2005; Girardin et al., 2008; Qiu, 2009; Truong et al., 2008; PPF, 2012; IBM Global Business Services, 2013). Therefore, in addition to conducting the needed surveys, supporting tools facilitating *ad hoc* processes for fostering either collaboration (e.g., IBM Jazz, Action Log, or Caramba) (Dustdar, 2004; Jazz, 2012) or better bookkeeping should be utilized to enhance the execution of virtual team operational practices. As a result, in the VPDS example, more and better quality data can be gradually derived and mined from its IT-enabled service system.

6.3.3.4 Systems Data and Process Logs, and External Data Sources As mentioned earlier, data and information related to all the activities (tasks, communications, systems dynamics, and so on and so forth, either automated or manual) related to service engineering, operations, and management should be timestamp logged. However, it is well understood that in reality the collected and aggregated data and information for any today's service system are typically overwhelming, in which a high percentage of the collections could be nonassociated, redundant, and context-insensitive (Dunham, 2003; Qiu, 2006). Discussions on how to design, develop, and implement systematic or standardized methods to perform data, information, and knowledge integration led to the constructs of consistent, relevant, and sound data, information, and knowledge bases are out of scope of this book. Some good technical references on big data technologies including data mining are provided in Chapter 9.

As a supplementary measure, periodic surveys using different questionnaires can be used throughout the development duration of each project. To make data processing efficient, Likert scales should be mainly used in each of the conducted surveys. Rad and Levin (2003) provide numerous instruments, covering member attributes, motivation study, leadership, success factors, and team maturity measures. Duarte and Snyder (2001) show different set of instruments to understand virtual team success factors and IT setting for better collaborations. Runde and Flanagan (2008) provide the appropriate measures of team conflicts in project management. To ensure the presented approach to meet different objectives, the instruments can be significantly revised to fit the needs of other corresponding investigations in practice.

We all know the saying, "garbage in, garbage out". Regardless of the goal of a specific investigation, systems data, process logs, and external data sources play a key role in analytics. We must always pay much attention to data integrity and quality. With the appropriate support of systems data, process logs, and external data sources, we can have a better understanding of the VPDS, and refine the presented approach to modeling, monitoring, and managing the defined and *ad hoc* processes within the VPDS.

6.4 SERVICE SYSTEMS' PERFORMANCE: METRICS AND MEASUREMENTS

As discussed earlier, to stay competitive, it might make more sense to measure the viability rather than simple profitability of a service system. At a given time and market for a given service, the viability of a service system might rely on a combination of both profit and user satisfaction as shown in Figure 6.2(b). To retain its competitive advantage, it is obvious that the service system's trajectory should be well controlled throughout the service lifecycle. Adequate metrics with applicable measures to be used for evaluating how a service system performs from time to time are indispensable for service providers to make swift and informed decisions, so that the service system can successfully navigate throughout uncertainties and be viable in the long run.

Modeling using computational thinking focuses on exploring, capturing, understanding, and managing systemic behaviors, interactions, connections, complex relations, and interdependencies of a service system. However, resources, largely people—the main focus of a service system, are more complex to model and study as people participating in service production and consumption have physiological constraints, psychological traits, cognitive capability, and sociological obligations, etc. Therefore, measurements of a service system should be collected using methods and means capable of capturing the insights and dynamic social-technical behaviors of the service system, directly and indirectly. Ultimately, through qualitative and quantitative analysis, a service system can successfully navigate throughout a variety of uncertainties and stay competitive (Figure 6.11).

In focus on systemic behavior, structure equation modeling (SEM) has been widely used to study social and/or economic behavior of organizations. By carefully designing the indicators (i.e., measurements) from both social and technical

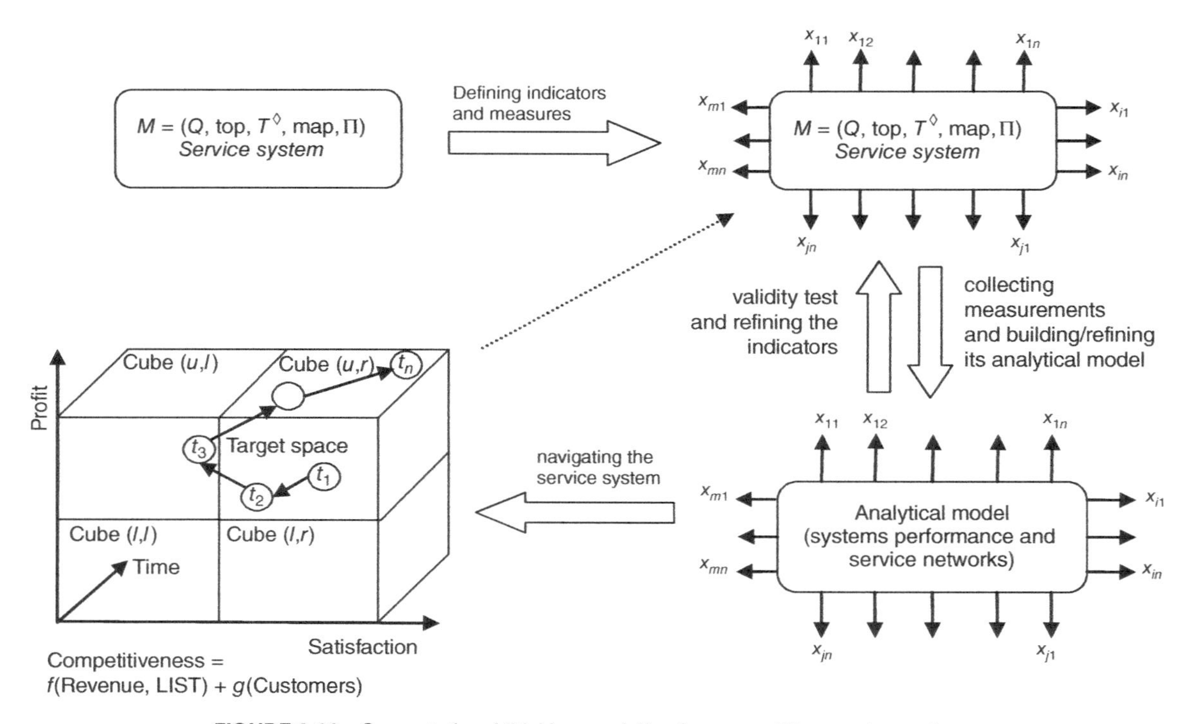

FIGURE 6.11 *Computational thinking modeling for competitive service systems.*

perspectives of a service system, improved SEM can be effectively applied in this interdisciplinary field. As a measurement example that is further enriched from the discussion in the previous section, the dynamics of the presented VPDS can be essentially described using the following latent model constructs/variables: (i) conflicts (indicated by culture issues, managerial styles, personalities, tasks, etc.); (ii) communication effectiveness (indicated by language barrier, customs, infrastructures, meeting setting, etc.); (iii) size of the project (indicated by geographical locations, the number of members, etc.); (iv) project management (indicated by project management method, tools, cost, commitment, risk matrices and control, etc.); (v) project goal (indicated by targeted marketplace, timeframe, etc.); and (vi) quality and satisfaction (Figure 6.12).

As compared to many covariance-based modeling approaches, the partial least squares approach to structural equation modeling (PLS SEM) is a soft modeling with relaxation of measurement distribution assumptions. In addition, PLS SEM requires only a small size of measurement samples and tolerates data collection errors in a more amicable manner. Indeed, this is a very good start point for us to analyze the systems performance of a social-technical service system. We focus on the development of real-time and closed-loop framework to help service organization engineer and manage their service systems. Developing an approach to model service systems while allowing performing continual improvements is unique, differentiating this book from others.

We give a brief introduction to SEM. More discussions will be provided in later chapters. For a reflective measurement model, measurement variables are a linear function of their latent variables ξ plus a residual ε, and π is the corresponding loading set, that is,

$$x_h = \pi_{h0} + \pi_h \xi + \varepsilon_h$$

$$E(x|\xi) = \pi_{h0} + \pi_h \xi \tag{6.7}$$

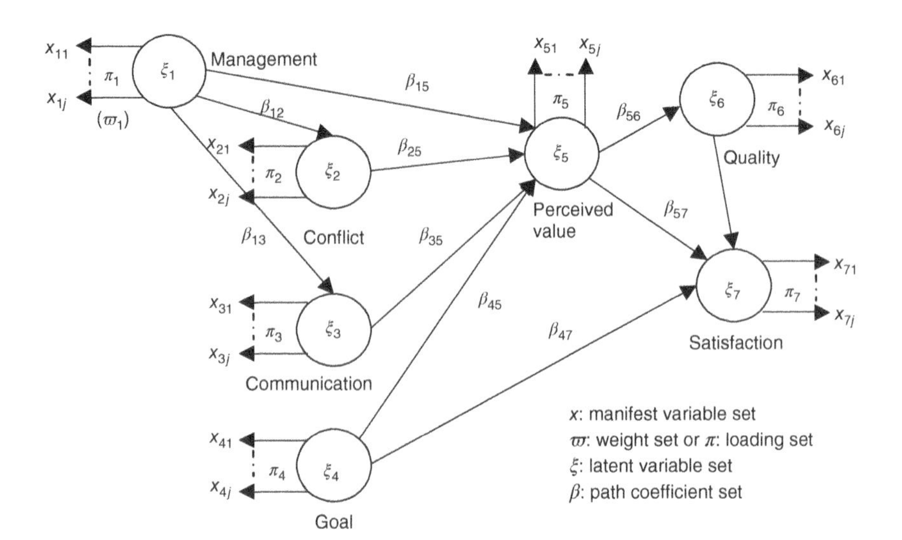

FIGURE 6.12 *An SEM for the presented VPDS.*

The above-mentioned equation implies that the residual ε has a zero mean and uncorrelated with the latent variables ξ. For a formative measurement model, the latent variable ξ is a linear function of its measurement variables plus a residual δ, that is,

$$\xi = \sum_j \varpi_j x_j + \delta$$
$$E(\xi|x) = \sum_j \varpi_j x_j \tag{6.8}$$

For the structural model, the path coefficients between latent variable ξ is given by

$$\xi_j = \sum_i \beta_{ij}\xi_i + \zeta_j \tag{6.9}$$

where ζ is the vector of residual variance.

Note that the operational dynamics of a service system can be easily and well explored and studied using a PLS SEM. To briefly illustrate the insights from the proposed modeling, this example simply shows the following analyses about the service system (e.g., the presented VPDS) at the point of need through the PLS SEM approach:

- *Performance Index.* Which level of the business operational objectives was achieved (e.g., the level of customer satisfaction for a given engineered and delivered service)?
- *Impact Scores.* How did an individual measured factor affect the performance index (e.g., the adopted communication media's impact on the final customer satisfaction level)?

As an example, Figure 6.13 shows an estimated model for the presented VPDS. The performance index for a latent variable is estimated at a 1-to-100 scale basis for easy human interpretation. Thus, a weighted average of scores from corresponding measurement variables is used by converting the original 7-point (X) scale to a 100-point scale (V) (Martensen and Gronholdt, 2003).

More specifically, Figure 6.13 gives some detailed analytical results from a set of collected measurements used in the presented VPDS:

- *Performance Index.* The service satisfaction is at the performance index of 78 out 100. As the model is able to explain 69% of what drives user satisfaction ($R^2 = 0.69$), the model indeed delivers a high and substantial explanatory level.
- *Impact Scores.* With regard to communication effectiveness in the VPDS, the effect of one-point increase in the communication effectiveness might result in 0.284 ($0.366 \times 0.362 + 0.366 \times 0.742 \times 0.562$) points' improvement in the satisfaction performance index. Note that the calculated value has a symbolic meaning rather than a real outcome that could be actually generated. In other

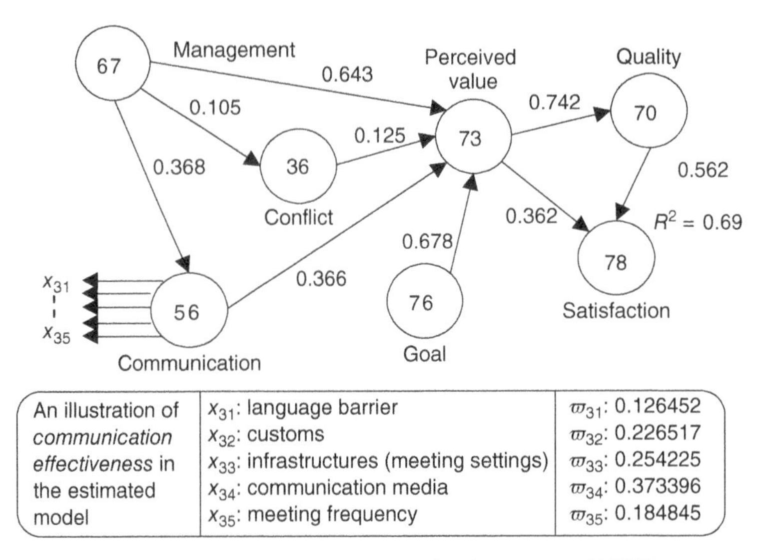

FIGURE 6.13 *An estimated model for the presented VPDS.*

words, the effectiveness ranking of changes can be identified. Therefore, it is appropriate for a comparison exploration, which helps service organizations understand the insights into possible changes and their effectiveness.

As schematically illustrated in Figure 6.11, once we can directly and indirectly collect adequate measurements of a service system that truly and timely capture the insights and dynamic social-technical behaviors of the service system, the organizational effectiveness, operational efficiency, and adaptiveness of the service system can be analyzed and evaluated in a quantitative and qualitative manner, which optimally ensure smarter decision making for business to stay competitive.

The provided insights into the systems performance of a service system, proper transformation can be pinpointed and then applied, resulting in the successful navigation throughout a variety of uncertainties in reality. With the advances of people-centric sensing technologies, methods, and tools, we can fully explore service encounters to strive for more profound understandings of services in real time. Discussions on service networks with a focus on service interactions will be introduced later. However, exploratory modeling of service networks in a comprehensive manner is provided in the later chapters of this book.

6.5 PDGroup AS AN EXPLORATORY EXAMPLE OF SERVICE SYSTEMS MODELING

PDGroup as a global project development cluster, which is formed on a project basis in an international bellwether service organization, was full introduced in Chapter 2. Here comes a quick review of the project background information. An international chemical company called ChemGlobalService has manufacturing facilities in Beijing, China (A), Prague, Czech (B), and Houston, United States (C) across three countries in order to serve its customers across different continents in a cost-effective and efficient way. However, certain materials that can be made only by one facility are required by two other facilities in order to make final products. Each of the three facilities produces different common materials that are needed by others. As a result, these regionally made chemical materials must be transported among three warehouses on a weekly basis. The ChemGlobalService contracted the PDGroup to help integrate three local warehouse systems applications to make sure that their operational coordination among three warehouses is conducted in a collaborative, timely, and effective manner.

More specifically, daily business activities related to the common materials should be monitored and coordinated across three facilities. The right information should be shared and made available to corresponding end users at the point of need. In light of integrating three existing warehouse systems to meet the identified needs, the project's main requirements can be summarized as follows:

- Business processes should be defined in support of fully coordinated operational activities within and across warehouses.
- Industry-specific specifications should be supported.
- All environmental protections, treaties, customs, and other related regional and international regulations and policies must be fully complied. Instructions should be provided at the point of need to all the internal and external personnel who are involved in the process of transporting the hazard components.
- User-friendly human interfaces should be provided to different warehouse employees who understand different languages and have different educational and cultural backgrounds.

On the basis of the general practice in software engineering and the design principles of BPM, we can easily create a task-based hierarchical structure for this service project (Figure 6.14).

The PDGroup includes six small groups of talents. Groups are located in different geographic areas, aimed at leveraging their strengths to meet the project needs. A top-level management unit (i.e., Team A) stays in New York City (NYC), United States. Team A, consisting of one team manager, one team architect, and one team business analyst, oversees and coordinates the project development and deployment across the entire virtual project team. Each of other groups has certain unique skill sets of from 5 to 15 talent employees, including a software designer, a group architect, programmers, quality assurance staff, business analysts, and a group manager. The following list reviews the above-mentioned individual group's respective and unique competency:

- *Team A Located at NYC, United States.* This team is essentially the administrative team, leading, overseeing, and coordinating the project development and deployment across the entire virtual project team.

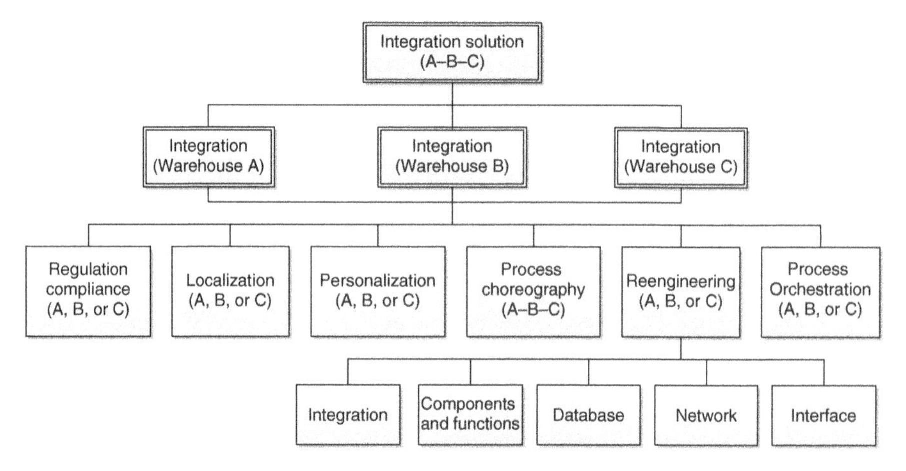

FIGURE 6.14 *Overview of the project's task-based hierarchical structure.*

- *Team B Located at San Jose, CA, United States.* This team has a group of persons of talent in human interface design and development.
- *Team C Located at Houston, TX, United States.* This team has a group of persons of talent in the field of warehouse systems. This team is local and close to the customer facility in Houston, United States. The team will be able to get familiar with the local warehouse system quickly and thus understand local warehouse operations and relevant managerial needs.
- *Team D Located at Prague, The Czech Republic.* Similar to Team C, this team has a group of persons of talent in the field of warehouse systems. This team is local and close to the customer facility in Prague, Czech Republic. The team will also be able to get familiar with the warehouse system in Prague quickly and understand local warehouse operations and relevant managerial needs.
- *Team E located at Beijing, China.* Just like Teams C and D, this team has a group of persons of talent in the field of warehouse systems. This team is local and close to the customer facility in Beijing of China. Hence, it will be convenient and easy for the team to get familiar with the customer warehouse system in Beijing and surely understand local warehouse operations and relevant managerial needs.
- *Team F Located at Bengaluru, India.* This team is a software outsourcing partner. This team has a group of persons of talent in the field of software design, development, integration, and systems test.
- *Team G Located at Sydney, Australia.* This team has a group of persons of talent in the field of enterprise application integration, business analytics, and international regulation and policy compliance.

Figure 6.15 shows the organizational chart for the PDGroup. Organizational charts at the group or virtual team level in the PDGroup is illustrated in Figure 6.16.

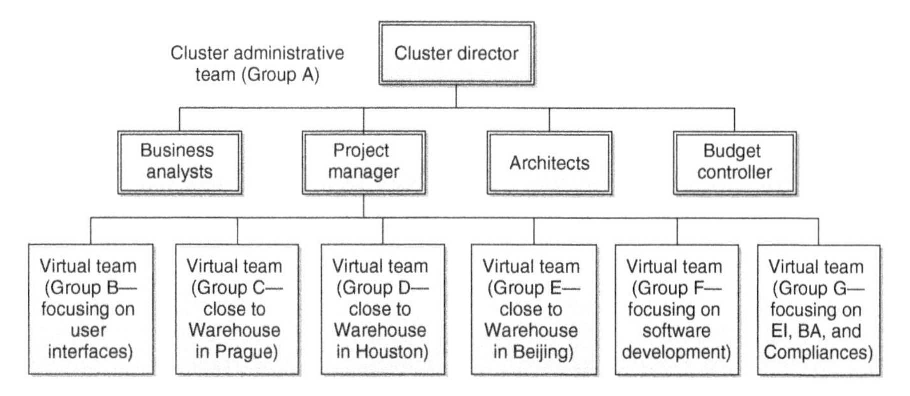

FIGURE 6.15 *Organizational chart at the cluster level.*

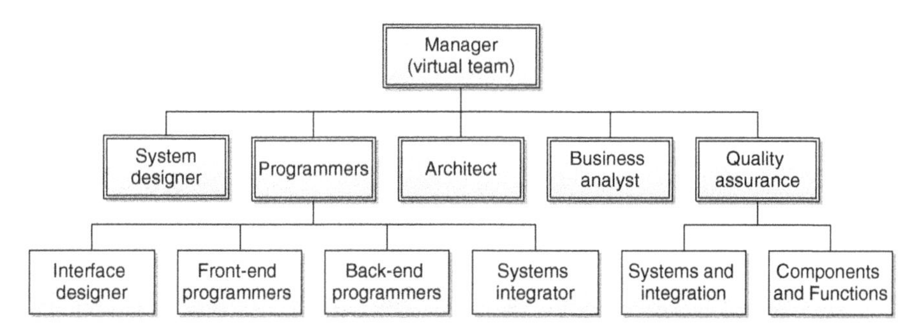

FIGURE 6.16 *Organizational chart at the group or virtual team level.*

Surely, it is challenging to coordinate the daily development activities across the cluster as groups are geographically dispersed around the world.

6.5.1 The Competitiveness of Service Systems: A Systems Approach

PDGroup as a service system can be easily modeled when it is compared to a large-scale service organization. However, the modeling of the PDGroup is never easy and straightforward because of its natural span over the time and space and the existing cultural barriers, in which coordination must be done across nations. It is a desirable goal for the PDGroup to have a way to manage and coordinate the development of an effective solution to solve the discussed ChemGlobalService's warehouse issue. Finding an approach to model the dynamics of the PDGroup surely is the first step to realize the goal.

Using computational thinking, we can have an approach to modeling of the PDGroup dynamics, focusing on collaboration, people-centric sensing, and viability. People-centric sensing facilitates and promotes social-sensing applications (Campbell et al., 2008). Figure 6.17 shows our exploratory roadmap for the focused model, highlighting modeling tasks in a stepwise manner. Through enabling mechanisms

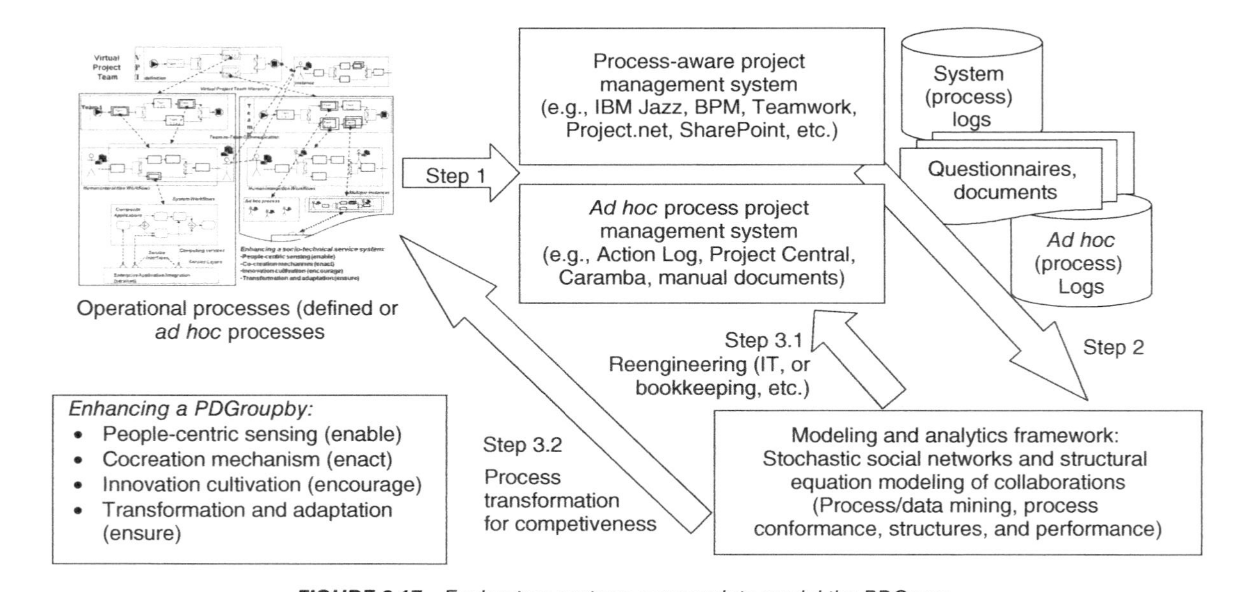

FIGURE 6.17 *Exploratory systems approach to model the PDGroup.*

of people sensing to capture their physiological and psychological issues, cognitive capability, and sociological constraints during the lifecycle of project services, the developed approach shall provide with the means (e.g., *Step 1*) to explore, model, capture, and manage systemic behaviors, interactions, connections, complex relations, and interdependencies of sociotechnical systems, resulting in a better understanding (e.g., *Step 2*) of what constitutes an effective and robust PDGroup and under what conditions and agile transformation (e.g., *Step 3*), the PDGroup's performance and potential would be more predictable, controllable, competitive, and sustainable from time to time.

Once again, at a given time and market for a given service, the competitiveness of a service system might rely on a combination of both profit and user satisfaction. In general, adequate service metrics with applicable measures to be used for evaluating how the PDGroup performs from time to time are indispensable to making swift and informed decisions, so that the PDGroup can successfully navigate throughout uncertainties and be viable in the long run (Qiu, 2009). To fully investigate the acceptance's constituents in any service question, we must explore how people (PDGroup's staff and ChemGlobalService's employees) interact with each other throughout the service lifecycle. For instance, we should understand well how cross-cultural management plays a critical role in ensuring virtual team's success (Figure 6.18). In the long run, we should explore collaborative mechanisms that can be applied in coordinating team works cost-effectively and efficiently. Therefore, social networks formed in business operations must be deeply studied. Because of the people-centered nature in modeling, SNA becomes an essential component of the proposed approaches in this book.

6.5.2 The Competitiveness of Service Systems: A Network Approach

In general, we understand that this global project development service surely requires a series of interactions and coordination, physically and/or virtually. Varieties of service encounters throughout the lifecycle of global project development service are collaborative in nature. The effectiveness of the project development service highly depends on interactions between PDGroup's staff and ChemGlobalService's employees. Only when fully understood, designed, and executed, interactions could be well managed and controlled so that each interaction would occur in a timely, efficient, and effective manner (Figure 6.19).

As indicated in Figure 6.19, a service interaction from a series of needed service encounters throughout the service lifecycle for a given end user or customer representative can start at any point and end at a point that is after the start point of the service. To an end user or customer, such an event-based series service-oriented interactions essentially constitute the customer's service lifespan, which largely depends on the role of the end user or customer representative from the ChemGlobalService. In other words, individual's service lifespan varies with his/her role at work. By the same token, a member of the PDGroup, including the personnel at the outsourcing group, will also take a role-based trajectory based on a series of service encounters. As a value cocreation service interactive activity, each service encounter surely makes

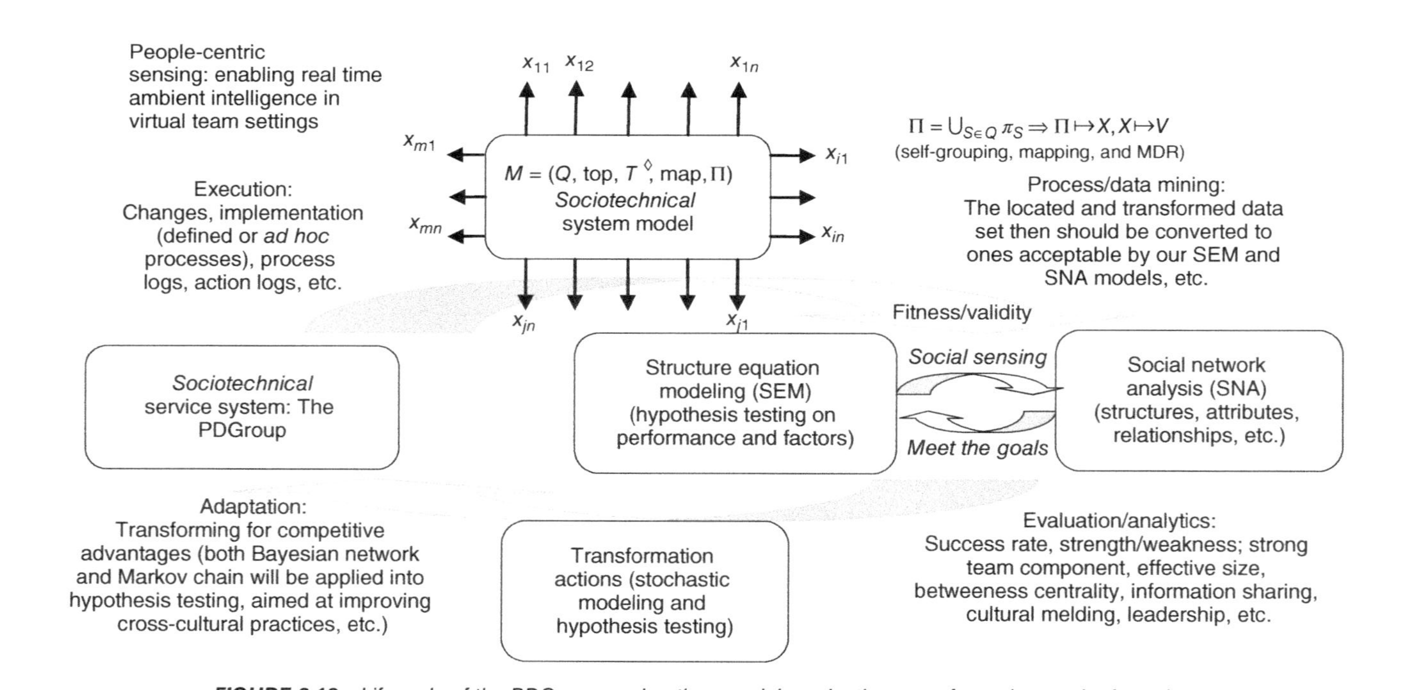

FIGURE 6.18 *Lifecycle of the PDGroup exploration: model, evaluation, transformation, and adaptation.*

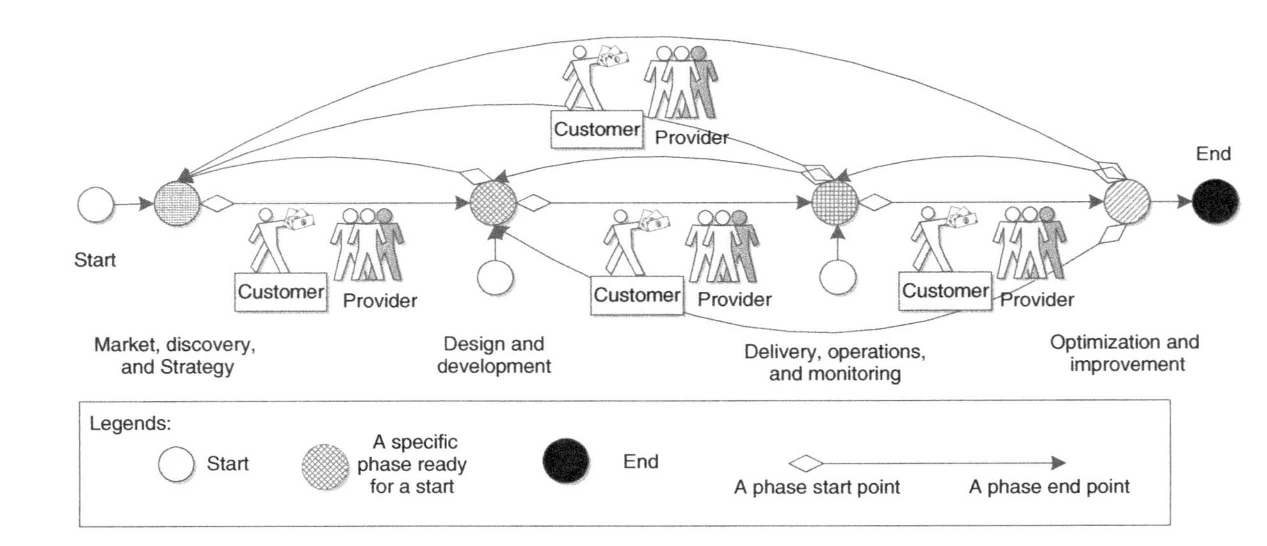

FIGURE 6.19 *Cocreation-oriented management in pursuit of a series of positive service interactions.*

a difference. To the service provider as a whole (i.e., PDGroup's perspective), the sum of all the series of service encounters essentially create a service encounter network as discussed in Chapter 4.

The efficacy and effectiveness of planning and design of the service encounter network throughout the service lifecycle will directly impact the value of the provided service. It is well understood that the execution of a service might require some changes at the point of delivery. Sometimes the execution could be substantially deviated from its original design and plan. Evidently, engineering, operating, and managing services in a world-class service organization are extremely complex and challenging. Therefore, managing and control of service (encounter) networks become necessary for service organizations to ensure that all the services will be designed, developed, delivered, and operated to meet the needs of both service providers and customers.

6.5.3 Market, Discovery, and Strategy

At the market, discovery, and strategy phase, a project draft specification might be brainstormed when the top-level management group meets with a group of customer representatives. The participations of managerial and operational personnel from the organization and unit levels at different facilities of the ChemGlobalService are necessary. Onsite visits might also be needed, aimed at collecting the requirements from daily operations by discussing with end users.

In addition to paying attention to collecting service requirements by fully understanding customers' operational and social-psychological needs, the PDGroup should also pay attention to its organizational supports. Figures 6.20 and 6.21 show how a variety of supports from the organization should be well defined and accordingly coordinated in a collaborative and effective manner.

More specifically, the PDGroup starts to work with customers to understand their needs and starts to lay out a plan for managing and controlling service networks in a comprehensive and holistic way by paying considerable attention to the following four flows on the service profit chain:

- *Customer Dynamics Flow*. Working with customers to understand both the utilitarian and psychological needs of customers; planning a chain of positive and interactive service encounters; and defining a contingency plan to recover service failures.
- *Organizational Behavior Flow*. Working on a plan to ensure that the project can also result in a high level of employees' job satisfaction while surely meeting the customers' utilitarian and psychological needs through enabling a chain of positive and interactive service encounters. Appropriate training for employees in the PDGroup might be necessary.
- *Physical Flow*. Planning the supports for service engineering, operations, and delivery, such as providing employees and customers with the right tools, servicescape, and other necessary resources.

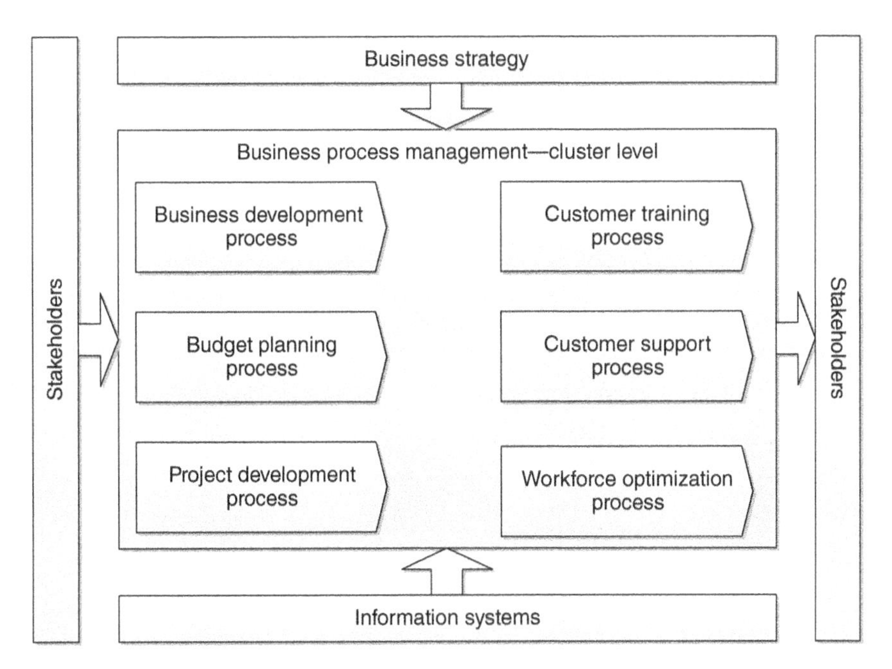

FIGURE 6.20 *Business process management at the cluster level.*

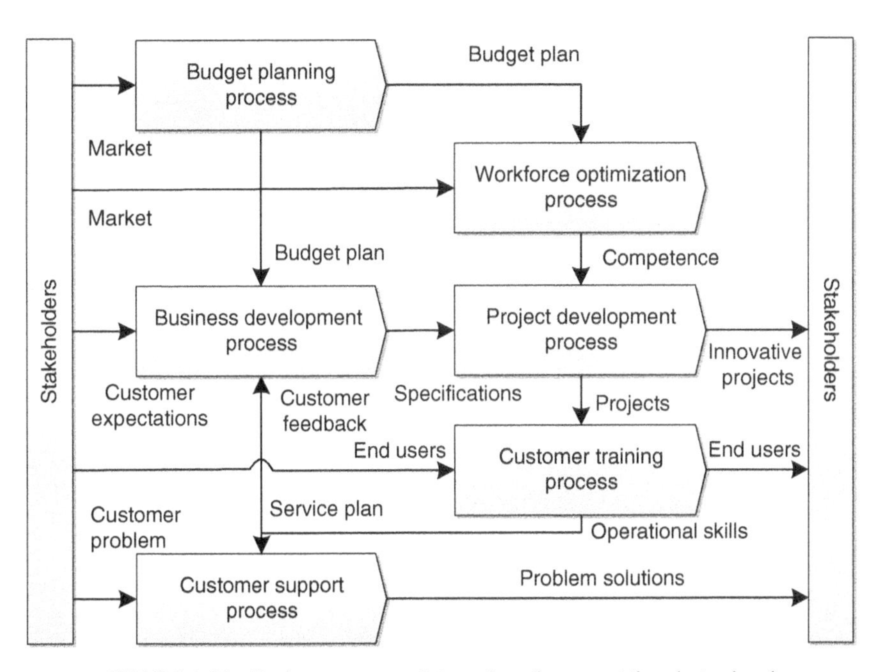

FIGURE 6.21 *Business process interaction diagram at the cluster level.*

- *Information Flow.* Facilitating the collection of service requirements; planning tools and IT services that can capture right data/information in a timely manner and support all operational and managerial needs in a more intelligent way across the service lifecycle.

6.5.4 Design and Development

At the design and development phase, the project specification is usually revised and significantly adjusted and enriched. This is particularly true when more technical functionalities become clearer as time goes. Unless the project is completed, it is typical that the specification will keep changing to some extent. Surely each revision will be the outcome of numerous onsite or virtual meetings among related representatives from the ChemGlobalService and all teams, that is, Team A to Team G. Changes are probably requested and then applied after each test. Customer representatives could be directly or indirectly contacted by group members if necessary. In addition, the capability of teams should be appropriately identified, developed, and aligned with the needs of development tasks.

Note that by relying on a generic computational model, a collaborative process-aware development and management framework can be applied to facilitate service execution across the service lifecycles (Figure 6.22). At this stage, the applied framework provides flexible and adaptable mechanisms to foster team member interactions. Figure 6.22 supports the collection of systems, human interaction, and consequence data when teams work on assigned tasks. All the activity data especially how members collaborate with each other can be appropriately recorded as needed.

At the second phase of the service lifecycle in this example, the PDGroup begins to carry out the plan developed at the market, discovery, and strategy phase. Engineering the needed IT solution is the focus at this stage. The PDGroup still has an opportunity to enhance the solution delivery and operational plan. As a key component of the value cocreation process in providing IT solution services, the PDGroup should also manage and control the formed service networks in a comprehensive and holistic way by paying considerable attention to the following four flows:

- *Customer Dynamics Flow.* Working with customers to cocreate the integration solution to address the needs of the ChemGlobalService. The PDGroup might start to provide appropriate training activities for customers.
- *Organizational Behavior Flow.* Continuously focusing on employees' job satisfaction by designing and preparing the chain of interactive service encounters necessary for ChemGlobalService's employees to run the integration solution. Engineering the needed solution in a satisfactory and competitive manner becomes the highest priority within the PDGroup.
- *Physical Flow.* Providing employees and customers with the right tools, servicescape, and other necessary resource supports to facilitate service encounters at this phase in meeting the needs of engineering the needed IT solution.
- *Information Flow.* Capturing right data/information in a timely manner and supporting the operational and managerial needs in a more intelligent way at this

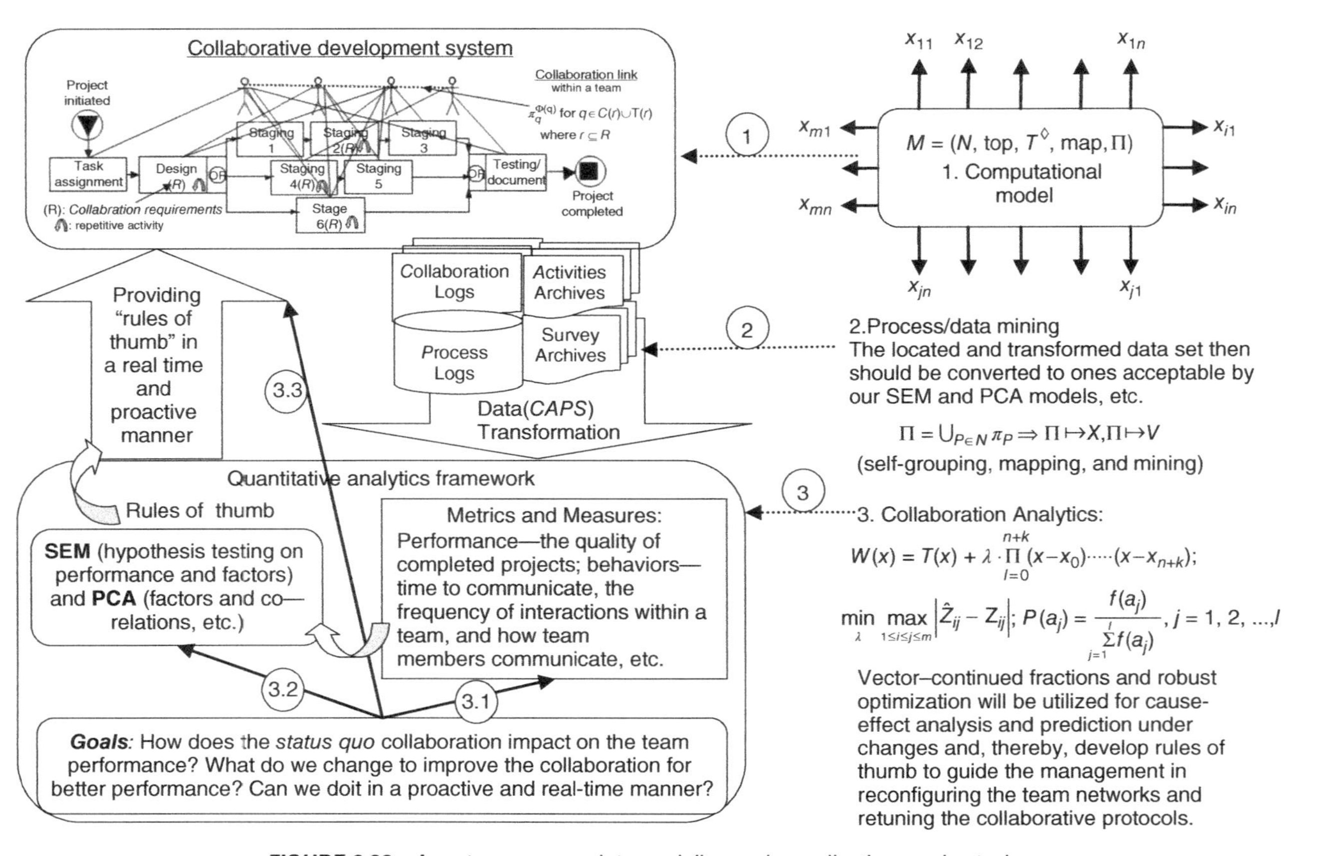

FIGURE 6.22 *A systems approach to modeling and coordinating service tasks.*

phase so that the trusted data, information, and knowledge can be provided at the point of need.

6.5.5 Delivery, Operations, and Monitoring

At the delivery, operations, and monitoring phase, most likely, people from Teams A, B, C, D, and E would have to be involved. PDGroup must make sure that end users will be well trained. The daily business operations can be well coordinated and monitored among three facilities. During operations, member profiles, communication timing and frequencies, adopted communication media, and delivery and operational data are critical to both the PDGroup and ChemGlobalService. Data transformation algorithms to support process/data mining and analytics in support of operations and management are extremely important at this stage.

The PDGroup continues to carry out the plan by delivering the engineered solution to the ChemGlobalService's facilities. Customers begin to really appreciate the value cocreated with the provider. The PDGroup should continue to manage and control service networks in a comprehensive and holistic way by paying much attention to the four flows so that the service value cocreation process can be carried out in a satisfactory manner:

- *Customer Dynamics Flow.* Meeting both the utilitarian and psychological needs of customers; delivering the solution by leveraging a chain of collaborative and interactive service encounters; and ensuring that the customers have the appropriate training, utilize the solution well, and get responsive helps if needed.
- *Organizational Behavior Flow.* Making sure that all field support employees are appropriately assigned and well trained; continuously focusing on employees' job satisfaction by meeting the customers' utilitarian and psychological needs through enabling a chain of interactive and positive service encounters.
- *Physical Flow.* Providing field employees and customers with the right tools, servicescape, and other necessary resource supports to support service encounters at this phase in meeting the needs of service delivery and operations.
- *Information Flow.* Capturing delivery and operational data/information in a timely manner; supporting the operational and managerial needs by delivering the right data, information, and knowledge service to end users at the point of need.

6.5.6 Optimization and Improvement

At the optimization phase, for this kind of IT solution service project, it might be true that teams are reorganized after the engineered IT solution is successfully deployed on a customer site. We assume that all the teams will remain although they might be scaled down somewhat. If the service is not phased out yet, all teams should be involved to some degree. However, Teams A, B, and G would have more interaction with related representatives from the ChemGlobalService, aimed at understanding the weakness of the deployed solution and new additions to the changes of business operations. As indicated in Figure 6.22, vector-continued fractions, robust optimization, or other optimizations can be utilized for cause-effect analysis and prediction under changes (Qiu et al., 2011). The PDGroup can focus on refining and delivering rules of thumb to guide the participating teams to retune their collaborative protocols in a proactive and real-time manner to improve team performances.

Like other phases, the four flows should also be comprehensively and real-time explored to ensure that all aspects of the deployed solution under operation are understood. Issues and areas for improvements are identified as time goes. Ultimately, the optimal solution can be delivered, operated, and managed through well-executed collaborations between the service provider and the customers.

6.5.7 Final Remarks on This Exploratory Example

This example provides an overview discussion on how computational thinking can be well applied to exploring service systems. With the support of computing, networking, and service science, a service system can be managed and controlled with scientific rigor. A service system must be managed and controlled in a comprehensive and holistic way by simultaneously paying significant attention to the following four flows on the service profit chain:

- *Customer Dynamics Flow*. Meeting both the utilitarian and psychological needs of customers; focusing on a chain of interactive service encounters. Customer dynamics flow must be explored with the support of behavioral science, consumer behavior, and cognitive science, to offer and truly create an excellent customer experience.
- *Organizational Behavior Flow*. Focusing on employees' job satisfaction by meeting the customers' utilitarian and psychological needs through enabling a chain of interactive and positive service encounters. Organizational behavior flow must be explored with the support of behavioral science, cognitive science, individual and group dynamics, organizational dynamics, operations management, and workforce optimization to improve job satisfaction and organizational behavior.
- *Physical Flow*. Focusing on the conduits of service provision. An efficient and effective physical flow should provide employees and customers with the right tools, servicescape, and other necessary resource supports to facilitate service encounters in meeting both utilitarian and psychological needs of customers while improving job satisfaction.
- *Information Flow*. Capturing right data/information in a timely manner and supporting the operational and managerial needs in a more intelligent way across the service lifecycle. An optimal information flow should enable the right data, information, and knowledge service for employees and customers at the point of need.

Indeed, the above-mentioned four flows shift their priorities as services progress throughout their lifecycles. From this exploratory example, we understand that systems performance and network analytics can be applied to services in a complementary manner. We provide much more detailed discussions with analytics examples in Chapters 7 and 8, which will help readers understand that the presented science to service in this book, as one approach to help service organizations discover, design, engineer, and manage their competitive services, can be scientifically applied in practice.

6.6 CONCLUSIONS

In today's globalized economy, enterprises are keen on building highly profitable service-oriented businesses by taking advantage of their own unique engineering and service expertise, aimed at shifting gears toward creating superior outcomes to best meet their customer needs in order to stay competitive. However, there is a lack of full-fledged sciences that could systematically guide the plan, design, marketing, engineering, and delivery of services to meet the challenges highlighted by changes, complexity, and dynamics from political, social, and economic aspects; thus, we are in great need of the theory and principles toward engineering, operating, managing, and evolving service systems in the service-led economy.

By fully leveraging today's ubiquitous digitalized information, computing capability, and computational power, this Chapter presented an approach to model the dynamics and adaptiveness of service systems, enabling mechanisms of people-sensing to capture their physiological and psychological issues, cognitive capability, and sociological constraints during the lifecycle of service. The presented approach has more potential, including the enablement of system configurability. As we fully took into account service system's characteristics (e.g., people-sensing, cocreation values, human interactions, and consequences), this Chapter should lay out a solid foundation for the later chapters.

As discussed earlier, the following mechanisms should be developed to implement and enhance service systems:

- A mechanism to timely capture end users' requirements, changes, expectation, and satisfaction in a variety of technical, social, and cultural aspects
- A mechanism to efficiently and cost-effectively provide employees right means and assistances to engineer services while promptly responding the changes
- A mechanism to allow involved people consciously infuses as much intelligence as possible into all levels and aspects of decision making to assure necessary system adaptiveness for smarter operations.

As these mechanisms vary with service systems, they must be developed and enhanced over time. Ultimately, with the support of service science, service systems will then be operated in a smarter, competitive, and satisfactory manner.

ACKNOWLEDGMENT

A significant portion of this Chapter is derived from a paper entitled "Computational Thinking of Service Systems: Dynamics and Adaptiveness" that was published in an INFORMS' journal—*Service Science*. The paper received the best paper award in 2013 from the INFORMS Service Science Section, recognizing publication excellence in the Section's journal *Service Science* from 2009 to 2012. Reprinted by permission of the publisher.

REFERENCES

Buchanan, L. (2013). Between Venus and Mars: 7 traits of true leaders. *Inc. Magazine*, June, 64–74+130.

BusinessWeek. (2009). When service means survival. *BusinessWeek*, March 2, 26–40.

Campbell, A., Lane, N., Miluzzo, E., Peterson, R., Lu, H., Zheng, X., Musolesi, M., Fodor, K., Eisenman, S., & Ahn, G. (2008). The rise of people-centric sensing. *IEEE Internet Computing*, July–August, 12–21.

CMU. (2009). What is computational thinking? *Center for Computational Thinking at Carnegie Mellon*, Retrieved on 2009 from http://www.cs.cmu.edu/~CompThink/index.html.

Duarte, D., & Snyder, N. (2001). *Mastering Virtual Teams*. San Francisco, CA: Jossey-Bass, A Wiley Imprint.

Dunham, M. (2003). *Data Mining: Introductory and Advanced Topics*. Upper Saddle River, NJ: Prentice Hall.

Dustdar, S. (2004). Caramba—a process-aware collaboration system supporting ad hoc and collaborative processes in virtual teams. *Distributed and Parallel Databases*, 15, 45–66.

Dustdar, S., Hoffmann, T., & van der Aalst, W. (2005). Mining of ad-hoc business processes with Teamlog. *Data & Knowledge Engineering*, 55, 129–158.

Gerzema, J., & D'Antonio, M. (2010). *Spend Shift: How the Post-crisis Values Revolution is Changing the Way We Buy, Sell, and Live*. San Francisco, CA: Jossey-Bass.

Girardin, F., Blat, J., Calabrese, F., Fiore, F., & Ratti, C. (2008). Digital footprinting: uncovering tourists with user-generated content. *Pervasive Computing*, October–December, 36–43.

IBM Global Business Services. (2012). Analytics: the real-world use of big data—how innovative enterprises extract value from uncertain data. *IBM Global Business Services Business Analytics and Optimization Executive Report*. Retrieved on Mar. 5, 2013 from http://www-935.ibm.com/services/us/gbs/thoughtleadership/ibv-big-data-at-work.html.

IBM Global Business Services. (2013). Analytics: a blueprint for value: converting big data and analytics insights into results. *IBM Global Business Services Business Analytics and Optimization Executive Report*. Retrieved on Mar. 5, 2013 from http://public.dhe.ibm.com/common/ssi/ecm/en/gbe03575usen/GBE03575USEN.PDF.

Jazz. (2012). About Jazz—Overview. Retrieved on May 10, 2012 from https://jazz.net/learn/.

Kanellos, M. (2004). Perspective: IBM's service science. *CNET News*. Retrieved on Dec. 10, 2012 from http://news.cnet.com/2010-1008_3-5201792.html.

Kotler, P. (2010). Forward in *Spend Shift: How the Post-crisis Values Revolution is Changing the Way We Buy, Sell, and Live*, ed. by J. Gerzema and M. D'Antonio. San Francisco, CA: Jossey-Bass.

Martensen, A., & Gronholdt, L. (2003). Improving library user's perceived quality satisfaction and loyalty: an integrated measurement and management system. *The Journal of Academic Librarianship*, 29(3), 140–147.

Partnership for Public Service. (2011). From data to decisions: the power of analytics. IBM Center for the Business of Government. Retrieved on Dec. 6, 2012 from http://www.businessofgovernment.org/report/data-decisions-power-analytics.

Partnership for Public Service. (2012). From data to decisions II: building an analytics culture. IBM Center for the Business of Government. Retrieved on June 6, 2013 from http://www.businessofgovernment.org/report/data-decisions-ii.

PPF. (2012). *Putting People First*. Retrieved online Nov. 28, 2012 from http://www.experientia.com/blog/.

Qiu, R. G. (2006). Towards ontology-driven knowledge synthesis for heterogeneous information systems. *Journal of Intelligent Manufacturing*, 17(1), 109–120.

Qiu, R. G. (2009). Computational thinking of service systems: dynamics and adaptiveness modeling. *Service Science*, 1(1), 42–55.

Qiu, R. G., Tang, Y., & Joshi, S. B. (2008). A process-driven computing model for reconfigurable semiconductor manufacturing. *Robotics and Computer-Integrated Manufacturing*, 24(6), 709–721.

Qiu, R., Wu, Z., & Yu, Y., (2011). A tractable approximation approach to improving hotel service quality. *Journal of Service Science Research*, 3(1), 1–20.

Rad, P., & Levin, G. (2003). *Achieving Project Management Success Using Virtual Teams*. Fort Lauderdale, FL: J. Ross Publishing.

Runde, C., & Flanagan, T. (2008). *Building Conflict Competent Teams*. San Francisco, CA: Jossey-Bass, A Wiley Imprint.

Schroeck, M. (2013). How to make real-world use of big data. *The Huffington Post*, A post entry by Michael Schroeck on April 3, 2013. Retrieved on June 6, 2013 from http://www.huffingtonpost.com/michael-j-schroeck/.

Truong, H., Dorn, C. Casella, G. Pollleres, A., Reiff-Marganiec, S., & Dustar, S. (2008). inConText: On coupling and sharing context for collaborative teams. *Proceedings of the 14th International Conference on Concurrent Enterprising : a New Wave of Innovation in Collaborative Networks*, 225–232.

Van Der Aalst, W. M., & Ter Hofstede, A. H. (2005). YAWL: yet another workflow language. *Information Systems*, 30(4), 245–275.

Weske, M. (2007). *Business Process Management: Concepts, Languages, Architectures*. Berlin: Springer-Verlag.

7

Education as a Service and Educational Service Systems

According to the American Heritage® Dictionary of the English Language (http://www.ahdictionary.com), education essentially is defined as follows:

- The act or process of educating or being educated when it is referred to a series of activities for imparting or acquiring knowledge.
- The knowledge or skill obtained or developed through a learning process when it is suggested as learning outcomes.
- A program of instruction of a specified kind or level, such as driver training, and K-12 and college education, when it is implied to types or purposes of learning;
- The field of study that is concerned with the pedagogy of teaching and learning if it indicates a discipline.
- An instructive or enlightening experience when it is meant as human cognitive development that improves one's ability or competency over the period.

Evidently, the implied meaning of education surely varies with the context in which a given type of education takes place or the subject that one genuinely wants to learn at a given time. In this book, education mainly implies the act or process of imparting or acquiring knowledge through school systems. In particular, we are interested in education in schools where formal educational service is offered.

More specifically, at a given education institution, instructors are education service providers and learners are education service consumers. Administrators and many

Service Science: The Foundations of Service Engineering and Management, First Edition. Robin G. Qiu.

daily support staff manage and assist instructors and learners to carry out a series of activities relevant to imparting and acquiring knowledge. The marketed programs and courses are the service products that will be offered by the institution. The learning outcomes might be accessed by items, such as how much the learners will learn and what impact the acquired knowledge can have on their lives at work or home in the short term or the long run.

In this chapter, we will first review systems of schooling from the Service Science's perspective. Then, we use two different examples to show how education service systems can be explored by applying the principles of Service Science. In the first example, we apply a systems performance approach to understand the dynamics of an educational school system as a whole. We focus on the mechanisms that can be practically and effectively utilized to adjust school policies and administrative practices for the purpose of improving the effectiveness of education service systems. In the second example, we use a high school off-campus education enrichment study to show how a systems approach can be adopted to reengineer off-campus learning processes to improve students' aptitude to science, technology, engineering, and mathematics over time.

7.1 SYSTEMS OF SCHOOLING: SERVICE SCIENCE'S PERSPECTIVE

Regardless of the scope and goal of a school system, the school system is typically institutionalized with an array of curricula with the support of a variety of teaching and learning resources. At a given school, administrators, instructors, and staff are the people who provide educational services; learners then are the people who consume the provided educational services. Surely, individual learners' purposes for pursuing education vary, so does the school's. However, pursuit of effective learning should be the common goal for both the learners and the school.

Education surely is a service. Schools are educational service systems. On the basis of the discussion on Service Science we have had so far, the systems and holistic perspective of education service must include the following fundamental understandings (Figure 7.1):

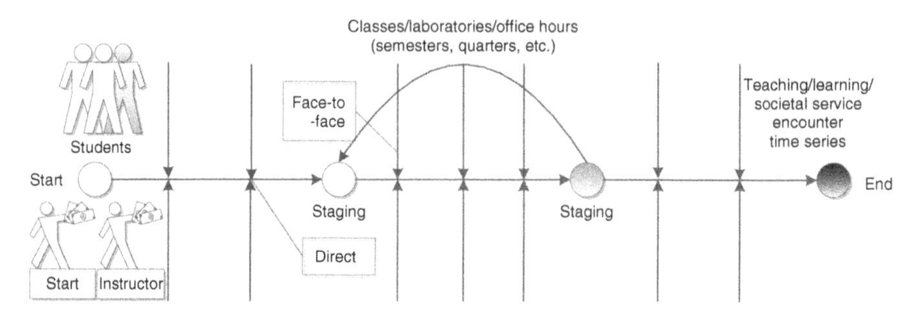

FIGURE 7.1 *Teaching/learning service encounters in knowledge transformation processes.*

- Education service is indeed a knowledge transformation process, delivering the values that are beneficial for both service providers and learners.
- Education service through a knowledge transformation process is centered at people rather than service products themselves.
- An education service provision system is a sociotechnical service system. It might consist of a number of interrelated and interacting domains systems empowered by a variety of resources, which are coordinated in a collaborative manner to help realize their common goal.
- The realized value of an education service is the total value that is perceived from learning outcomes derived from the education service. Note that the perception of education service by a learner is cumulative. In other words, the realized value of an education service is cumulatively perceived from a series of service encounters throughout the service lifecycle.

Regardless of private or public schools, systems of schooling should have no exception in organizational and operations management when compared to any other service systems. By complying with the principles of Service Science, we understand that systems of schooling should be managed and operated in delivering the daily educational needs to the students we serve. Hereinbelow, using a public high school as an example, we identify service encounters that are essential throughout the education service lifecycle.

At the market, discovery, and strategy phase, a list of courses might be developed based on the current and future market (i.e., students) needs. The participants in the discussion on new courses should include administrators, teachers, students, and parents. In addition to traditional ways of collecting the references from certain general guidelines provided by federal and state educational agencies and some course works available from the colleagues, peer-school websites, and well-recognized teaching materials in the field, the inputs that should be brought in could also come from market surveys and analyses, which could be well aggregated through online social media, such as school blogs, Facebook, or Twitter. Understanding and discovering the market and changing needs play a critical role in identifying the courses that should be developed and offered to students. A school's board and its administrators surely and significantly contribute to the strategic development for all the new courses that are planned to be developed and offered.

At the design and development phase, in general, course curricula will be revised and enriched as time goes. Unless a course is completely delivered, it is typical that its curriculum will keep changing to some extent. Surely, each revision will be the outcome of numerous discussions and meetings among related course stakeholders. In other words, although teachers are most likely the main contributors to the design and development of new courses, they revise and enrich contents and learning materials for each course by enhancing the adopted teaching pedagogies through collecting more suggestions and feedbacks from course stakeholders on an ongoing basis. Current and prospective students' viewpoints, if they can be collected in any way, should be thoroughly considered.

At the delivery, operations, and monitoring phases, courses are offered and delivered face-to-face, virtually, or in a hybrid manner. It takes time for the covered knowledge to get imparted and acquired. To the learners, the process of knowledge transformation consists of many interactive learning activities. On the basis of the principles of Service Science, knowledge acquired by students is essentially cocreated by the instructor and students. In other words, effective interactions within the community of the instructor and students substantially impact the learning outcomes. Figure 7.2 shows the classic residential teaching/learning service encounters experienced throughout a given learning period. The interactions include class discussions, questions and answers during after-class office hours, and other learning engagement activities. Students always appreciate these interactions with the instructor and other students on campus.

At the optimization and improvement phase, all participants should be involved to some degree. However, the instructor and administrators mainly contribute to this stage although suggestive and constructive feedbacks could be further collected from students and parents. The instructor and administrators must analyze all collected data, understanding the weakness of the delivered course and identifying mechanisms to make changes in organizational structures and service operations for improved learning outcomes and stakeholders' satisfaction. Of course, they should take actions on the identified changes, monitor the changes, and ensure that the implemented changes indeed improve the offered service.

Residential instructor-led courses have proven effective in education (Figure 7.2). With the fast development of Internet and communication technologies, online courses (Figure 7.3) emerge as they provide a new learning model, called any time, any place, and any pace education paradigm. Indeed, by leveraging the strengths of both instructor-led and virtual courses, the blended approach (Figure 7.4) gradually becomes a popular educational model on campus.

When we study a specific topic in education, we must stay focused and pay much attention to the areas we can deal with as the available resources might be limited for the time being when we conduct the study. Surely, each study will have its unique objective. Education service delivery mechanisms change as time goes, which

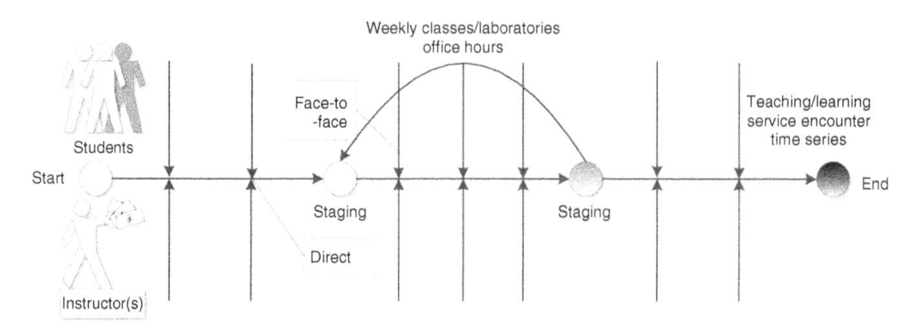

FIGURE 7.2 *Classic residential teaching/learning service encounters.*

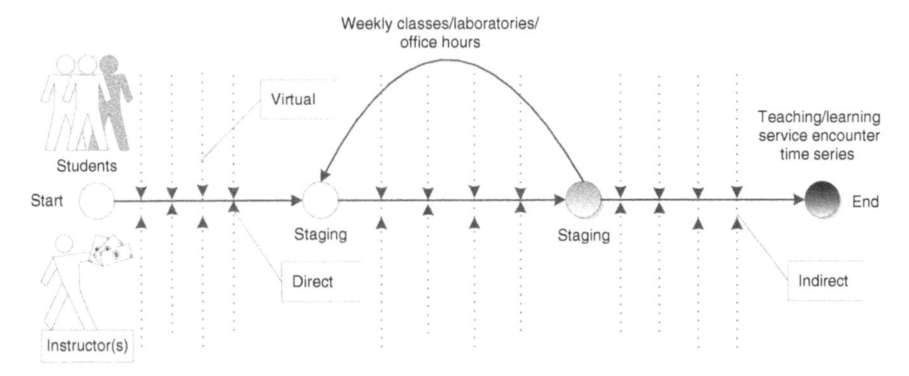

FIGURE 7.3 *Online-based teaching/learning service encounters.*

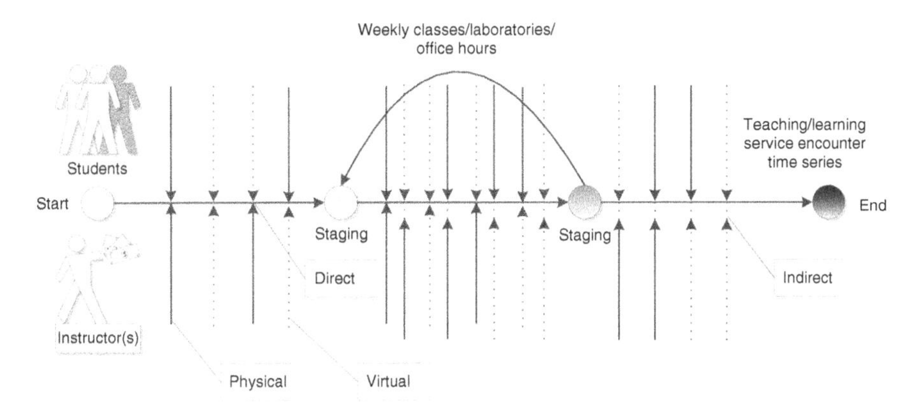

FIGURE 7.4 *Blended teaching/learning service encounters.*

certainly vary with the service contexts. However, service encounters are always centered regardless of the changes of service delivery mechanism. Face-to-face education paradigms are essential for the resident pedagogical engagement. When the traditional pedagogical engagement is weakening in the virtual learning setting, we must explore state-of-the-art alternatives to compensate for the missing essential components and ensure that we can deliver education service that continuously meets the needs of learners.

7.2 A QUALITY CONTROL AND MANAGEMENT CASE STUDY FOR RESIDENT EDUCATION: THE SYSTEMS PERSPECTIVE

Whether a service is attractive and sustainable highly depends on a variety of factors, including what the service is in terms of what it offers to meet the needs of prospective customers, how it is delivered, and whether it meets the needs of customers during its service transformation process. In general, education is a service. College education surely is a service. Compared to a traditional resident college education, education service focusing on professionals is more challenging, given that professional students have varieties of backgrounds, work experiences, and expectations. In other words, for a given education program, the backgrounds of the student body in a school change fast, so do their expectations. Therefore, cocreation processes in building and managing a competitive program play a critical role in ensuring educational service offered and delivered to be competitive and satisfactory.

Empirical studies of education service quality using exploratory and confirmatory analyses have been conducted in academia for over several decades (Hill, 1995; Gruber et al., 2010; Duarte et al., 2012), aimed at gaining insights of higher education management through understanding the education market changes and identifying factors that influence education service quality and students' successes during their studies in college. The following example is an empirical study of the education quality for a resident professional school, aimed at identifying administrative actions for the potential of improving students' satisfaction.

7.2.1 A Typical System-Based Empirical Approach to Explore a Service System

Structural equation models (SEMs) have been widely used to study the social and/or economic behavior of organizations (Bollen, 1989; Kline, 2005; Martensen and Gronholdt, 2003; Raykov and Marcoulides, 2006). As compared to many covariance-based modeling approaches, the partial least squares approach to SEM (PLS SEM) is a soft modeling approach with relaxation of measurement distribution assumptions (Chin et al., 2003). PLS SEM requires only a small size of measurement samples and tolerates certain measurement errors. It is a perfect analytical model for this case study because our sample size can be small when compared to many other types of empirical studies adopted in the commercial world (Wang et al., 2009).

By focusing on providing customer (i.e., students) satisfaction excellence, the following SEM model is mainly applied to understand the operational and managerial structure of the service system with a variety of latent variables (i.e., constructs) (Figure 7.5). Manifest and latent variables for analyzing education programs in school vary with service contexts. Table 7.1 gives relevant indicators for manifest and latent variables used in this professional program exploratory example. This exploratory example aims at illustrating how an SEM can be applied to the modeling of an education service system. In particular, we analyze education service quality at the institutional level of a graduate professional study school. Through pinpointing weak areas across the studied service system, we recommend appropriate actions to school's administrators. Ultimately, after the suggested actions are properly and promptly carried out, the program quality and students' satisfaction can be significantly improved over time.

Formative and reflective measurements as two different measurement models have been widely used to construct SEMs (Fornell and Bookstein, 1982; Bollen and Lennox, 1991; Coltman et al., 2008; Diamantopoulos et al., 2008; Edwards, 2011).

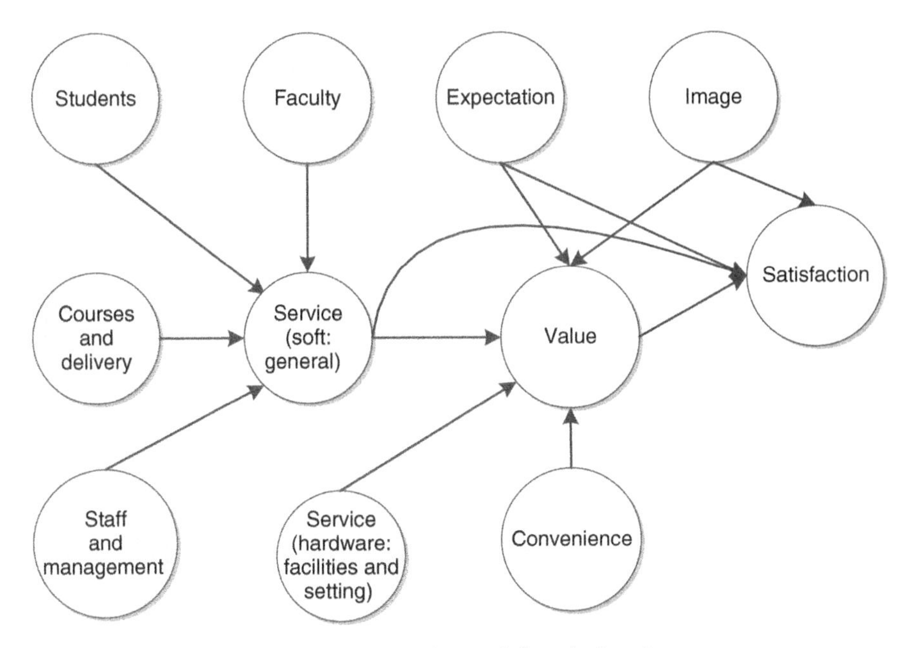

FIGURE 7.5 *Structural equation modeling of education programs.*

If measures are treated as causes of constructs, we apply formative measurements in modeling; if the constructs are viewed as causes of measures, we apply reflective measurements in modeling. Reflective measurement modeling is popularly used in developing a latent measurement model, which essentially is the principal factor model. In a reflective measurement model, covariation among the measures or indicators is caused by, and therefore reflects, variation in the underlying latent construct (Jarvis et al., 2003). The choice of formative and reflective measurements has been never easy as it depends on the dimensionality, internal consistency, identification, measurement error, construct validity, and causality of the subjects under study. Reflective measurement models surely have been applied widespread in practice over decades, while formative measurement models begin to receive more and more attention from researchers and practitioners.

By merging studies in the psychology, social science, management, and marketing fields to examine a variety of issues concerning the conceptualization, estimation, and validation of formative measurement models, Diamantopoulos et al. (2008) advocate that formative measurement models are advancing over the years. Coltman et al. (2008) demonstrate the viability of formative measurement modeling of two applications. However, Edwards (2011) criticizes that the presumed viability of formative measurement is a fallacy and concludes that formative measurement is not a viable alternative to reflective measurement.

Regardless of types of measurement models, it is essential for all researchers to justify their choice of measurement models (Jarvis et al., 2003). Coltman et al. (2008) articulate that using an incorrect measurement model for a service system undermines

TABLE 7.1 Manifest and Latent Variables for Analyzing Education Programs in School

Latent Variables	Mediators	Indicators (Formative/Reflective)
Image		• Reputation (ranking) • Education • Research • Community service • Sports and alumni
Expectation		• Career enhancement • Change of career
Service	Student	• Background • Experience
	Faculty	• Background and knowledge • Industrial experience • Advice • Availability • Adjunct
	Staff and administration	• Assistance • Availability
	Courses and delivery	• Challenges • Course projects • Team works • Professional classmates • Quality of courses • State of the art • Hands-on • Taught by full-time or part-time faculties
	(In general)	• Diversity of courses • Web-based course management system (e.g., Angel) • Tuition • Variety of programs/tracks • Advisors • Orientation • Information sessions • 7 weeks versus 14 weeks • Course description and prerequisites • Guest speakers from the fields • Degree requirement changes • Availability of courses • MS thesis • Social networks on campus

TABLE 7.1 (*Continued*)

Latent Variables	Mediators	Indicators (Formative/Reflective)
Facility and settings		• Parking • Access to the library • Safety • Computer laboratories • On-campus bookstore • On-campus ATM • Cafeteria • Access to campus (weekend and late night) • State-of-the-art equipped classrooms • Access to networks
Convenience factors		• Location • Schedule of classes (7 weeks) • Online courses • Financial aid
Perceived value		• Meeting the expectation • Promotion effect • Impacts on responsibility at work
Satisfaction		• Satisfied with this school • Would choose this school again if satisfied with the quality of education • Willing to recommend to friends

the content validity of constructs, which could misrepresent the structural relationships between constructs and lower or even put doubts on the usefulness of the identified insights of the explored service system. Therefore, the validity of models should be confirmed so that the insights of service systems in light of the system's principal factors and relationships, behavior, and performance can be meaningful and thus can be practically applied to the decision making in business operations and management by a service organization.

The modeling validity should be applied to both measurement and structural models. There are a variety of discussions on how to validate and report constructed system models. The following validations are well utilized in the literature (Henseler et al., 2008). In particular, they are required as essential reports for publications when PLS SEMs are applied to model the dynamics of service systems:

- Evaluating reflective measurement models
  - *Factor Loadings*. Factor loadings should be higher than 0.7. Otherwise, either the factors might be insignificant or the method of collecting the relevant data set of the factors should be revised.

- *Composite Reliability* ρ_c. $\rho_c = \left(\sum \lambda_i\right)^2 / \left(\sum \lambda_i\right)^2 + \sum_i \text{var}(\varepsilon_i)$, where λ_i is the component loading to an indicator and $\text{var}(\varepsilon_i) = 1 - \lambda_i^2$. The component reliability as a measure of internal consistency should be higher than 0.6.
- *Average Variable Extracted (AVE).* $\text{AVE} = \sum \lambda_i^2 / \sum \lambda_i^2 + \sum_i \text{var}(\varepsilon_i)$, where λ_i is the component loading to an indicator and $\text{var}(\varepsilon_i) = 1 - \lambda_i^2$. The average variable extracted should be higher than 0.5.
- *Discriminant Validity.* The extracted average variances of the latent variable should be greater than the square of the correlations among the latent variables. In other words, more variance should be shared between the latent variable component and its block of indicators than with another block representing a different block of indicators. Cross-loadings can be another test for discriminant validity, which will testify that each block of indicators loads higher for its respective latent variable than as indicators for other latent variables. If an indicator has a higher correlation with another latent variable, then the appropriateness of the model may be reconsidered.

• Evaluating formative measurement models

- *Significance of Weights.* Estimates for formative measurement models should be at significant levels. This significance can be evaluated using the bootstrapping procedure.
- *Multicollinearity.* Manifest variables in a formative block must be tested for multicollinearity. The variance inflation factor can be used for such test. Values that are higher thus reveal a critical level of multicollinearity, thereby indicating that the measurement model must be reconsidered.

• Evaluating the structural models

- *R^2 of Latent Endogenous Variable.* R^2 – results of 0.67, 0.33, and 0.19 for latent endogenous variable in the structural model are described as "substantial," "moderate," and "weak" (Chin, 1998, p. 323).
- *Estimates for Path Coefficients.* The estimated values for path relationships in the structural model should be at significant levels. This significance can be evaluated using the bootstrapping procedures.

Note that there are many other empirical methods that can be adopted in modeling people-centric service systems. In the following sections, we present step by step how PLS SEM is adopted to investigate the education service quality at the institutional level of a graduate professional study school (i.e., an education service system). We focus on pinpointing weakness across the studied service system, and aim at identifying appropriate actions for recommendation to school's administrators.

7.2.2 Questionnaires and Responses

As indicated earlier, education service in the professionals market is more challenging than focusing on traditional resident education service in which students are most likely engaged in full-time studies. In general, a professional school that serves part-time adult students faces more challenges in addition to typical ones faced by full-time resident college programs. For example, students in the professionals market typically have a variety of backgrounds, work experiences, and expectations. It could be even more challenging when students are financially sponsored by their employers because of the fact that their expectations vary with factors including their employers' reimbursement policies. Hence, the enrollments or attendances in the professional school could be substantially influenced by their employers' reimbursement policies, business travel schedules, and even their families' matters.

We know that a competitive and sustainable service of a service system mainly relies on customers' satisfaction, which highly depends on how effective the service system meets their expectations in a timely manner. Education in the professionals market is of no exception. To act on identified issues to meet students' needs at a graduate professional school, we, the school as a service provider, should have an effective way to answer the following questions from time to time:

- Do we really know how we perform in terms of meeting their needs?
- What are the main obstacles in achieving students' satisfaction excellence, so we can transform and outperform our competitors?
- Given the limited resources (which restrict us from having more programs/tracks), what will be the best approach to transform our educational service system (i.e., institution) into a more adaptable, responsive, and competitive one?

To analyze whether the offered education programs meet the students' needs and understand what they want over time, the best approach will be to capture what the served students think in real time. Although online social media using Facebook, blogs, Twitter, Google+, and others are attractive, it is still risky of missing the real voices from students as the technologies in support of mining the real insights of the served students' dynamics and thinking are not mature. Social media will be surely the best way to use down the road when the related technologies are fully enabled. At present, cross-sectional and longitudinal surveys using well-defined questionnaires prove effective. As a result, the appended questionnaire in this chapter is used, which helps us to collect sufficient data from time to time. It indeed demonstrates that well-executed surveys and responses allow us to analyze what the status of our educational service system is in light of meeting our students' needs in a timely manner.

Student participations play a key role in completing this investigation. A web site was designed and deployed. Current students are asked to fill in the web page questionnaire during the middle of each class. As usual, the questionnaire is designed to be volunteer based. No identity and confidential data are required. We clearly indicate that collected data are solely used for a research and/or internal decision-making reference purpose.

This conducted study aims at investigating a scientific approach to model the effectiveness of such an education service system. By focusing on finding ways to provide customer (i.e., students) satisfaction excellence, we identify certain factors substantially contributing to adversely affecting the quality of education and students' satisfaction and then determine how those priority issues should be addressed in order to transform the service system for competitive advantage. We also investigate certain measures and mechanisms to ensure that the proposed approach indeed works in terms of meeting our students' needs. Ultimately, as time goes and the job market and business environment change, we can always take a smarter action so that we can keep this graduate professional school as an education system more competitive.

7.2.3 Modeling and Analytics

Validity and reliability of instruments are critical in questionnaire-based empirical studies. Validity of instruments refers to whether our questionnaire that is used will actually measure what we want to measure. Reliability refers to two things. First, reliability means we would get similar results if we repeat our questionnaire soon afterwards with the same class. The "repeatability" of the questionnaire should be high. This is called test–retest reliability. The other aspect of reliability concerns the consistency among the questions. Because all the questions relate to the program quality in the school, we would expect all the answers to be fairly consistent when answered by numerous different groups of students.

Because we sent the request to a small group of students at the very beginning, we collected only 58 responses. The sample size surely was too small such that no meaningful insights could be concluded and then delivered to the administrators. The first step indeed was to check the validity and reliability of instruments and the validity of our proposed systems dynamics and behavioral model. We use SmartPLS tool to run all SEMs in this book (Ringle et al., 2005). Using these 58 responses, we create our first SEM that models the program quality of the graduate professional school under study (Figure 7.6).

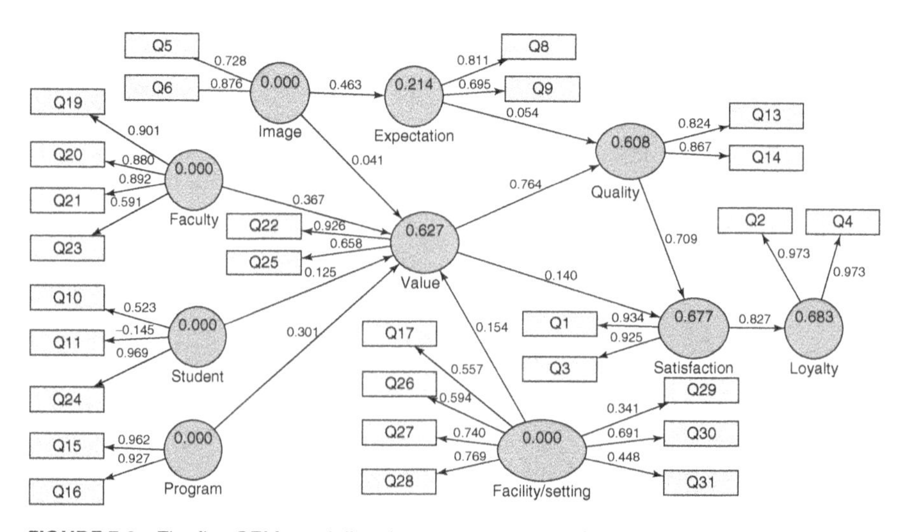

FIGURE 7.6 The first SEM: modeling the program quality of a graduate professional school.

TABLE 7.2 The Assessment Data of the Measurement and Structural Models (Round 1)

	AVE	Composite Reliability	R^2	Cronbach's Alpha
Expectation	0.569852	0.724811	0.213610	0.248591
Facility/setting	0.372048	0.796720		0.721068
Faculty	0.682682	0.893514		0.836747
Image	0.648871	0.785741		0.469362
Loyalty	0.946254	0.972385	0.683233	0.943203
Program	0.892106	0.942960		0.881620
Quality	0.715262	0.833900	0.597523	0.603405
Satisfaction	0.864042	0.927061	0.675709	0.842846
Student	0.411014	0.511824		0.011536
Value	0.651490	0.785893	0.616824	0.491644

Note that we have to assess the models, including both the measurement and structural models, before we can really use the interpreted insights from the modeling to arrive at a conclusion on the health of the service system. Table 7.2 provides the assessment data of the measurement and structural models during the first round of modeling using the collected 58 responses. It shows that both the measurement and structural models seem marginally appropriate.

We resent the request to more groups of students after we finished our first round of modeling and we collected 79 more responses. Figure 7.7 shows the second SEM when we model the program quality of the graduate professional school using the second round of responses. Table 7.3 then presents the assessment data of the measurement and structural models during the second round of modeling using the collected 79 responses. Similar to the round 1 modeling, it shows that both the measurement and

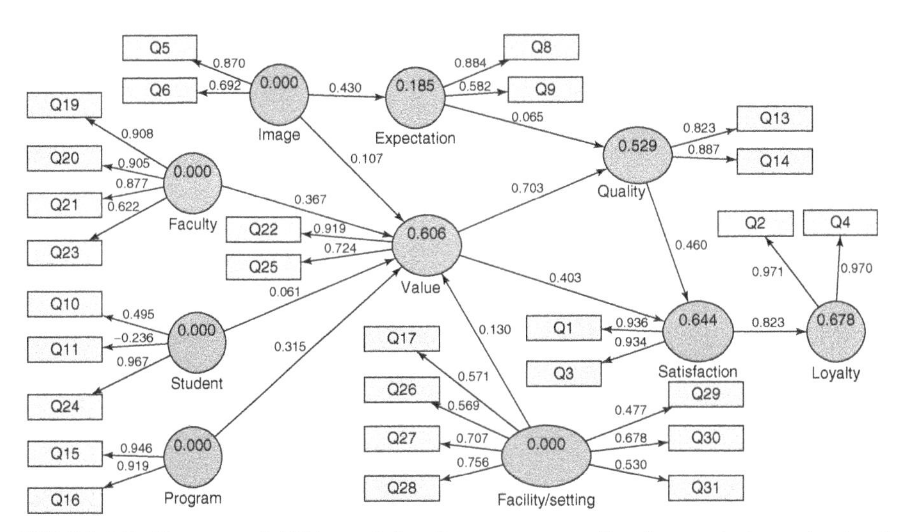

FIGURE 7.7 The second SEM: modeling the program quality of a graduate professional school.

TABLE 7.3 The Assessment Data of the Measurement and Structural Models (Round 2)

	AVE	Composite Reliability	R^2	Cronbach's Alpha
Expectation	0.560510	0.709870	0.184996	0.238394
Facility/setting	0.384121	0.810046		0.740291
Faculty	0.699809	0.901327		0.849513
Image	0.618270	0.761790		0.395998
Loyalty	0.942398	0.970345	0.677691	0.938885
Program	0.870564	0.930790		0.852700
Quality	0.731662	0.844860	0.528686	0.636544
Satisfaction	0.874561	0.933084	0.643735	0.856577
Student	0.411889	0.459742		0.145019
Value	0.684238	0.810396	0.606353	0.564130

structural models again prove marginally appropriate. However, the outcomes highlighting service quality and students' satisfaction seem consistent, which indicates that the validity and reliability of instruments are well checked.

To increase the sample size of the collected responses, we combine the two rounds of surveys. Then, we use all the responses to develop our overall model for the program quality study, which is illustrated in Figure 7.8. Table 7.4 gives the assessment data of the measurement and structural models using all the collected responses. Note that it is necessary for researchers to continuously refine the instruments and models until the adopted instruments and models are completely validated if the study is really used to serve organizations and assist administrators to make informed decisions in facilitating organizational business development and improving operations management. Here, we simply use this example to demonstrate the applicability of the presented approach to model the dynamics and behavior of service systems.

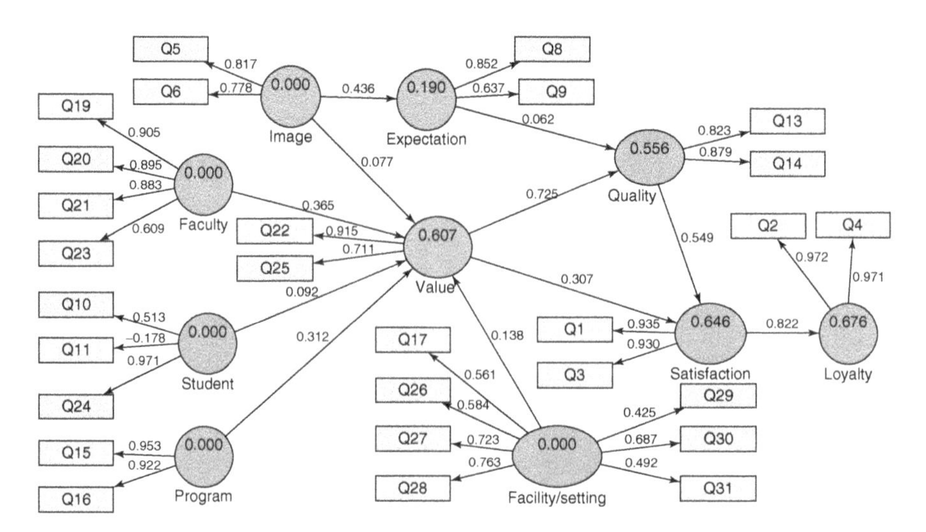

FIGURE 7.8 The overall SEM: modeling the program quality of a graduate professional school.

TABLE 7.4 The Assessment Data of the Measurement and Structural Models (Overall)

	AVE	Composite Reliability	R^2	Cronbach's Alpha
Expectation	0.565523	0.718315	0.190307	0.243258
Facility/setting	0.379210	0.804911		0.732110
Faculty	0.692748	0.898164		0.844313
Image	0.635863	0.777301		0.428032
Loyalty	0.944123	0.971259	0.676171	0.940820
Program	0.879225	0.935716		0.864524
Quality	0.724474	0.840083	0.555590	0.622111
Satisfaction	0.868986	0.929901	0.646310	0.849278
Student	0.412880	0.492218		0.096273
Value	0.670964	0.800600	0.606728	0.535843

7.2.4 Analytics and Decision-Making Supports

Given that a competitive and sustainable service relies on customers' satisfaction, we must keep our service system responsive and effective in order to meet students' expectations from time to time. As indicated earlier, the purpose of this modeling is to assist the administration to identify the overall campus-wide strategic and tactic actions to further improve school's program quality on an ongoing basis. Capturing systems' dynamics and ensuring the good understandings of the insights should help answer the following questions discussed earlier in a timely manner:

- Do we really know how we perform in terms of meeting their needs?
- What are the main obstacles in achieving students' satisfaction excellence, so we can transform and outperform our competitors?
- Given the limited resources (which restrict us from having more programs/tracks), what will be the best approach to transform our educational service system (i.e., institution) into a more adaptable, responsive, and competitive one?

To better understand the behavior and systems performance of this explored school, we remove those indicators, either those that have little impact on the desirable outcomes or the ones there is nothing the school can do about at the time of investigation. As a result, by simply focusing on highly influential factors or potential actions that can be done to improve the levels of program quality and student satisfaction, we thus develop a new SEM as illustrated in Figure 7.9. Table 7.5 shows the assessment data of the measurement and structural models, in which constructs are mainly reflected by influential or actionable indicators. Table 7.6 lists the results of latent variable correlations in the focus model illustrated in Figure 7.9.

To pinpoint how individual factors impact the outcomes of a modeled service system, we must estimate path coefficients in the developed SEM. The estimated

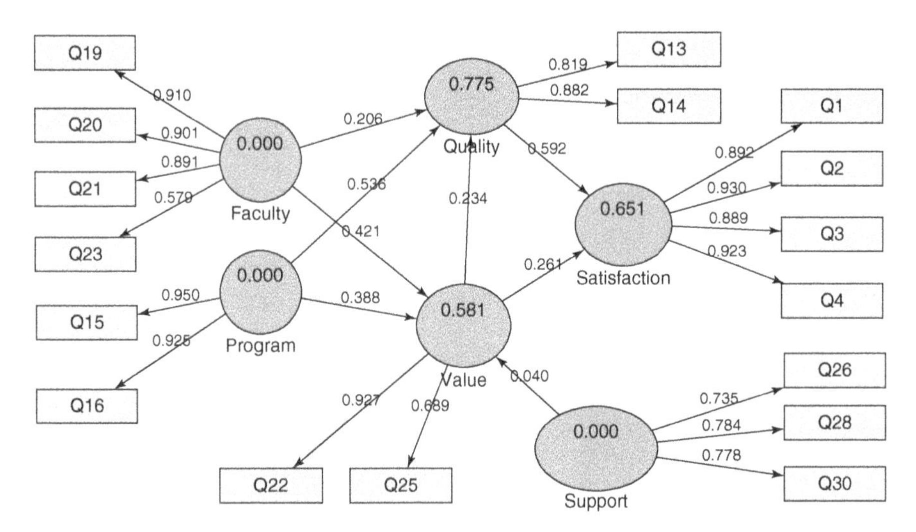

FIGURE 7.9 A focus SEM that models the program quality in a graduate professional school.

TABLE 7.5 The Assessment Data of the Measurement and Structural Models (A Focus SEM)

	AVE	Composite Reliability	R^2	Cronbach's Alpha
Faculty	0.692209	0.897357		0.844313
Program	0.879668	0.935971		0.864524
Quality	0.724094	0.839783	0.774870	0.622111
Satisfaction	0.825928	0.949928	0.650854	0.929633
Support	0.586977	0.809888		0.654831
Value	0.666706	0.796586	0.581142	0.535843

TABLE 7.6 Latent Variable Correlations in the Focus SEM

	Faculty	Program	Quality	Satisfaction	Support	Value
Faculty	**0.692209**					
Program	0.673252	**0.879668**				
Quality	0.731086	0.836926	**0.724094**			
Satisfaction	0.699606	0.756430	0.787987	**0.825928**		
Support	0.447598	0.510850	0.523520	0.560359	**0.586977**	
Value	0.700608	0.692432	0.749747	0.705272	0.427105	**0.666706**

values for path relationships in the structural model should be at significant levels. As discussed earlier, this significance can be evaluated using the bootstrapping procedures that are provided in SmartPLS (Ringle et al., 2005). We run bootstrapping with 100 cases and 300 samples. The corresponding bootstrapping results are shown

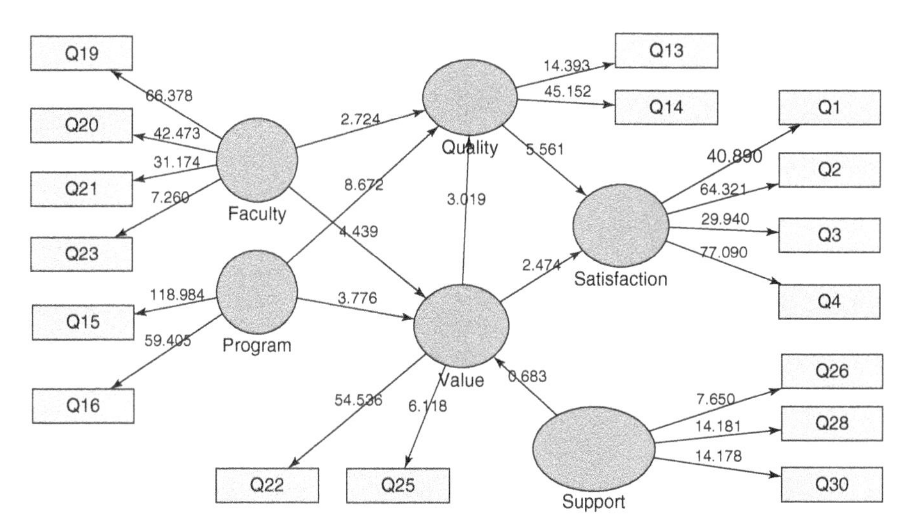

FIGURE 7.10 *The bootstrapping results of path relationships in the focus SEM.*

in Figure 7.10. Table 7.7 lists the path coefficients (Mean, STDEV, *T*-values) for the developed SEM.

From Table 7.7, we find that we fail to pass the test as the path coefficient Value to Quality lacks the necessary confidence level. Therefore, we must revise our model to ensure that the path coefficient can pass the *t*-test. The final SEM is shown in Figure 7.11. Table 7.8 shows the assessment data of the final measurement and structural models, in which constructs are mainly reflected by influential or actionable indicators. Table 7.9 lists the results of latent variable correlations in the final model illustrated in Figure 7.11.

Cross-loadings are often used to show the discriminant validity of a reflective measurement model. Table 7.10 lists the results of cross-loadings in the final model illustrated in Figure 7.11. It is confirmed that each block of indicators loads higher for its respective latent variable than indicators for other latent variables.

We run bootstrapping with 100 cases and 300 samples using the final SEM. The corresponding bootstrapping results are shown in Figure 7.12. Table 7.11 lists the path coefficients (Mean, STDEV, *T*-values) for the final SEM. This bootstrapping test shows that the estimated values for path relationships in the final structural model are indeed at significant levels. Therefore, both the final measurement and structural models are valid and reliable.

To assist decision makers with insights of service operations and management, we can convert the performance index of a latent variable using estimates on a 1-to-100 scale basis for easy human interpretation. Thus, a weighted average of scores from corresponding measurement variables is used by converting the original 7-point scale to a 100-point scale. By converting the results shown in Figure 7.11, we show the converted results of the final SEM in Figure 7.13, which are illustrated using estimates on a 1-to-100 scale basis. In addition to the

TABLE 7.7 Path Coefficients (Mean, STDEV, *T*-Values) Report for the Focus SEM

	Original Sample (O)	Sample Mean (M)	Standard Deviation (STDEV)	Standard Error (STERR)	*T*-Statistics (IO/STERRI)
Faculty → Quality	0.206184	0.199749	0.075704	0.075704	2.723537
Faculty → Value	0.421154	0.415019	0.094873	0.094873	4.439138
Program → Quality	0.535961	0.541161	0.061804	0.061804	8.672009
Program → Value	0.388348	0.399482	0.102851	0.102851	3.775838
Quality → Satisfaction	0.591970	0.586952	0.106458	0.106458	5.560609
Support → Value	0.040210	0.079948	0.058911	0.058911	**0.682567**
Value → Quality	0.234177	0.235352	0.077571	0.077571	3.018858
Value → Satisfaction	0.261444	0.269461	0.105657	0.105657	2.474450

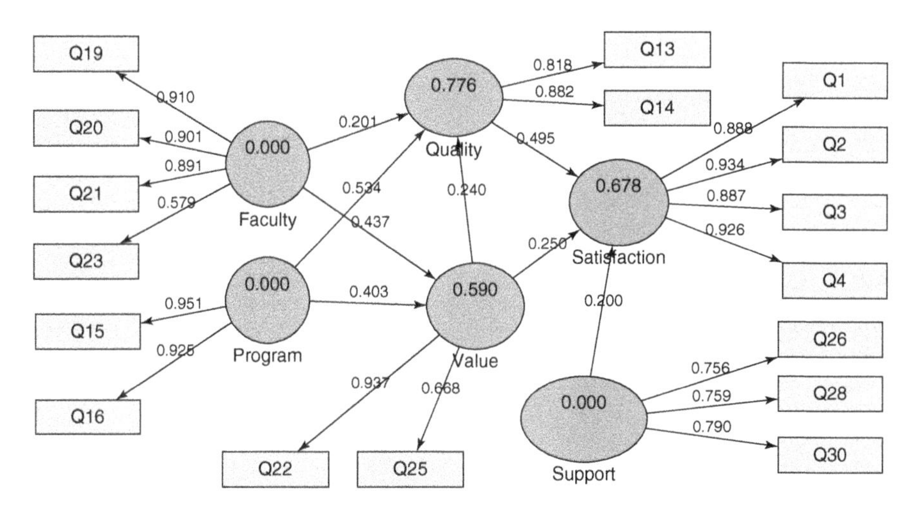

FIGURE 7.11 *The final SEM that models the program quality in a graduate professional school.*

TABLE 7.8 The Assessment Data of the Measurement and Structural Models (The Final SEM)

	AVE	Composite Reliability	R^2	Cronbach's Alpha
Faculty	0.692226	0.897372		0.844313
Program	0.879634	0.935952		0.864524
Quality	0.724072	0.839767	0.775709	0.622111
Satisfaction	0.826129	0.949988	0.678209	0.929633
Support	0.590186	0.811982		0.654831
Value	0.662208	0.792283	0.590154	0.535843

TABLE 7.9 Latent Variable Correlations in the Final SEM

	Faculty	Program	Quality	Satisfaction	Support	Value
Faculty	**0.692226**					
Program	0.673304	**0.879634**				
Quality	0.731001	0.837117	**0.724072**			
Satisfaction	0.697838	0.755955	0.786953	**0.826129**		
Support	0.438959	0.512730	0.516582	0.560036	**0.590186**	
Value	0.708142	0.696983	0.755148	0.707349	0.418563	**0.662208**

path coefficients and performance indices, Figure 7.13 also provides the weights of individual indicators. For the sake of convenience, we have all the weights shown in Figure 7.13 listed in Table 7.12. Total effects in the final SEM is then provided in Table 7.13, which essentially present how much one perceived construct might impact another perceived construct in the investigated service system.

TABLE 7.10 Cross-Loadings in the Final SEM

	Faculty	Program	Quality	Support	Value	Satisfaction
Q1	0.660608	0.683556	0.732008	0.428316	0.703793	**0.888106**
Q13	0.593550	0.544060	**0.818422**	0.409559	0.593901	0.645482
Q14	0.648777	0.854105	**0.882230**	0.466578	0.685985	0.693108
Q15	0.677687	**0.950655**	0.844523	0.454466	0.728331	0.722632
Q16	0.576694	**0.924945**	0.714606	0.514630	0.564128	0.694342
Q19	**0.909940**	0.706526	0.755019	0.379925	0.709701	0.697847
Q2	0.592926	0.688390	0.700471	0.608285	0.616795	**0.933561**
Q20	**0.900549**	0.529666	0.645550	0.310709	0.597193	0.603205
Q21	**0.891449**	0.582194	0.638540	0.416317	0.580035	0.612982
Q22	0.753876	0.704682	0.770847	0.329745	**0.936786**	0.708534
Q23	**0.579003**	0.361575	0.283688	0.398547	0.433970	0.334659
Q25	0.281010	0.355895	0.369957	0.412342	**0.668468**	0.375278
Q26	0.249149	0.428591	0.279704	**0.755613**	0.263120	0.384230
Q28	0.484586	0.372700	0.500332	**0.758757**	0.382257	0.473021
Q3	0.672891	0.687021	0.717865	0.461890	0.649054	**0.887431**
Q30	0.252472	0.385911	0.387386	**0.789870**	0.306630	0.423995
Q4	0.611125	0.689025	0.710416	0.535686	0.601131	**0.925588**

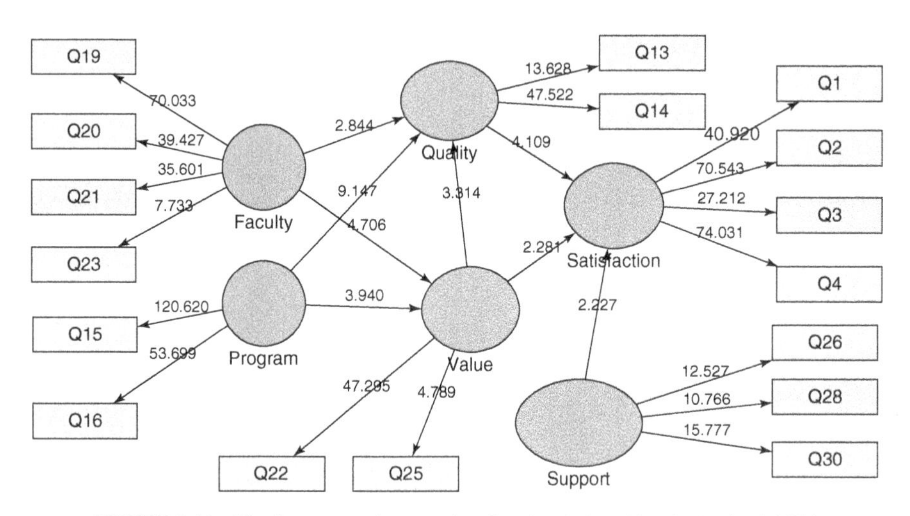

FIGURE 7.12 The bootstrapping results of path relationships in the final SEM.

Apparently, we can directly and indirectly collect adequate measurements of a service system that truly and timely capture the insights and dynamic behaviors of the service operations in a school. If appropriate information systems and services are well deployed (Figure 7.14), the presented approach should provide real-time business analytics to the administration in a school, resulting in that the administration can make informed decisions and take prompt and optimal actions in keeping the provided education service attractive, competitive, and satisfactory.

TABLE 7.11 Path Coefficients (Mean, STDEV, *T*-Values) Report for the Final SEM

	Original Sample (O)	Sample Mean (M)	Standard Deviation (STDEV)	Standard Error (STERR)	*T*-Statistics (IO/STERRI)
Faculty → Quality	0.201083	0.199764	0.070711	0.070711	2.843728
Faculty → Value	0.436945	0.440188	0.092844	0.092844	4.706211
Program → Quality	0.534113	0.532635	0.058390	0.058390	9.147273
Program → Value	0.402786	0.407524	0.102226	0.102226	3.940169
Quality → Satisfaction	0.495082	0.487465	0.120478	0.120478	4.109310
Support → Satisfaction	0.199684	0.217621	0.089650	0.089650	2.227361
Value → Quality	0.240485	0.243602	0.072562	0.072562	3.314193
Value → Satisfaction	0.249909	0.246934	0.109539	0.109539	2.281449

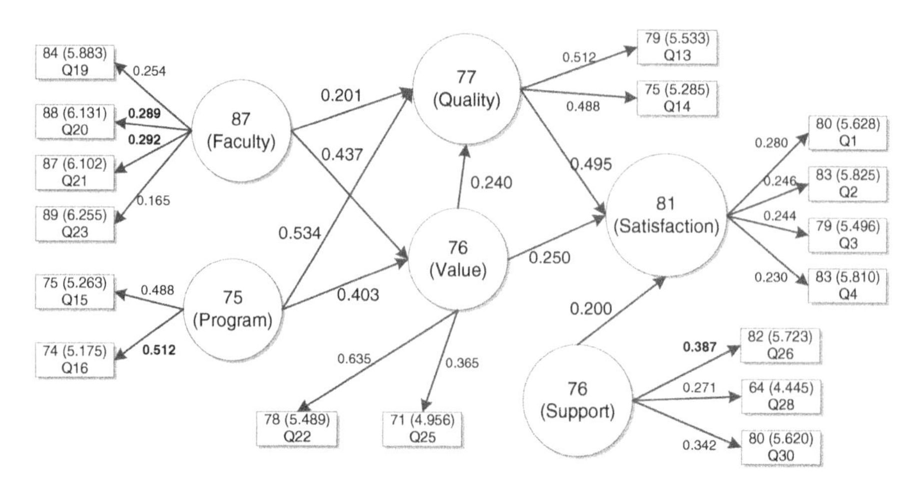

FIGURE 7.13 *The final SEM that is illustrated using estimates on a 1-to-100 scale basis.*

TABLE 7.12 Measurement Model Index (Weight)

	Faculty	Program	Quality	Satisfaction	Support	Value
Q1				0.279928		
Q13			0.512377			
Q14			0.487623			
Q15		0.488399				
Q16		0.511601				
Q19	0.253861					
Q2				0.245576		
Q20	0.289077					
Q21	0.292138					
Q22						0.635362
Q23	0.164923					
Q25						0.364638
Q26					0.387226	
Q28					0.270559	
Q3				0.243989		
Q30					0.342215	
Q4				0.230506		

TABLE 7.13 Total Effects in the Final SEM

	Value	Quality	Satisfaction
Faculty	0.436945	0.306162	0.260771
Program	0.402786	0.630977	0.413045
Quality			0.495082
Support			0.199684
Value		0.240485	0.368968

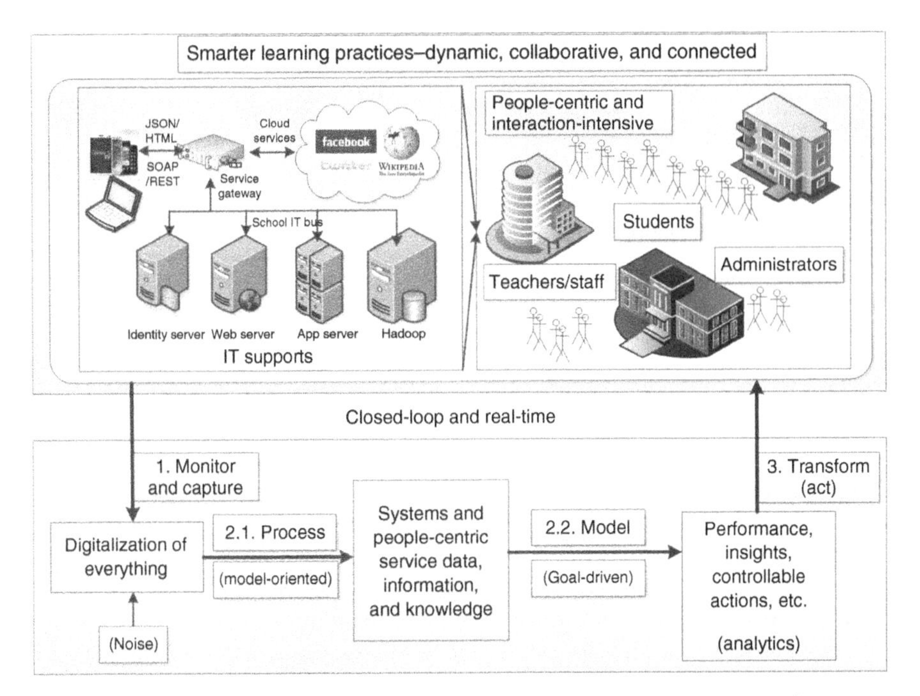

FIGURE 7.14 *Retuning of a school service system in real time for competitive advantage.*

Apparently it is not easy for us to analyze a whole service system like a school in a book. Without loss of generality, we look into a small-scale service system in our next example so that we can easily present approaches to make informed decisions and take prompt and optimal actions in adjusting the service system's dynamics for improved outcomes.

7.3 OFF-CAMPUS LEARNING: AN EXAMPLE OF HIGH SCHOOL STEM EDUCATION ENHANCEMENT

Without question, science and technology has played a more and more critical role in the world economy than ever before. However, the yearly percentage of students in the United States who choose science, technology, engineering, and mathematics (STEM) careers after their graduations from high schools continues to decline. Statistics shows that the United States now lags behind much of the world in terms of the percentage of students who want STEM careers when they join the workforce.

Mathematics is the foundation of the STEM fields as it essentially provides the knowledge and skills necessary for advancement and innovation in every other STEM field. However, the overall math competence of *K-12* school students in the United States is worrisome, because many high school graduates in the United States do not possess good preparation in math to enter the workforce (Texas Instruments, 2006).

For example, Saad (2005) reports, in a 2004 Gallup poll, that 37% of teens regarded math as their most difficult subject, while only 23% considered math as their favorite one. The 2006 Program for International Student Assessment (PISA) corroborated this, showing that the performance of 15-year-old students in math literacy in the United States was 24 points lower than the global average. At the 90th percentile, the United States was 22 points below the global average. These scores ranked the United States below all but six participating jurisdiction countries tested (Baldi et al., 2007).

On the basis of this observation, math performance among teens in the United States is inferior to that among teens in many other countries. This phenomenon can be attributed to many different reasons. In terms of broad sociocultural causes, a study of students with advanced math proficiency, over several decades, found that the US society does not accord mathematics with much recognition. For instance, social ostracism and ridicule make male and especially female students stay away from math as a career. Consequently, currently 80% of female math professors and 60% of male math professors hired in the past few years were born outside of the United States (Andreescu et al., 2008). In addition, family expectations, peer influences, and teacher qualifications all have a substantial impact on students' attitude to math and their math performance in schools. Interestingly, studies show that family expectations have the strongest impact on all ethnic groups when it comes to math achievement (National Center for Education Statistics, 2004; Wenglinsky, 2000; Yan and Lin, 2005; Matthews, 2007). Hence, an approach to positively changing family expectations could make a difference in improving students' overall attitude to the STEM.

Note that the US federal and state agencies and organizations have launched a variety of initiatives in the STEM education to change this situation. More specifically, many research projects on STEM education in both college and *K-12* school have been conducted, and hundreds of school districts across the United States have established STEM programs since the early 1980s (NGA, 2009; NSB, 2010). More promisingly, significant attention has recently been directed to encourage and develop the talents of adolescents with interests in STEM, aimed at boosting the domestic STEM talent supply from the root (Schaefer et al., 2003; Miller and Barney Smith, 2006; Holmes et al., 2007; Atkinson et al., 2007; Qiu and Doris, 2013).

Academically, suggested solutions for improving students' math performance include a communal structure within schools, more cohesive curriculum, smaller class sizes, cooperative learning, and strategies that help students overcome math anxiety. Lee et al. (1997) discuss the results of a study focused on the elements of high school organizations that are directly related to the improved learning of mathematics and science as well as the "equity" between the first 2 years and the past 2 years of high school. The term "a mile wide and an inch deep," first used by William Schmidt of Michigan State University, describes the US math curriculum as a vast series of topics that are covered in each grade. However, each topic is not covered in detail and little connection is made among topics. According to Roger Bybee, head of a Colorado Springs-based nonprofit science curriculum development organization, this leads to a curriculum that has no coherency. He and other educators advocate a national curriculum standard that determines core skill sets and promotes textbooks that establish connections between math topics. With a national curriculum, students will be pushed to master the core content and be evaluated as per established standards (Brown and Brown, 2007).

In addition, smaller class sizes, cooperative learning, and less math anxiety all play a role in improving individual student performance. Rice (1999) conducts a study on class sizes and analyzed their impacts on students' performance. She unveils that class size had a greater impact on math classes than science classes, inversely affecting three instructional variables, namely, small group time, innovative instructional practices, and whole group discussions. For cooperative learning, Slavin et al. (2009) review effective programs in middle school and high school mathematics and found that math curricula and the use of existing computer-based instruction had little effect on achievement, while there was significant positive effect for cooperative learning programs. Programs emphasizing teaching quality and student interaction along with textbooks and technology have more positive results than those that emphasize textbooks and technology only. Finally, Tobias (1993) defines math anxiety as a conditioned emotional response to participate in a math class and/or talking about math. This response produces a fear that precludes students from maximizing their performance in the subject at school and from pursuing possible careers in the future. Curtain-Phillips (1999) indicates that such anxiety results from three common classroom practices, namely, imposed authority, public exposure, and deadlines. Teachers and parents can also heighten the anxiety by forcing their perspective of math on the students. Rossman (2006) suggests that math teachers should encourage their students to be active learners, make the math relevant to the children's world, and promote collaborative learning.

In summary, improving math performance in the United States requires providing the necessary resources and making positive changes for students from different social and economic strata, building a better communal structure within schools, creating more cohesive math curricula, offering smaller classes, encouraging cooperative learning, and helping students overcome math anxiety. As teens spend significant time outside school, their interests toward a STEM field would be substantially influenced by their after-school activities. Therefore, when we address STEM education at school, it is enticing and effective if we can promote STEM education in an after-school setting by fully leveraging certain after-school advantages (e.g., any time, any pace, relaxing and less anxiety, family's involvement, etc.).

Currently, even though many web sites with online *K-12* learning materials exist (OnlineK12, 2013) (for instance, a Google search will return hundreds of web sites), it remains difficult to locate pertinent literature for quality STEM education outside the school setting. It is equally difficult to find literature that focuses on incorporating the state-of-the-art cyber-based self-service technologies and uncovered in-school STEM education knowledge for effective off-campus STEM education, although "open educational resources" initiatives (e.g., The William and Flora Hewlett Foundation at http://www.hewlett.org/programs/education-program/open-educational-resources) focusing on improving students learning experience while leveraging external resources are gaining momentum (OER, 2010).

Furthermore, few effective approaches that were derived from empirical research are available to substantiate the success of STEM programs (Cavanagh, 2008). Brody (2006) notes that most measures regarding STEM programs in place focus on "inputs," which refer to the number of participating teachers or attitudinal changes among students. She points out that there lacks a systematic approach in the current array of research, resulting in nebulous findings regarding the effectiveness. Clearly, there is an array of research questions that must be well addressed, including what short-term and long-term outcomes should be measured, what are the strengths and weaknesses of various outcome metrics, and what other success definitions should be studied. We are indeed concerned about the absence of a sound knowledge base for determining the effectiveness of a STEM program in light of meeting the nation's STEM talent growth objectives (Subotnik et al., 2007; Subotnik et al., 2009). However, we believe that an approach to retune a service system in a real-time and gradual manner for competitive advantage, as shown in Figure 7.14, can be viable in practice.

More specifically, the emergence of Service Science brings in a promising approach to facilitate STEM program offering in meeting the needs of education. By applying the concept of Service Science to the development process of a web-based education service system, we understand that the learning experiences that public school students have outside of the classroom positively impact on their learning outcomes (Qiu and Doris, 2013). We fully understand that we should avoid one-size-fits-all thinking. In other words, different paradigms might be more appropriate for certain circumstances. Therefore, "What kind of approach and proactive guidance should be provided to effectively customize and enrich such an outside school learning experience for improved impact on STEM education?" becomes an obvious question in the long run. The solution to this question with scientific rigor is truly critical for the continuous improvement and widespread adoption of the discussed model on a large-scale basis.

As indicated earlier, when we study a specific topic in education, we must stay focused and pay much attention to the areas in which we are interested. On the basis of the early discussion, we are motivated by the fact that effectively creating more interest among high school students and providing an outside-school, self-learning experience, through optimally leveraging state-of-the-art cyber-based technologies, will significantly enrich and enhance our nation's STEM education. By having a strong connection with the fast-changing and digitalized living environments, the potential influence on high school students' attitudes and perceptions is enormous.

To essentially enrich the knowledge base of STEM education, we present how a transformative education service system (TESS) can significantly address the defined challenges, namely, "Do the learning experiences that K-12 students have outside of the classroom contribute significantly to their choice of a STEM career?," "How does the effectiveness of an off-campus STEM education module get evaluated quantitatively?," and "What kind of approach and proactive guidance should be provided to effectively improve and enrich such a learning experience for improved impact on STEM education?" By taking the above-mentioned challenges into consideration during our TESS prototyping, we can ensure that the developed TESS meets the needs of the school, students, and their parents. Therefore, the desirable off-campus STEM education service can be well offered.

The detailed explanation of this example will be presented in the remaining parts of this section. More specifically, we thus discuss a Service-Science-based approach to enrich this off-campus high school STEM education program. The adopted approach includes the following three focused areas:

- *Web 2.0 System Prototyping*. We fully utilize the discovered knowledge on how to improve STEM education at school. Appropriate instruments and measures are developed, implemented, and enriched to assess engagement, persistence, and other relevant constructs of student motivation and learning.
- *Systems Performance Analysis*. We then use SEMs to answer the exploratory questions. As tests, data collection, and analyses are conducted semester by semester, an effective and applicable off-campus STEM education approach can evolve and mature over time.
- *Optimization and Improvement*. By using goal-driven tractable approximations and stochastic modeling, a guideline for potentially improving STEM education can be synthesized and compiled (Qiu, 2009; Qiu et al, 2011).

7.3.1 A Transformative Education Service System (TESS)

Given the increased dynamics and complexity of this fast-changing world, it could be extremely difficult to fully address the challenges. As a breakthrough exploration, the study explores an innovative STEM program with an initial focus on math education in a high school off-campus setting (Figure 7.15). Using the latest Web 2.0 technology, an off-campus, self-learning STEM service system will be easy and fun to use, encouraging teens to appreciate math and math learning. Visualization, interaction, collaboration, knowledge sharing, and intuitive programming will help to create

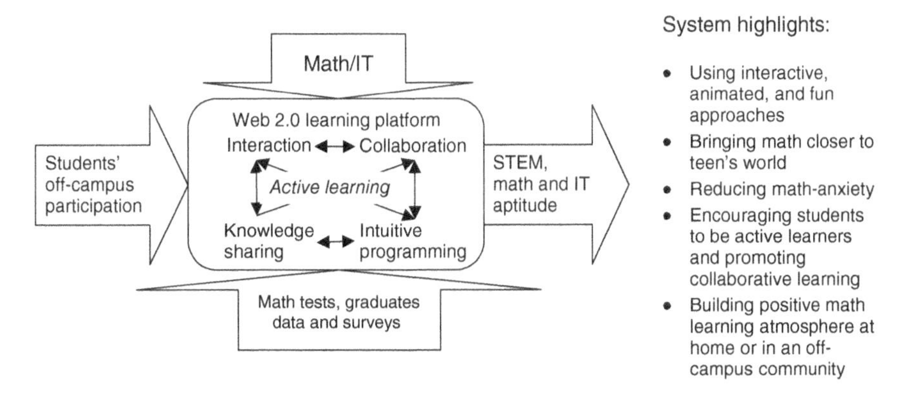

FIGURE 7.15 *Overview of TESS with a focus on off-campus STEM education service (Qiu and Lee, 2013).*

and nurture an active learning environment, keeping students pedagogically engaged. This service system is designed in a way that promotes a cohesive and hierarchical arrangement for anxiety-free math learning and gaining positive IT experience. High school students shall exit with greater proficiency in math, scoring higher on standardized testing and in school math programs. As a result, more students will choose a college major in a STEM field,

As discussed earlier, this example shows how to design, implement, test, and synthesize an innovative and systemic approach for enriching and improving math education (Figure 7.16). Using well-defined longitudinal studies, we focus on the mechanisms of retuning the service system in a comprehensive, quantitative, and qualitative manner. To further prove and demonstrate this example's exploratory potential, an approach to identify appropriate guidance for further improving STEM education, synthesized from longitudinal analytical studies, is provided in Section 7.3.3.

By collaborating with high schools, we design and develop a Web 2.0-based system to provide students with interactive math learning materials and some intuitive programming skills in an off-campus setting. For example, by closely working with math teachers in a high school, we can first identify 500 math cases that are difficult for high school students to learn in class. These cases can be evenly distributed across different math subjects that are offered in school (Figure 7.16). As an example, many students make mistakes in answering a question related to the object dilation of a *xyz* factor (e.g., 2, 0.2). We can create an animation page that is linked to this question as shown in Figure 7.17. By interacting with the object, students can directly watch its animated dynamic change with the change of the factor value or types of objects.

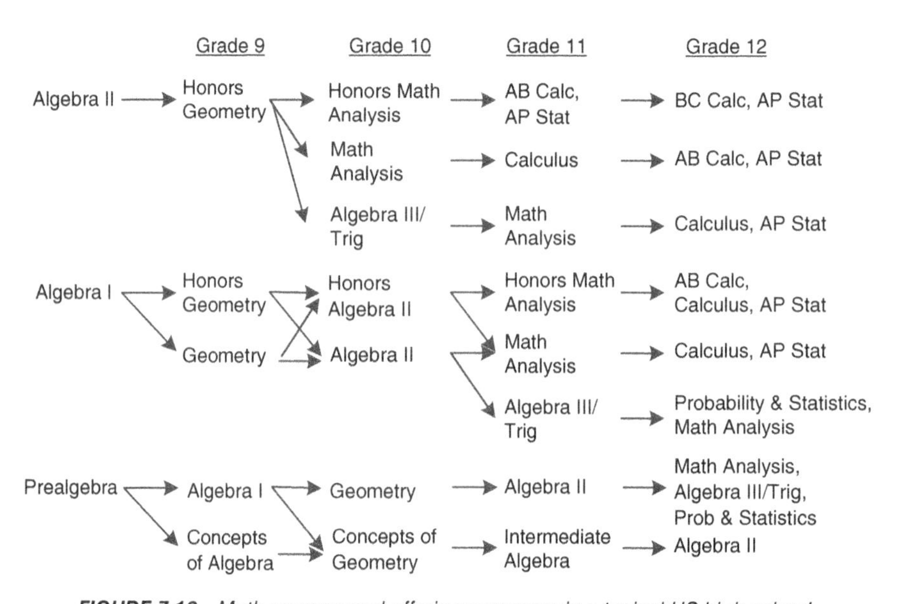

FIGURE 7.16 *Math courses and offering sequence in a typical US high school.*

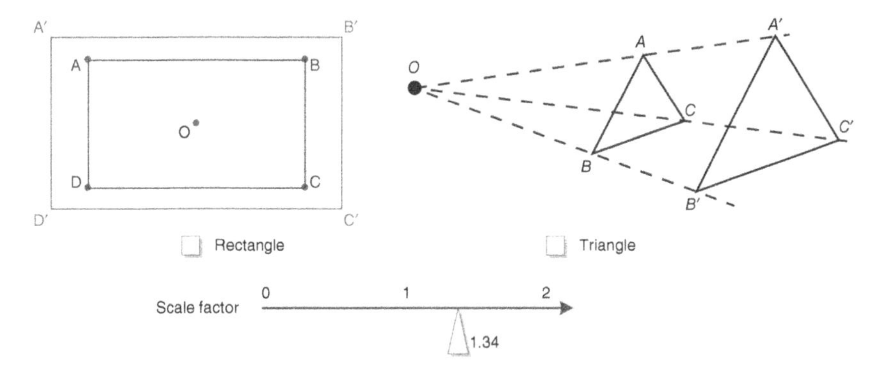

FIGURE 7.17 *Learning through visualization and interaction.*

Features in support of visualization (3D), interaction, collaboration, knowledge sharing, and intuitive programming can also be fully integrated, which create and nurture an active learning environment that helps students learn off-campus while having a lot of fun. This will promote a cohesive and hierarchical arrangement for anxiety-free math learning, within an open, positive, constructive, and collaborative atmosphere (Gokhale, 1995), while gaining state-of-the-art IT experience (Cuff and Molinaro, 2005). Essentially, we design and develop a Web 2.0-based animated learning system (Figure 7.18), focusing on keeping students engaged in a positive way.

7.3.1.1 Web 2.0 Services in a Real-Time and Collaborative Manner

Today's high school students are frequently called generation next or the Millennial Generation in a broad sense. Five important characteristics, "tech-savvy," "impatient," "fickle," "ambitious," and "communicators," have been used to describe them. Indeed, significant advancements in the computing, networking, and telecommunication technologies have been made since the turn of this new millennium. As a result, they grow up essentially with constant connection to the Internet. They like to use smart phones and tablet computing devices or the like, enjoying the

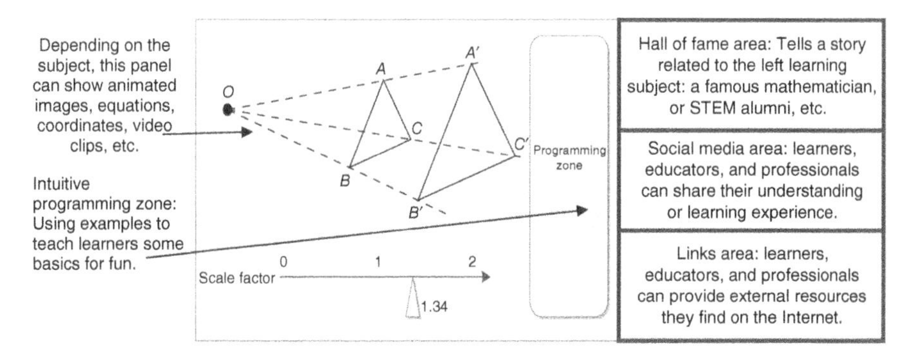

FIGURE 7.18 *Web 2.0-based animated learning system: an interface view.*

power, convenience, and connections enabled by the mobile computing devices and technologies. Indeed, "they're accustomed to using technology in every part of their lives and fully believe in its power to make their lives easier" (Ippolito, 2012). With the future becoming ever-more computer-oriented, network-connected, they have a stronghold on how to function in the digital age. Therefore, a learning service system should be well equipped with supports, meeting the needs of their interests and passions.

By allowing the individual users to interact and collaborate with each other, a Web 2.0 site focuses on leveraging end users' collective contributions to its served community. The quick advances in online social media have made possible interactions among end users in real time and at ease while engendering more fun and better personalization. Figure 7.19 shows how the TESS leverages the Web 2.0 services and takes advantage of the rich information over the Internet to deliver modules on a daily basis to subscribers. More importantly, online chats and instant messages are enabled, which allow students to interact and learn in a collaborative manner.

7.3.1.2 Big Data: Architecture and Transformation To provide effective and customized learning practices for students, we must know how they perform and what they like. To enable real-time guidance in support of their daily learning activities, we then have to analyze their learning practices as time goes. Capturing the dynamics of students as time goes becomes critical for us to develop and operate the Web 2.0 system, so that better guidance for individual learners over time can be truly provided.

TESS is designed with the capability of collecting daily learning activities for all the individuals. In other words, the activities of accessing the system are fully logged by TESS. As the system has links to many external sources, the data becomes big and overwhelming. For instance, the discussions on the provided forums, communications over Facebook and Twitters, and online chatting are most likely unstructured

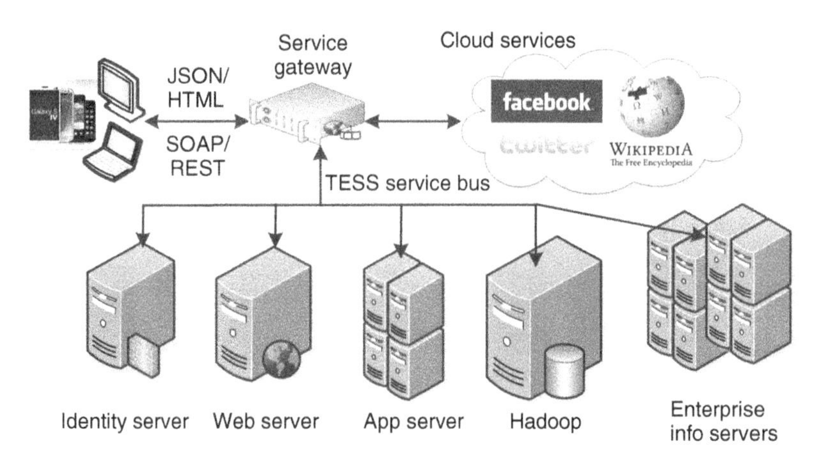

FIGURE 7.19 Web 2.0 services in a collaborative learning setting.

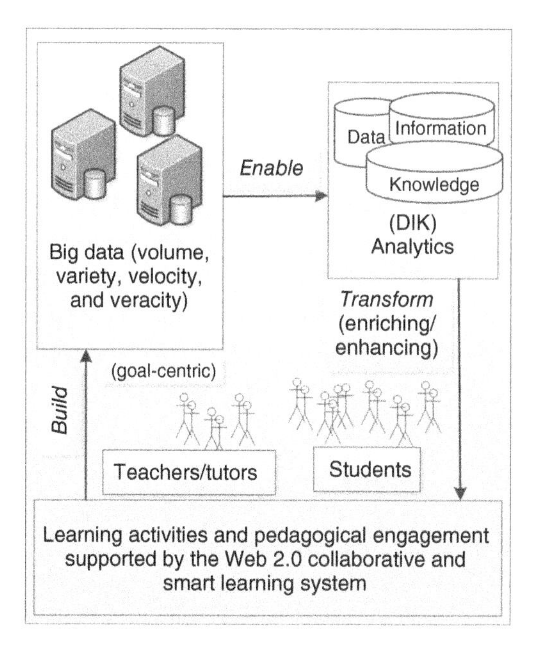

FIGURE 7.20 *Big data architecture applied to TESS.*

data. That is to say, the majority of data coming from disparate and heterogeneous sources are semistructured or unstructured (e.g., text and media data); thus, conventional approaches and tools that were designed to work with large structured data sets simply cannot handle this big data if learning analytics is the goal for the collected-data.

The prototype of TESS does not include the big data component. Figure 7.20 shows the overall big data architecture that can be used in the TESS when data for all learning activities managed by external sites can be well retrieved and logged. Figure 7.21 illustrates how we can appropriately deal with the voluminous data transformation services in support of Web 2.0-based collaborative learning. More importantly, new analytic methodologies and frameworks must be explored and introduced to the market to help the TESS bring order to the big data from diverse sources and thus harness the power of the big data. By doing so, we can further glean the insights and values of the application of TESS by conducting learning analytics to meet the needs of TESS in the long run (IBM Global Business Services, 2012, 2013).

7.3.2 Systems Performance Analysis

The goal of this TESS is to help high school students gain more interests in STEM-related subjects and careers through active participations in an out-of-school, self-learning program. In practical terms, the study will attempt to prove STEM's practical value by implementing a Web 2.0 learning system and investigating whether students using the learning system will produce a better math performance.

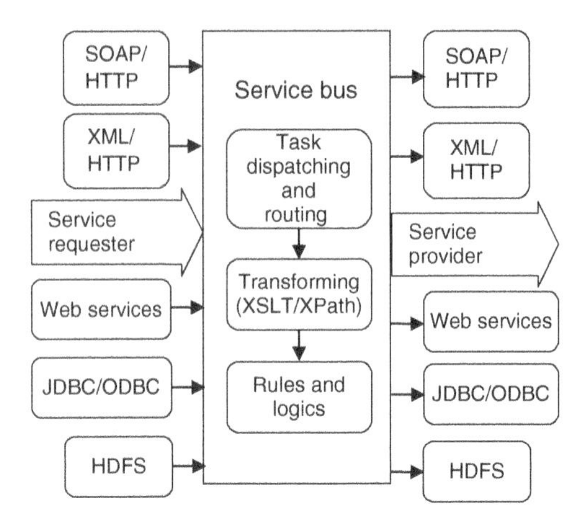

FIGURE 7.21 *Integration service bus used in TESS (Qiu, 2013).*

Furthermore, it will investigate whether the learning system leads them to choose a college major in a STEM field, meeting the growing demand for professionals and information technology workers in the US STEM workforce (NGA, 2009; NSB, 2010).

By incorporating the concept and principles of Service Science into this study, we show an innovative approach to encourage and improve high school STEM education through well-defined longitudinal studies. This study explores how our off-campus, self-learning module (i.e., a scaled-down version of the STEM program) will function in today's digitized world from a comprehensive, quantitative, and qualitative perspective. The developed approach is generic so that any STEM program in the United States can apply it to analyze challenges within its operations. On the basis of the analysis, a STEM program can then identify optimal countermeasures and priorities that will facilitate transforming learning processes in an effective manner.

The early stage of TESS development and relevant experiments can be used as the foundation for the study with a focus on the test of hypotheses. For example, before we fully design and develop the TESS, we should initially answer this question: "How much will the impact on high school students choosing STEM careers be after they receive enriched, cyber-based Math/IT education in an outside-school setting?" Using a cross-sectional empirical study that is simply based on SEM and principal component analysis (PCA) methods, we can conduct a full model fitting study, aimed at improving the adopted questionnaires and identifying features and functionalities capable of collecting the needed data from the TESS for further analytics.

Students in high schools can be invited to use the TESS. Depending on the course a student is taking, he/she will have different privileges to access relevant course materials. On the basis of the defined math course sequence (Figure 7.16), a student will be allowed to access the materials for all the courses that he/she already completed. By doing so, students will provide more comprehensive feedback information to this study over time. By working with teachers, administrators, and some volunteer students, numerous sets of questionnaires for students based on groups and grades can

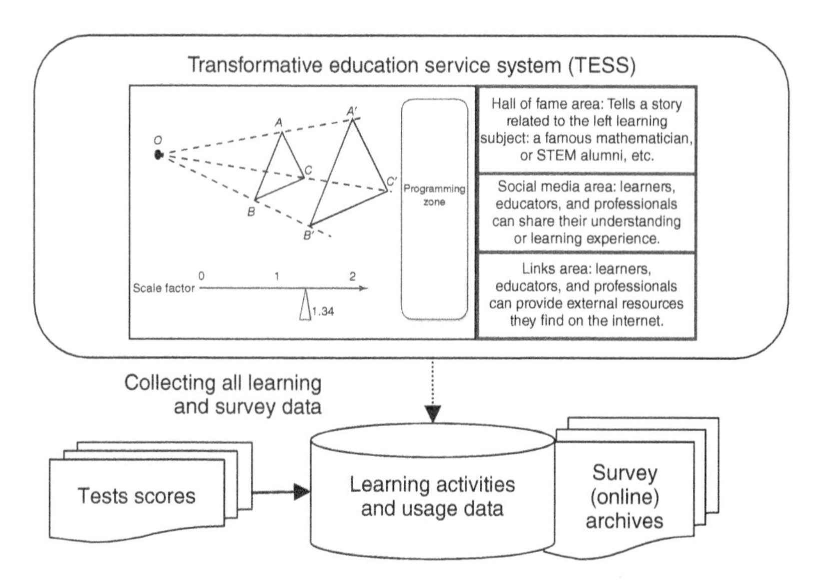

FIGURE 7.22 *Data collection mechanisms used in the Web-based interactive learning system.*

be developed. As a result, by fully leveraging the data collection mechanisms embedded in the system, data measuring the learning process, materials, interactions, usage, and outcomes (e.g., tests scores) can be collected in a timely manner (Figure 7.22).

Voluminous data and rich information on students' learning practices can be collected as the experiments continue year by year (Figure 7.23). Promisingly, the conducted study will go beyond a cross-sectional empirical one; a longitudinal study shall result, which can provide much better understanding of learning practices in an online and off-campus setting. More importantly, the learned insights of students'

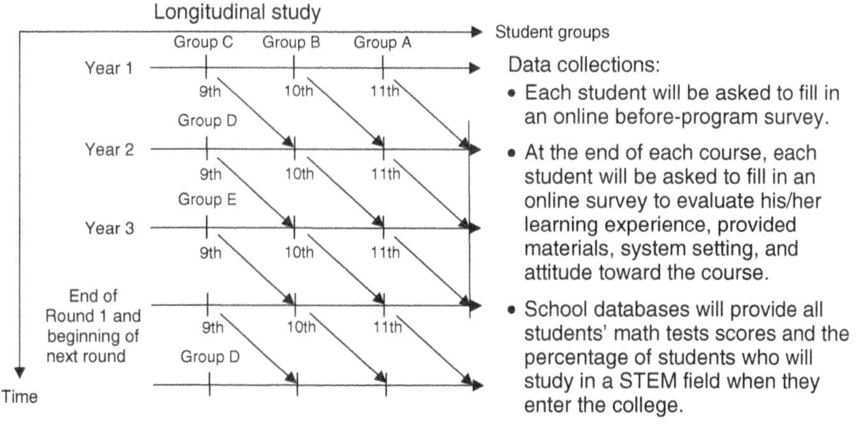

FIGURE 7.23 *Tests and data collections for a period of 3 years.*

learning behavior and needs can help us improve TESS system designs (e.g., interfaces, features, and interactions) and enrich learning materials (e.g., more examples and cases if possible). Ultimately, students' off-campus learning experience can be substantially impacted in a positive way.

Note that data preprocessing is critical for SEM and PCA analyses (Wang et al., 2010). Most of the logged TESS data will be in standard XML formats, with the exceptions of the survey data that are mainly in Likert scale. Data collection and transformation algorithms are essential for discovering and extracting the needed data. Transformed data and texts are mapped as the required formats such as Likert-scale data, integers, or binary numbers can be properly loaded into SEM and PCA models (Agrawal et al., 1993; Zeng et al., 2007; Wang et al., 2009). Essentially, data from individuals' learning practices, different learning modules, and collaborative learning activities through the TESS must be fully collected and appropriately transformed, and then made ready for conducting learning analytics.

7.3.2.1 Learning Analytics in the TESS Engendering more interest among high school students, providing an outside-school active learning experience, and nurturing a societal culture of highly respectful STEM recognition substantially contribute to the teen development. As discussed earlier, we focus on the fundamental understanding of how we can create and maintain active and self-paced STEM learning atmospheres in off-campus settings, resulting in positive STEM workforce development, participation, and improvement. As it takes place in the context of the family, the peer group, and the neighborhood or community with the support of the advanced cyber technologies, the learning relying on the TESS becomes essentially a contemporary and dynamic sociotechnical phenomenon.

By referring to a variety of SEM-based STEM studies such as ones conducted by Dauphinee et al. (1997), Bernold et al. (2007), and Tempelaar et al. (2007), we develop an SEM based on indicators covering areas from study help, students' attitude, engagement, persistence, motivation, and other relevant latent constructs that are known to impact off-campus learning processes one way or another. From the constructed SEM illustrated in Figure 7.24, these latent variables are measured and analyzed using the following measurable variables in groups.

- *Value.* A list of questions is used to understand students' attitudes regarding the usefulness, relevance, and worth of math and IT through their personal observations and life experience.
- *Difficulty.* A list of questions is used to measure students' attitudes toward the difficulty level of math or IT as a learning subject.
- *Motivation.* A list of questions is used to measure societal recognition and respect in students' mind.
- *Effort.* A list of questions is used to measure the amount of work that a student expends to learn math and IT over the exploratory period.
- *Study Help.* A list of questions is used to measure the learning setting and available learning assistance and supports: availability of online help, easiness of getting the learning subjects, appropriateness of learning materials, etc.

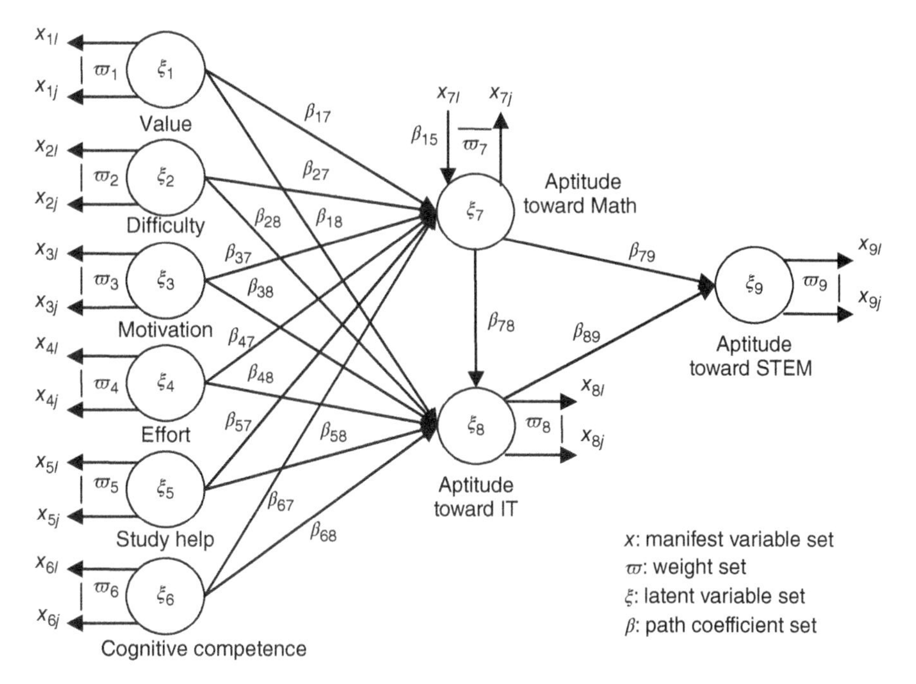

FIGURE 7.24 *An SEM model for the needed cross section and longitudinal empirical study.*

- *Cognitive Competence.* A list of questions is used to measure students' intellectual competence, subject knowledge, and skills applied to the process of learning math and IT.
- *Aptitude Toward Math.* A list of measures is used to gauge the enjoyment aspect of intrinsic value and the acquired knowledge of learned relevant math subjects.
- *Aptitude Toward IT.* A list of measures is used to gauge the student's perceptions, state of mind, or feelings toward IT: *I have fun using state-of-the-art web technologies; I enjoy the visualization and interactions, etc.*
- *Aptitude Toward STEM.* A list of questions is used to measure the changes concerning STEM improvement: *improved percentage of students choosing STEM career, better a variety of testing results, etc.*

The perceived learning outcomes are extremely subjective and also vary with groups of students and schools. Hence, a suite of priority measures (i.e., the main factors influencing students' attitude toward STEM) must be developed for evaluating the effectiveness of the TESS and conducted experiments. Although it is clear that the yearly percentage change in students' attitude toward STEM is what we will use to measure the effectiveness of this study, a generic effectiveness measure can be more appealing in practice. This generic effectiveness measure might shed a light of establishing better quantitative measures for achieving different objectives when circumstances change over time or similar explorations are conducted in different schools:

$$\Omega(\text{Effectiveness}) \propto \bigcup_{i=1}^{n} \lambda_i f_i(\xi_i) \tag{7.1}$$

where n is the number of constructs considered at the time when such a study takes place, $f(\xi)$ is a mapping function for a given objective latent variable, and λ is the weight defined by a given school district and can be adjusted as time goes. The defined weight can be determined using analytic hierarchy process, analytic network process, or other decision-making methodologies (Ahmad and Qiu, 2009).

In this analytical study, by relying on PCA and SEM mediator and moderator effect analyses, we can analyze how certain factors (i.e., more visual geometry models, hands-on programming, or more interactions) highly correlated, and understand how an action will quantitatively impact the high school students' inclination toward STEM as time goes. As a result, rules of thumb for students and tutors can be developed, which can be then applied to guide teaching and learning practices to improve students' inclination toward math, in particular, and STEM, in general.

7.3.3 A Goal-Driven Learning System: Optimization and Improvement

"The only constant is change." Student inclination toward STEM could vary with changes (Figures 7.15 and 7.24): new IT technologies, changed user profiles because of improved individual's learning and cognitive ability, etc. To tackle changes and assure the desirable trend of STEM education in the focused group, stochastic processes and algorithms, to ensure tractable approximations, should be incorporated into the study. Once again, we fully understand that we should avoid one-size-fits-all thinking as different paradigms might be more appropriate for certain circumstances. Therefore, one way to gain a quantitative understanding of off-campus math learning by relying on the TESS is what we mean in the following discussion.

A goal-driven approach to optimize and improve the TESS essentially focuses on answering the following question, "*What kind of approach and guidance should be used to effectively enhance and enrich such a learning experience for improved impact on STEM education?*" We discuss one possible way to meet the defined needs. In this subsection, we will focus on (i) defining a vector-continued fractions interpolation method for trajectory approximations (Qiu et al., 2011); (ii) developing a robust optimization approach to address changes as time goes; and (iii) further enhancing the developed TESS by fully incorporating the research findings. Ultimately, the developed mechanisms support retuning of the learning guidance in a proactive and real-time manner, so that students would have tractable and desirable performances (see Figures 7.23 and 7.25). The developed Service-Science-based framework essentially could lay out a solid foundation for a future scale-up study regionally or nationally.

7.3.3.1 *Cause–Effect Analysis and Prediction* For a given n-dimension-based performance trajectory (Figure 7.25), vector-continued fractions can be used to

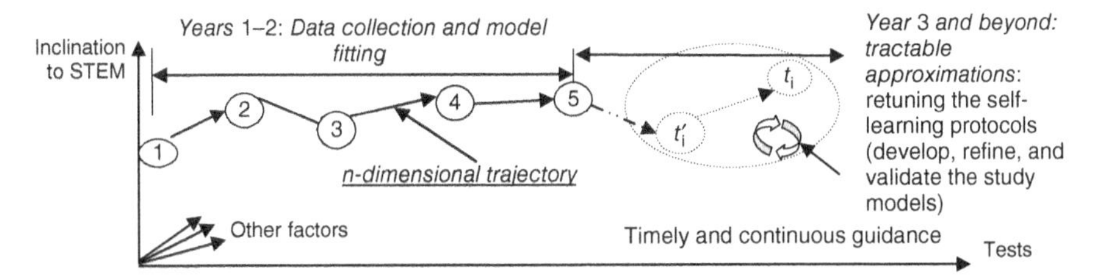

FIGURE 7.25 *Test, effectiveness metrics, and tractable trajectory approximations.*

find its fitting or path approximation equation (Roberts, 1999; Haydock et al., 2004). We use the equation to approximate the student inclination toward STEM by predicting its next dynamic behavior. Assume that there is a data set from 1 to t (e.g., each test period derives one set in the TESS), which is denoted as $H^j (j = 1, 2, \ldots, t)$. Each H^j has $m + n$ vectors $\varpi_1, \varpi_2, \ldots, \varpi_m, \beta_1, \beta_2, \ldots, \beta_n$ (see Figure 7.24). The dynamic trajectory of the TESS can be approximately described by the SEM data set of loadings and coefficients. By introducing an interpolation method, we aim to improve the precision of the trajectory approximations, which was successfully done in our hotel service quality project (Qiu et al., 2011).

Definition 1 Suppose that $A = (a_{ij})_{m \times n}$ is a $m \times n$ matrix. Using continued fractions, we can construct $\mathrm{Vec}(A) = (a_{11}, \ldots, a_{1n}, a_{21}, \ldots, a_{2n}, \ldots, a_{m1}, \ldots, a_{mn})$, $\mathrm{Vec}(A)^{-1} = \mathrm{Vec}(A)/|\mathrm{Vec}(A)|^2$, $A^{-1} = A/|A|^2$, where $|A|^2 = \sum_{i=1}^{m} \sum_{j=1}^{n} a_{ij}^2$ and A^{-1} is the pseudo inverse of vector A, and $A^{-1}A \neq E$.

Definition 2 Assume $V_0, V_1, \ldots, V_m$ are vectors, we define the following matrices,

$$\varphi[x_i] = V_i, i = 0, 1, 2, \ldots, m$$

$$\varphi[x_p, x_q] = \frac{x_q - x_p}{\varphi[x_q] - \varphi[x_p]}; \quad \ldots$$

$$\varphi[x_i, \ldots, x_j; x_k, x_l] = \frac{x_l - x_k}{\varphi[x_i, \ldots, x_j; x_l] - \varphi[x_i, \ldots, x_j; x_k]}$$

Therefore, $\varphi[x_0, \ldots, x_l]$ is the φ-order vector-valued inverse difference of vector set $V^n = (V_0, V_1, \ldots, V_m)$ in $x_0, \cdots, x_l$. We can have

$$R_n(x) = \varphi[x_0] + \frac{x - x_0}{\varphi[x_0, x_1]} \ddots + \frac{x - x_{n-1}}{\varphi[x_0, x_1, \ldots, x_n]}, \text{ where } R_n(x_i) = V_i (i = 0, 1, 2, \ldots, n) \tag{7.2}$$

Goal-Driven Tractable Approximation Algorithm. On the basis of the definitions given earlier, we construct the following interpolation function sets, based on $X_{n+k} = \{x_i; i = 0, 1, \ldots, n+k\}$ and $V_{n+k} = \{v_i; i = 0, 1, \ldots, n+k\}$, for improved path approximations:

Step 1. Divide X_{n+k} and V_{n+k} into two sets: $X_1 = \{x_0, x_1, \ldots, x_{k-1}\}$, $V_1 = \{v_0, v_1, \ldots, v_{k-1}\}$, and $X_2 = X_{n+k} - X_1$, $V_2 = V_{n+k} - V_1$. On X_1 and V_1, we construct a $k-1$-order Newton interpolation polynomial $P_{k-1}(x)$, or a vector-continued fraction interpolation $R_{k-1}(x)$, $P_{k-1}(x_i) = v_i, i = 0, 1, 2, \ldots, k-1$; $R_{k-1}(x_i) = v_i, i = 0, 1, 2, \ldots, k-1$.

Step 2. For $i = k, k+1, \ldots, n+k$, define $U_i = v_i - P_{k-1}(x_i)/\prod_{l=0}^{k-1}(x_i - x_l)$ or $U_i = v_i - R_{k-1}(x_i)/\prod_{l=0}^{k-1}(x_i - x_l)$, we have $S_n(x) = \varphi[x_k] + \frac{x-x_k}{\varphi[x_k, x_{k+1}]} \ddots + \frac{x-x_{k+n-1}}{\varphi[x_k, \ldots, x_{k+n}]}$, where $\varphi[x_i] = U_i, i = k, k+1, \ldots, k+n$.

Step 3. Consider $T(x) = P_{k-1}(x) + S_n(x)\prod_{l=0}^{k-1}(x - x_l)$ or $T(x) = R_{k-1}(x) + S_n(x)\prod_{l=0}^{k-1}(x - x_l)$.

Step 4. Construct continued fraction interpolation function:

$$W(x) = T(x) + \lambda \cdot \prod_{l=0}^{n+k}(x - x_0) \cdots (x - x_{n+k}) \tag{7.3}$$

where $W(x_0) = v_0, W(x_1) = v_1, \cdots, W(x_{n+k}) = v_{n+k}$.

Equation 7.3 essentially can be used to predict how a student will be performing next (e.g., t_i' if point 5 indicates the current status in Figure 7.25 when no guidance will be provided proactively and in real time. Genetic algorithm, using iterative processes, can be applied to find the value of λ for optimal approximation based on the longitudinal study (Qiu et al., 2011).

7.3.3.2 *Robust Optimization to Develop STEM Education Guidance* If the prediction model shown in Figure 7.25 demonstrates that a student trajectory, in terms of meeting the objectives, is heading to t_i' (not likely to meet goals), the question becomes what actions can be undertaken with confidence to guide the student toward t_i (likely to meet the goals).

Stochastic optimization is one of the most popular approaches used to address data uncertainty in operation research (Dantzig, 1955). We used a Markov Chain method to find optimal policies for group decision making in a changing IT project management circumstance (Shen et al., 2008). However, the underlying probability distribution of the data is, in general, impossible to know. Robust optimization circumvents this difficulty and has been used as an alternative to stochastic optimization since it was introduced (Soyster, 1973; Goldfarb and Iyengar, 2003; Bertsimas et al., 2004; Ben-Tal et al., 2005; Bertsimas and Thiele, 2004). A budget of uncertainty on the data is an efficient way to measure the trade-off between conservativeness and performance (Bertsimas and Sim, 2003, 2004). The budget of uncertainty represents the overall cumulative amount of variation away from nominal values that must be shared among uncertain data. Yamashita et al. (2007) discuss a robust optimization model for a project scheduling problem. Adida and Joshi (2009) then present a robust optimization approach to project scheduling and resource allocation problems.

By analyzing all the collected measurable data from the TESS, a budget of uncertainty can be determined to reflect the overall cumulative amount of variation away from the nominal values in this study. On the basis of this established budget, a robust optimization approach can be defined to address a variety of uncertain circumstances. As a result, a list of self-learning guidelines for students and administrators can be developed, which can help to guide their teaching/learning practices in a proactive manner.

7.3.4 Continuously Enhancing STEM Education

Through iterations of design, implementation, and synthesis, the discussed approach can surely help high school students to learn STEM subjects in off-campus online learning settings. However, as a laboratory research project, the implemented framework will truly require more tremendous efforts than we have done to make this TESS full-blown operational in practice. We understand that the applicability of the discussed approach might also be limited because of some restricted accessibility for certain groups of students, which must be further addressed in practice.

Note that although this study initially focuses on enriching math education for high school students, it will create a solid foundation for developing an integrated and systematic approach and framework to enrich STEM education in general. The presented approach and framework can then be applied for *K-12* education on a large-scale basis. With the guidance of Service Science principles, the developed models, algorithms, and Web 2.0-based system can also be significantly revised and enhanced and then integrated with evolving in-school education systems for general educational improvement.

As there is lacking sufficient data for us to present detailed analytics as of the time we are writing this chapter, we cannot generate a concrete set of rules of thumb that can guide schools' administrators and teachers, students, and parents to carry out best practices in off-campus education. To show how Service Science modeling and analytics help to improve education management and operations, we provided a data-rich analytics earlier using an example of education program quality control and management study. More concrete and detailed analytics examples are provided in Chapter 8.

7.4 A LIFECYCLE AND REAL-TIME-BASED APPROACH TO SERVICE ENGINEERING AND MANAGEMENT

As compared to approaches taken in the fields of psychology, social science, and marketing, the discussed systems approach in this chapter is not focusing on hypothetic tests. Instead, we focus on methodologies of enhancing the effectiveness of business practices through a series of real-time and proactive guidance to offer and deliver satisfactory services. Using computational and transformative thinking, (i) a suite of mathematical models in the form of computation, data mining, and integrated structural equation modeling and PCA are applied to describe a variety of service circumstances; (ii) vector-continued fractions and robust optimization are then utilized for cause–effect analysis and prediction over time or on a daily operation basis when enabled technologies are fully implemented; accordingly, (iii) rules of thumb are identified to assist stakeholders to manage and govern service systems to meet their needs in a satisfactory manner. Ultimately, by responding to the principles of Service Science and putting them into use, we can act to retune management and operational practices in a proactive and real-time manner, resulting in further improving the performance of service systems as illustrated in Figure 7.14.

For instance, in the first example, the administrative team at a school frequently question themselves:

- Do we really understand our campus as an education service system?
- What can we change in school so that we can achieve student satisfaction excellence?
- What will be the best approach to transform our education service system into a more adaptable, responsive, and competitive one?

Note that these questions surely have no easy answer, given that the changes in the education market are becoming faster than ever before. The answers also certainly vary with the schools. With the fast advancing technologies and accelerating globalization, it becomes significantly challenging because viable solutions can also change and be substantively different over time.

For example, we know that providing state-of-the-art programs matters significantly from the analysis that is clearly indicated in Figure 7.13. However, to make a change in the offer programs would take considerable efforts and lot of investments over a long period. In a real-life situation, the school surely should have strategic and tactic plans in place to have practical and applicable approaches to achieve the goal. The enriched process must also be monitored and adjusted during the implementation. In other words, when we determine an action for education program improvement that is strategically viable and tactically implementable, we must take into consideration all the inputs from all the stakeholders. Therefore, we must adopt a lifecycle and holistic approach that is essentially promoted and supported in Service Science.

To briefly illustrate an example in applying closed-loop and real-time analytics to enhance service quality and students' satisfaction, let us use the construct called "support" in the above-discussed school service system. In Figure 7.13, Q26 (i.e., "whether the classrooms are appropriately equipped for classes offered on campus") carries the highest weight (i.e., 0.387) in reflecting students' perceived learning support on campus. Surely, we know that students appreciate investment on retaining classrooms well-equipped and keeping up campus learning supports with top-notch information, computing, and mobile technologies. However, given the constrained IT budget on campus, an effective and practical teaching equipment and IT transformation or investment plan should be first created.

The literature suggests surveys to collect the inputs from stakeholders (i.e., employees and students). The framing questions on IT investment needs and directions from a website at Penn State (http://www.psu.edu/dept/it/strategies/strategicplan/openforums-alt.php) could be asked in the questionnaire that focuses on how IT could facilitate and transform services in a school system. Similar to any investment made (or simply an action taken) on campus, its plan, design, delivery, operation, and improvement comprise its lifecycle. Integrating qualitative and quantitative approaches to define and carry out the investment should be an effective way to improve our education service in overall and ensure our service system's sustainability and competitiveness. In other words, the adopted approach must be executed throughout the service lifecycle:

- Market, discovery, and strategy
- Design and development
- Delivery, operations, and monitoring
- Optimization and improvement

Only if we could carry out closed-loop and real-time explorations of our service systems, we would keep the service systems competitive and retain in the world-class club.

7.5 SUMMARY

In this chapter, we first reviewed systems of schooling in the Service Science's perspective. Then, we used two different examples to show how education service systems can be explored by applying the principles of Service Science. In the first example, we applied a systems performance approach to understand the dynamics of an educational school system as a whole. To introduce the concept of real-time and proactive actions for the purpose of guiding systems dynamics trajectory, in the second example, we used a high school off-campus learning system to show how a systems approach could be adopted to reengineer the off-campus learning system to help improve students' aptitude to science, technology, engineering, and mathematics over time.

Using systems thinking, this chapter focused on retuning a service system in real time for competitive advantage. To capture and understand people-centric service management and operations, we must explore what kind of service products people as individuals want, how they participate in the process of transformation of needs, and how they change their expectations during the service transformation. A social network approach to look into the insights of service systems is discussed with examples in our next chapter, which further enrich our explorations of Service Science in service engineering and management.

ACKNOWLEDGMENT

Part of Section 7.3—Off-campus learning example was delivered as a keynote speech by the author at *2013 IEEE International Conference on Software Engineering and Service Science*, May 23–25, 2013, Beijing, China. A brief paper was also published in the conference proceedings, entitled "Transformative education Web 2.0 systems for enriching high school STEM education." Dr. Doris Lee, Professor of Education at Penn State, significantly contributed to the literature review of the STEM education in the United States in this chapter.

REFERENCES

Adida, E., & Joshi, P. (2009). A robust optimization approach to project scheduling and resource allocation. *International Journal of Services Operations and Informatics*, 4(2), 169–193.

Agrawal, R., Imieliński, T., & Swami, A. (1993). Mining association rules between sets of items in large databases. *ACM SIGMOD Record*, 22(2), 207–216.

Ahmad, N., & Qiu, R. G. (2009). Integrated model of operations effectiveness of small to medium-sized manufacturing enterprises. *Journal of Intelligent Manufacturing*, 20(1), 79–89.

Andreescu, T., Gallian, J. A., Kane, J. M., & Mertz, J. E. (2008). Cross-cultural analysis of students with exceptional talent in mathematical problem solving. *Notices of the AMS*, 55(10), 1248–1260.

Atkinson, R., Hugo, J., Lundgren, D., Shapiro, M., & Thomas, J. (2007). *Addressing the STEM Challenges by Expanding Specialty Math and Science High Schools*. The Information Technology and Innovation Foundation.

Baldi, S., Jin, Y., Skemer, M., Green, P., Herget, D., & Xie, H. (2007). *Highlights from PISA 2006: Performance of U.S. 15-Year Old Students in Science and Mathematics Literacy in an International Context*. U.S. Department of Education.

Ben-Tal, A., Golany, B., Nemirovski, A., & Vial, J. P. (2005). Retailer-supplier flexible commitments contracts: a robust optimization approach. *Manufacturing & Service Operations Management*, 7(3), 248–271.

Bernold, L. E., Spurlin, J. E., & Anson, C. M. (2007). Understanding our students: a longitudinal-study of success and failure in engineering with implications for increased retention. *Journal of Engineering Education*, 96(3), 263–274.

Bertsimas, D., Pachamanova, D., & Sim, M. (2004). Robust linear optimization under general norms. *Operations Research Letters*, 32(6), 510–516.

Bertsimas, D., & Sim, M. (2003). Robust discrete optimization and network flows. *Mathematical Programming*, 98(1–3), 49–71.

Bertsimas, D., & Sim, M. (2004). The price of robustness. *Operations Research*, 52(1), 35–53.

Bertsimas, D., & Thiele, A. (2004). A Robust Optimization Approach to Supply Chain Management in *Integer Programming and Combinatorial Optimization*, 86–100. Berlin, Heidelberg: Springer.

Bollen, K. (1989). *Structural Equations with Latent Variables*. John Wiley & Sons, Inc.

Bollen, K., & Lennox, R. (1991). Conventional wisdom on measurement: a structural equation perspective. *Psychological Bulletin*, 110(2), 305–314.

Brody, L. (2006). Measuring the Effectiveness of STEM Talent Initiatives for Middle and High School Students in *Identifying and Developing STEM Talent: A Planning Meeting*. September, 2006. Washington, DC: National Academy of Sciences.

Brown, A. S., & Brown, L. L. (2007). What are science and math test scores really telling US. *The Bent of Tau Beta Pi*, 13–17.

Cavanagh, S. (2008). Federal projects' impact on STEM remains unclear: few U.S. programs have studies evaluating their effectiveness. *Education Week*, 27(30), 20–21.

Chin, W. (1998). The Partial Least Squares Approach for Structural Equation Modeling in *Modern Methods for Business Research*, 295–336, ed. by George A. Marcoulides. Lawrence Erlbaum Associates.

Chin, W., Marcolin, B., & Newsted, P. (2003). A partial least squares latent variable modeling approach for measuring interaction effects: result from a Monte Carlo simulation study and an electronic-mail emotion/adoption study. *Information Systems Research*, 14(2), 189–217.

Coltman, T., Devinney, T. M., Midgley, D. F., & Venaik, S. (2008). Formative versus reflective measurement models: two applications of formative measurement. *Journal of Business Research*, 61(12), 1250–1262.

Cuff, K. E., & Molinaro, M. (2005 December). Improving attitudes toward STEM by providing urban-based environmental science research opportunities. *AGU Fall Meeting Abstracts*, Vol. 1, 0327.

Curtain-Phillips, M. (1999). *Math Attach: How to Reduce Math Anxiety in the Classroom, at Work, and in Everyday Personal Use.* Marilyn Curtain-Phillips.

Dantzig, G. B. (1955). Linear programming under uncertainty. *Management Science*, 1(3–4), 197–206.

Dauphinee, T. L., Schau, C., & Stevens, J. J. (1997). Survey of attitudes toward statistics: factor structure and factorial invariance for women and men. *Structural Equation Modeling: A Multidisciplinary Journal*, 4(2), 129–141.

Diamantopoulos, A., Riefler, P., & Roth, K. P. (2008). Advancing formative measurement models. *Journal of Business Research*, 61(12), 1203–1218.

Duarte, P. O., Raposo, M. B., & Alves, H. B. (2012). Using a satisfaction index to compare students' satisfaction during and after higher education service consumption. *Tertiary Education and Management*, 18(1), 17–40.

Edwards, J. R. (2011). The fallacy of formative measurement. *Organizational Research Methods*, 14(2), 370–388.

Fornell, C., & Bookstein, F. L. (1982). Two structural equation models: LISREL and PLS applied to consumer exit-voice theory. *Journal of Marketing Research*, 19(Nov.), 440–452.

Gokhale, A. (1995). Collaborative learning enhances critical thinking. *Journal of Technology Education*, 7(1), 22–30.

Goldfarb, D., & Iyengar, G. (2003). Robust portfolio selection problems. *Mathematics of Operations Research*, 28(1), 1–38.

Gruber, T., Fuß, S., Voss, R., & Gläser-Zikuda, M. (2010). Examining student satisfaction with higher education services: using a new measurement tool. *International Journal of Public Sector Management*, 23(2), 105–123.

Haydock, R., Nex, C. M. M., & Wexler, G. (2004). Vector continued fractions using a generalized inverse. *Journal of Physics A: Mathematical and General*, 37(1), 161–172.

Henseler, J., Hubona, G., & Ringle, C. (2008). Structural equation modeling using SmartPLS. *SmartPLS Workshop Lecture Notes*, George State University, Atlanta, GA, January 2008.

Hill, F. M. (1995). Managing service quality in higher education: the role of the student as primary consumer. *Quality Assurance in Education*, 3(3), 10–21.

Holmes, M., Rulfs, J., & Orr, J. (2007). Curriculum development and integartion for K-6 engineering education. *Proceedings of the 2007 American Society for Engineering Annual Conference & Exposition.*

IBM Global Business Services. (2012). Analytics: the real-world use of big data—how innovative enterprises extract value from uncertain data. *IBM Global Business Services Business Analytics and Optimization Executive Report.* Retrieved on Mar. 5, 2013 from http://www-935.ibm.com/services/us/gbs/thoughtleadership/ibv-big-data-at-work.html.

IBM Global Business Services. (2013). Analytics: a blueprint for value: converting big data and analytics insights into results. *IBM Global Business Services Business Analytics and Optimization Executive Report.* Retrieved on Mar. 5, 2013 from http://public.dhe.ibm.com/common/ssi/ecm/en/gbe03575usen/GBE03575USEN.PDF.

Ippolito, E. (2012). Meeting millennials: 5 characteristics that define your newest customers. *Plume & Post.* Retrieved June 19, 2012 from http://www.matchstickstrategies.com/.

Jarvis, C. B., MacKenzie, S. B., & Podsakoff, P. M. (2003). A critical review of construct indicators and measurement model misspecification in marketing and consumer research. *Journal of consumer research*, 30(2), 199–218.

Kline, R. (2005). *Principles and Practice of Structural Equation Modeling.* New York, NY: The Guilford Press.

Lee, V. E., Smith, J. B., & Croninger, R. G. (1997). How high school organization influences the equitable distribution of learning in mathematics and science. *Sociology of Education*, 70, 128–150.

Martensen, A., & Gronholdt, L. (2003). Improving library user's perceived quality satisfaction and loyalty: an integrated measurement and management system. *The Journal of Academic Librarianship*, 29(3) 140–147.

Matthews, C. M. (2007). *CRS Report for Congress/Science, Engineering, and Math Education: Status and Issues.* Washington, DC: U.S. Congress, Resource, Science, and Industry Division, Congressional Research Service.

Miller, R., & Barney Smith, E. H. (2006). Education by design: connecting engineering and elementary education. *Proceedings Hawaii International Conference on Education*, 4539–4550.

National Center for Education Statistics. (2004). *Qualifications of the Public School Teacher Workforce: Prevalence of Out-of-Field Teaching 1987-88 to 1999-2000.* U.S. Department of Education.

NGA. (2009). *Innovation America: A Final Report.* Retrieved Oct. 10, 2009 from http://www.nga.org.

NSB. (2010). *Science and Engineering Indicators.* National Science Foundation. Retrieved Oct. 10, 2012 from http://www.nsf.gov.

OER. (2010). (1) OER Commons, http://www.oercommons.org/. 2) MIT, http://ocw.mit.edu/OcwWeb/web/home/home/index.htm, (2) Carnegie Mellon, http://oli.web.cmu.edu/open learning/forstudents/freecourses.

OnlineK12. (2013). (1) Study Island, http://www.studyisland.com, (2) http://edhelper.com/, (3) American Education Corporation http://www.amered.com/, (4) K 12 Software http://www.k12software.com/, (5) PLATO Learning http://www.plato.com/, (6) Engineering, Go For It!. http://teachers.egfi-k12.org/, (7) http://www.hsalliance.org/stem/index.asp, (8) Google search "k-12 STEM education" to get many others.

Qiu, R. G. (2009). Computational thinking of service systems: dynamics and adaptiveness modeling. *Service Science*, 1(1), 42–55.

Qiu, R. G. (2013). *Business-Oriented Enterprise Integration for Organizational Agility*. Hershey, PA: IGI Global.

Qiu, R. G., & Doris, L. (2013). Transformative education Web 2.0 systems for enriching high school STEM education. *2013 IEEE International Conference on Software Engineering and Service Science*, May 23–25, 2013, Beijing, China, 352–355.

Qiu, R. G., Wu, Z., & Yu, Y. (2011). A tractable approximation approach to improving hotel service quality. *Journal of Service Science Research*, 3(1), 1–20.

Raykov, T., & Marcoulides, G. (2006). *A First Course in Structural Equation Modeling*, 2nd ed. Mahwah, NJ: Lawrence Erlbraum Associates.

Rice, J. K. (1999). The impact of class size on instructional strategies and the use of time in high school mathematics and science courses. *Educational Evaluation and Policy Analysis*, 21(2), 215–229.

Ringle, C. M., Wende, S., & Will, S. (2005). *SmartPLS 2.0 (M3) Beta*, Hamburg, Germany. Retrieved Oct. 10, 2010 from http://www.smartpls.de.

Roberts, D. (1999). On a representation of vector continued fractions. *Journal of Computational and Applied Mathematics*, 105, 453–466.

Rossman, S. (2006). Overcoming math anxiety. *Mathitudes*, 1(1), 1–4.

Saad, L. (2005). *Math Problematic for U.S. Teens*. Retrieved January 13, 2010 from Gallup.com: www.gallup.com/poll/16360/math-problematic-us-teens.aspx?version=print.

Schaefer, M. R., Sullivan, J. F., Yowell, J. L., & Carlson, D. W. (2003). A collaborative process for K-12 engineering curriculum development. *Proceedings of the 2003 American Society for Engineering Education Annual Conference & Exposition*.

Shen, H., Zhao, J., & Qiu, R. G. (2008). A group decision-support method for IT project management based on Markov chain. *2008 IEEE International Conference on Systems, Man and Cybernetics*, Singapore, October 12–15, 862–866.

Slavin, R. E., Lake, C., & Groff, C. (2009). Effective programs in middle and high school mathematics: a best-evidence synthesis. *Review of Educational Research*, 79(2), 839–911.

Soyster, A. (1973). Convex programming with set-inclusive constraints and applications to inexact linear programming. *Operations Research*, 21(5), 1154–1157.

Subotnik, R., A. Edmiston, K. Rayhack. 2007. Developing National Policies in STEM Talent Development: Obstacles and Opportunities in *Science Education: Models and Networking of Student Research Training under 21*, 28–38, ed. by P. Csermely, K. Korlevic, and K. Sulyok. Amsterdam, The Netherlands: IOS Press.

Subotnik, R., Orland, M., Rayhack, K., Schuck, J., Edmiston, A., Earle, J., Crowe, E., Johnson, P., Carroll, T., Berch, D., & Fuchs, B. (2009). Identifying and Developing Talent in Science, Technology, Engineering, and Mathematics (STEM): An Agenda for Research, Policy, and Practice. Chapter 69 in *International Handbook on Giftedness*, ed. by L. V. Shavinina. Springer.

Tempelaar, D. T., Schim van der Loeff, S., & Gijselaers, W. H. (2007). A structural equation model analyzing the relationship of students' attitudes toward statistics, prior reasoning abilities and course performance. *Journal of Statistics Education Research*, 6(2), 78–102.

Texas Instruments. (2006). *Texas Instruments National Math Month Survey Finds Teens Lack Math Requirements for Hottest Careers*. Retrieved January 13, 2010 from Texas Instruments Press Center: http://focus.ti.com/pr/docs/preldetail.tsp?sectionId=594&prelId=et06 0019.

Tobias, S. (1993). *Overcoming Math Anxiety*. New York: W.W. Norton Company.

Wang, J., Qiu, R. G., & Mi, C. (2009). The sources of customer's satisfaction and dissatisfaction. *2009 INFORMS International Conference on Service Science*, Hong Kong, China, Aug. 6–8, 231–235.

Wang, G., Wang, J., Ma, X., & Qiu, R. G. (2010). The effect of standardization and customization on service satisfaction. *Journal of Service Science*, 2(1), 1–23.

Wenglinsky, H. (2000). *How Teaching Matters: Bringing the Classroom Back into Discussions of Teacher Quality*. Educational Testing Service.

Yamashita, D. S., Armentano, V. A., & Laguna, M. (2007). Robust optimization models for project scheduling with resource availability cost. *Journal of Scheduling*, 10(1), 67–76.

Yan, W., & Lin, Q. (2005). Parent involvement and mathematics achievement: contrast across racial and ethnic groups. *The Journal of Educational Research*, 99(2), 116–127.

Zeng, L., Zhang, Y., & Qiu, R. G. (2007). Adaptive user profiling in enhancing RSS-based information services. *2007 IEEE International Conference on Service Operations and Logistics and Informatics*, Philadelphia, USA, Aug. 27–29, 63–67.

APPENDIX A

Questionnaire that is Used to Evaluate the Quality of Education in a Graduate Professional School

This group questions are related to your personal satisfaction with your enrolled program at this graduate professional school. Please choose a number from 1 to 7 for each question: 1 indicates the least satisfaction and 7 indicates the most satisfaction.

1. Satisfaction measures
   (a) To what extent is your graduate professional school experience meeting your expectations? Q1
   (b) If you had a chance to do it over again, would you choose to attend this graduate professional school? Q2
   (c) What is your overall impression of the quality of education at this graduate professional school? Q3
   (d) Would you recommend your friend or colleague to attend this graduate professional school if you had an opportunity? Q4
2. Expectation and school image measures, and tuition matters
   (a) How much did you know about this graduate professional school when you applied for your graduate study at Penn State? Q5
   (b) How much did Penn State's reputation impact your decision in your enrollment process? Q6

(c) When you applied for your graduate study, which of the following school reputation or characteristic made you choose this graduate professional school? (If you choose "other," your manual input is required.) Q7

   i. Program quality
   ii. Research
   iii. Outreach and community services
   iv. Sports and alumni
   v. Location
   vi. Other

(d) Please rate the extent to which you expected the quality of the program you applied for. Q8

(e) When you applied for your graduate study at this graduate professional school, to what extent did you expect that your graduate professional school experience would enhance your current or future career? Q9

(f) How do you rate yourself in light of your background, readiness, and commitment when you applied for your graduate study at this graduate professional school? Q10

(g) To what extent do you rely on your employer's tuition reimbursement to actively maintain your graduate student status at this graduate professional school? Q11

3. Academic experience measures

(a) What was your primary objective when you first chose to enroll in this program? (If you choose "other," your manual input is required.) Q12

   i. Career enhancement or promotion
   ii. Change of career
   iii. Other

(b) Has your graduate study at this graduate professional school been meeting your needs of your primary objective? Q13

(c) What is your general opinion on the quality of courses? In other words, are all courses truly and appropriately designed and taught at the graduate level? Q14

(d) How do you feel the challenges provided by the classes you have taken so far? Q15

(e) What do you think about the "state-of-the-art" of programs at this graduate professional school? In other words, has the program been well adapted to accommodate developments in the field or positioned to do so in the future? Q16

(f) How do you like the profession-kind atmosphere provided by this graduate professional school as the majority of students are professional on campus? Q17

(g) Which of the following change would you prefer to see in the near future? (If you choose " other," your manual input is required.) Q18

i. More full-time faculty
ii. More part-time professional faculty
iii. Program enhancement in terms of adding the "state-of-the-art"
iv. More challenging courses
v. More seminars by guest speakers from business and industry
vi. Other

4. Faculty and student measures
   (a) Please rate the general quality of the faculty in your program. Q19
   (b) What do you think about the general adequacy of instructors' knowledge in the field? Q20
   (c) Do you think that the instructors' background is qualified for offering the subject matter? Q21
   (d) Do you think that the amount of learned information and knowledge is adequate and meets your expectation? Q22
   (e) To what extent do you value instructors' industrial experience that impacts the effectiveness of the instructors in demonstrating the significance of the subject matter? Q23
   (f) Please try to accurately rate your general readiness in terms of your background and commitment when you take each class. Q24
   (g) How well did your graduate professional school experience prepare you for your current or future career? Q25
5. Campus services and facilities measures
   (a) Do you think that the classrooms are appropriately equipped for classes offered on campus? Q26
   (b) To what extent are you satisfied with the academic advising services? Q27
   (c) How do you feel about course descriptions provided by the university? Q28
   (d) Are the services from the library, computer labs, campus networks, bookstore, administration, and other logistics easily accessible and meeting your needs on campus? Q29
   (e) What is your overall impression of the ease of class scheduling at this graduate professional school? Q30
   (f) Is it important for the campus to provide assistance with your career development? Q31
   (g) What campus service change or addition you like to see in the near future? (input-based question) Q32

MISC (Choices or Input-Based Questions)

1. What is your major? Q33
   (a) INSC 1
   (b) SWENG 2

    (c) SYSENG 3
    (d) Other MBA 4
2. How many years have you been enrolled in the program? Q34
    (a) Less than 1 year 1
    (b) 1–2 years 2
    (c) 2–3 years 3
    (d) 3 or more years 4
3. When you were first enrolled, how many years had it been since the receipt of your baccalaureate degree? Q35
    (a) 0–5 years 1
    (b) 6–10 years 2
    (c) 11 or more years 3
4. When you were first enrolled, how many years had it been for you to work in the same field as your program? Q36
    (a) 0–5 years 1
    (b) 6–10 years 2
    (c) 11 or more years 3
5. Please provide the one aspect that you like most about this graduate professional school. Q37
    (a) Course scheduling (7 weeks)
    (b) Summer session classes
    (c) Faculty
    (d) Staff
    (e) Program quality
    (f) Location
    (g) Parking
    (h) Other
6. Please provide the one aspect that you like least about this graduate professional school. Q38
    (a) Course scheduling (7 weeks)
    (b) Summer session classes
    (c) Faculty
    (d) Staff
    (e) Program quality
    (f) Location
    (g) Parking
    (h) Other
7. What is your gender? Q39
    (a) Male 1
    (b) Female 2

8

Online Education Service and MOOCs

Radically, this chapter continues the discussion in Chapter 7, essentially serving as supplementary contents that complement the previous chapter. In Chapter 7, based on the known systems performance of a service system, we mainly focused on identifying some high priority factors that impacted the outcomes of consumed services and then recommended viable actions to retune the service system in real time for competitive advantage. To fully capture and understand people-centric service management and operations, we must explore what kind of service products people as individuals want, how they participate in the process of service transformation, and why they change their expectations throughout the service lifecycle. We understand that a service network approach must be investigated in order to accomplish this exploratory objective so that competitive services can be promptly discovered, personalized, designed and engineered, delivered, and improved in a repetitive and sustainable manner.

Approaches to improve education services in light of meeting different needs under different circumstances were the exploratory theme in Chapter 7. As we know, currently online teaching/learning is growing at a phenomenal rate. Indeed, online education is becoming a new trend, transforming the approach to higher education (Burnsed, 2011). The news on massive open online courses (MOOCs), starting from May 2012 when MIT and Harvard announced their launch of edX.com, has been overwhelming. MIT and Harvard operate edX (2012), Stanford and Berkeley run coursera (Coursera, 2013), and there are many others running as profit or nonprofit organizations that offer and deliver online higher education worldwide. Impressively, a growing number of elite private and public universities around the globe have joined

Service Science: The Foundations of Service Engineering and Management, First Edition. Robin G. Qiu.

these organizations to start to open their digital doors to the masses regardless of their locations, backgrounds, and educational purposes.

The participating universities are delivering some of their popular courses online for free. This new online education service engenders the opportunities to let anyone with an Internet connection learn from world-renowned professors and experts in the corresponding fields. Indeed, there is nothing new if the community is only interested in providing courses over the Internet. Distance education through broadcasting systems (e.g., TV-based course programs) has been widely adopted in China for over 30 years as of the fifth day of March in 2013. Distance education over the Internet has been particularly popular around the world since the Internet became pervasive in the 1990s. Many profit or nonprofit schools have offered degree-based courses for decades, and profit or nonprofit organizations around the world have been enjoying delivering their training programs to employees and customers by fully leveraging the Web technologies. However, the participation of top-tier research universities clearly marks a crucial milestone in the growing trend of digital learning. Disruptively, the proliferation of MOOCs "has the potential to transform higher education at a time when colleges and universities are grappling with shrinking budgets, rising costs and protests over soaring tuition and student debt" (Chea, 2012).

Although MOOC is still in its embryonic period and lacks sustainable business models, MOOC has notorious student retention rates. Massive opinions on MOOC in a negative way are centered at one that lacks the traditional pedagogies of engagement. Nevertheless, we will eventually find its business models that could be adapted, tailored, or customized by profit or nonprofit organizations. One potential use of MOOC surely is to help higher institutions improve their ongoing online and residential educations. In particular, degree-based online higher education can surely be improved by understanding the education market trend in terms of what individuals want and how they prefer to be pedagogically engaged over the virtual learning settings.

Truly online continuing higher education is quite different from residential college programs. Online degree students are typically part-time and much more diversified in both an educational background and work experience. Despite many higher education institutions providing online graduate professional studies, there is remarkably little published literature addressing scientific approaches that can be adopted to help online education service providers effectively apply pedagogies of engagement in an online higher education setting. By relying on the "known" knowledge in the pedagogies of engagement in college education, in this chapter we explore a system-based framework to acquire a better understanding of given online higher education settings. Through capturing the insights of formed learning networks, we discover methods and tools to positively leverage online students' engagement with a focus on improving learning effectiveness in online education in the long run.

More specifically, a suite of mathematical models in the form of integrated structural equation modeling (SEM) and social network analysis (SNA) approaches and corresponding relevant measures are defined and then applied to help evaluate and guide problem-based collaborative teaching/learning practices. As instructors and students are centered at this particular service context, we will focus on the

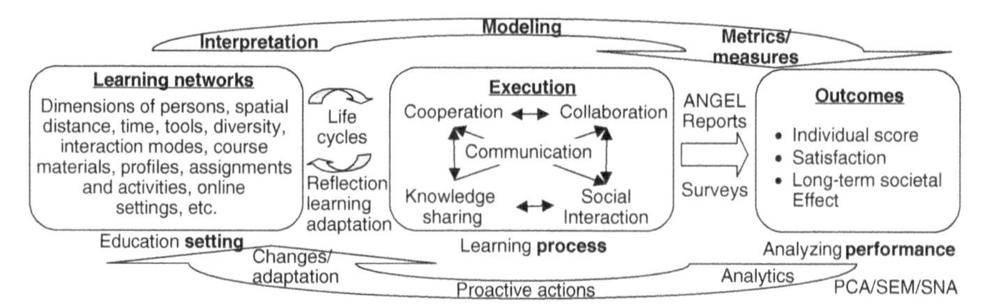

FIGURE 8.1 *An* integrative *and* interactive *approach to manage and engineer a learning service system.*

analysis of their pedagogic interactions and relevant dynamics in the service system, aimed at identifying mechanisms to influence their teaching and learning behaviors in a positive manner (Figure 8.1). In other words, rules of thumb are generalized to help students/instructors retune collaborative practices in a proactive manner whenever possible for retaining life-long effective learning engagement in an online education setting. Using the lifecycle perspective of service improvements, Figure 8.1 is enriched from Figure 4.4. Note that Figure 8.1 focuses on the dynamics and interactions in a learning service system.

The remaining chapter is organized as follows. After a short introduction in Section 8.1, Section 8.2 briefly discusses a computational model of collaborative learning, which essentially describes the dynamics of collaborative learning with a focus on harnessing collaboration in a virtual learning setting. Sections 8.3 and 8.4 present an introductory approach to evaluate collaborative learning performance and the formed learning networks, aimed at identifying best practices to help transform collaborative learning so that learning effectiveness can be significantly improved over time. Section 8.5 highlights the significance of the illustrated case study. Finally, Section 8.5 provides a brief conclusion for this chapter.

8.1 INTRODUCTION

Cooperative learning and face-to-face interactions are an essential part of pedagogies of engagement defined as a necessity in the best practices of classroom-based undergraduate and graduate college education. Pedagogies of engagement in a traditional college setting have been studied for decades. Many researchers have focused on three of the main principles, namely exemplary practices; they are cooperation/interaction among students, student–faculty interaction, and active learning. A large-scale correlational study of what matters in college (involving 27,064 students at 309 colleges) finds that interaction among students and interaction between faculty and students were the two learning practice factors best predictive of a positive change in college students' academic development, personal development, and satisfaction (Astin, 1997).

In the early years of this millennium, the project entitled "The National Survey of Student Engagement" (NSSE) (NSSE, 2003) furthered the understanding of how students perceive classroom-based learning in their college education. The NSSE project confirms the proposition that student engagement, the frequency with which students participate in a variety of learning activities, represents an effective educational practice and is a meaningful proxy for quality of education. After conducting a comprehensive classroom-based learning practice study, Smith et al. (2005) conclude that the faculty who create and maintain education programs must consider not only the content and topics that constitute the degree but also how students are engaged with these materials over time. Simply put, engaged and consistent interactions over the learning period are an essential part of pedagogies of engagement in higher education. Wyatt (2011) recently explores and confirms the best practices at University of Memphis for nontraditional student engagement.

In engineering education, according to Tryggvason and Apelian (2006), "Countless committees, task forces, panels, and commissions have already addressed the need and eloquently emphasized that the competitiveness of the country and thus the general standard of living hinges on the ability to educate a large number of sufficiently innovative engineers." Varieties of education research projects have been successfully completed, topics covering from pedagogy, policy, leadership, group learning, reengineering management for effectiveness, to new education paradigms (e.g., distance, blended/hybrid, self-learning) by leveraging the advancement of computing, networks, and telecommunication technologies (Gokhale, 1995; Garrison and Kanuka, 2004; Cheng, 2005; Moore, 2005; Tryggvason and Apelian, 2006; GHFP, 2006; NSB, 2007; NAP, 2009).

However, as the current way of educating the majority of engineers has changed remarkably little since the 1960s, there is a troubling lack of a good connection with the fast-changing reality of the global environment. It becomes necessary to transform the way we educate to produce the twenty-first century entrepreneurial engineer (O'Sullivan, 1999; Moore, 2005; Tryggvason and Apelian, 2006; Larson, 2009). Recently, many researchers have studied transformative education to address the twenty-first century phenomena of consumerism and globalization and have made progress in the ideological and pedagogical areas in a qualitative manner (O'Sullivan, 1999; Garison, 2000; Duerr et al., 2003; Mayo, 2003; Illeris, 2004; Moore, 2005; GHFP, 2006). One of the main challenges for transformative education is collaborative learning that receives much attention (Gokhale, 1995; Baker and Lund, 1997; Dillenbourg, 1999; Qiu, 2010). However, in the education community, little work has focused on promoting collaborative learning using a quantitative approach.

Collaborative learning is different from conventional structure-based cooperative learning. Collaborative learning as part of active learning emphasizes a group-based learning setting that is often complicated by cognitive, emotional, and social factors. It empowers members, and promotes the positive outcomes for both the group and individuals by motivating varieties of interaction and knowledge sharing (Cheng, 2005; Ebrahim et al., 2009; NAP, 2009; NSB, 2007; Thoms, 2011). Research in the classroom-based educational system has been done in promoting collaborative learning to address the twenty-first century phenomena of consumerism and globalization (Tryggvason and Apelian, 2006; Qiu, 2010). Little scientific research work on collaborative learning to improve students' learning engagement has been done for graduate professional studies, particularly in an online learning setting. Given that adult education is collaborative or participatory in nature (Dillenbourg, 1999; Marks et al., 2005; Fincher, 2010), collaborative learning should be fully leveraged for online and residential education.

Because of the quick advancements of digitalization and globalization, many scholars pay substantial attention to the descriptive and/or empirical study of virtual team-based project management (Andriessen and Verburg, 2004; Pauleen, 2004). Through an empirical study, Vickery et al. (1999) evaluate the inferred positions of virtual teams by examining the organizational structures' impact on team performance, and Lurey and Raisinghani (2001) focus on finding best practices in virtual teams. Shachaf and Hara (2005) then use an ecological approach to discuss team effectiveness in a virtual setting. A recent and quite comprehensive literature review on the study of virtual teams can be found in the work of Ebrahim et al. (2009), focusing on the analysis of innovation and activities and concluding that the knowledge and information that can be effectively captured, shared, and internalized is vital for innovation in a virtual R&D team.

A formal information-based network model of organizational performance in a distributed decision-making environment is presented by Carley and Lin (1997). Jin and Levitt (1996) develop a computational model of virtual design teams to understand the coordination needs derived from activity interdependencies. They model team members, activities, communications tools, and organizations; the developed model simulates the actions and interactions to evaluate organization performance by measuring project duration, cost, and coordination quality. Computational model approaches have also been used to study interventions, contingency, and cultural influences in virtual teams (Levitt et al., 1999; Wong and Burton, 2000; Thomsen et al., 2005). The question is if these newly explored approaches could be appropriately adopted in an online learning environment, in particular, for online engineering education.

To reduce the education cost by leveraging the advancement of e-learning, Goodwin et al. (2011) show the success of developing emulation-based virtual lab to give students an introduction to real-world control engineering design. By taking advantage of the convenience and richness of the Internet and multimedia technologies, Djenic et al. (2011) present an advanced and enriched variant of learning through delivering lessons over the Internet to enhance residential programs. As more and more higher education institutions offer online education to some extent, it becomes necessary for these higher institutions to understand how students perceive the delivery of e-learning. On the basis of a survey of 538 university students, Selim (2007) uses confirmatory factor models to reveal eight categories of e-learning critical success factors. These factors include instructor's attitude toward and control of the implemented e-learning technology, instructor's teaching style, student motivation and technical competency, student interactive collaboration, course coverage and structure, ease of on-campus Internet access, effectiveness of information technology (IT) infrastructure, and university support of e-learning activities. More recently, Orange et al. (2012) conclude a comprehensive evaluation of HigherEd 2.0—a Web 2.0-based pedagogical and technology framework for teaching STEM courses—in undergraduate mechanical engineering courses across four universities, aimed at identifying best practices in leveraging advanced online technologies to enrich and enhance higher education on campus.

However, online education is quite different from residential college programs. Quite often, they do not live on campus or in the same location. As they are geographically populated in different locations and around the world, the inherent on-campus types of interaction discussed earlier do not exist. Thus, the traditionally well-structured and mainly subject-based teaching/learning becomes less effective in an online education environment (Qiu, 2010). Unless online teaching/learning can be effectively adjusted, the same benefits of pedagogies of engagement will not result because of the absence of the needed learning engagement.

A pilot study concerned with the design, development, and evaluation of online courses for adult learners has been conducted by Antonis et al. (2011). They present a framework for the evaluation of three important educational issues, information, and support provided to learners at the start of and during their studies, learners' performance, and learners' satisfaction, involved in the process of online learning. They articulate the means to improve the proposed learning environment and the need for maintaining an optimal balance between synchronous and asynchronous activities, enhanced collaboration, and interactions among adult learners and instructors, aimed at optimally improving the effectiveness of online learning.

As an explanatory example for this book, we focus on online education. Relying on the concepts and principles of Service Science, this chapter presents a quantitative model and relevant measures to help evaluate and accordingly guide problem-based collaborative teaching/learning practices, aimed at positively leveraging online students' engagement to improve learning effectiveness in engineering degree-based education. A web-based course management system (eCMS) is used to manage the online courses under this study. The eCMS logs considerable data on how and when students have accessed learning materials and what kinds of and how teaching/learning interactions have occurred. The logged data in the eCMS, such as learning processes, teaching/learning activities, and student profiles and performance are used to evaluate and confirm the norm of pedagogies of engagement in an online learning setting. Similar to Chapter 7, by relying on the collected data, a suite of mathematical models—in the form of integrated SEM and SNA—is adopted to illustrate the dynamics of online learning circumstances. Rules of thumb are thus identified and generalized, aimed at guiding students/instructors to retune collaborative practices in a proactive manner for furthering the effective learning engagement.

8.2 A SYSTEMIC APPROACH TO ANALYZE COLLABORATIVE LEARNING

Although an offered online course in college is only an education system miniature on campus, it surely involves all the educational necessities, including an instructor, teaching assistants, students, learning materials, and IT support systems that are used to assist the teaching and learning. As discussed earlier, adult education is collaborative or participatory in nature (Dillenbourg, 1999). Therefore, collaborative and problem-based learning that essentially relies on teams and projects are assumed to be the most predictive of positive outcomes in graduate professional education, particularly in an online setting. However, little scientific research work has been done in analyzing and accordingly helping enrich collaborative learning in this presumably preferable learning setting for professionals. As opposed to the qualitative approach to explore project team dynamics proposed by Andriessen and Verburg (2004), an integrative and interactive approach in a quantitative manner (Thompson et al., 2009; Qiu, 2010) to achieve a *better understanding* of the online learning dynamics throughout semesters is necessary.

By relying on the "known" in the pedagogies of engagement in college education, we explore a systemic approach to guide online education practices. More specifically, quantitatively analyzing and accordingly facilitating collaborative learning in a project and team-based learning setting is the focus of this case study. As computational thinking can fully leverage today's ubiquitous digitalized information, computing capability, and computational power, it has evolved as one of the optimal ways of solving problems, designing systems, and understanding human behavior. Computational thinking promotes quantitative thinking in terms of abstractions, modeling, and understanding of the dynamics of a studied people-centric service system (Qiu, 2009). Derived from the discussion in the preceding chapters, and from Figure 5.12, Figure 8.2 gives a detailed view of the adopted computational thinking and sys-

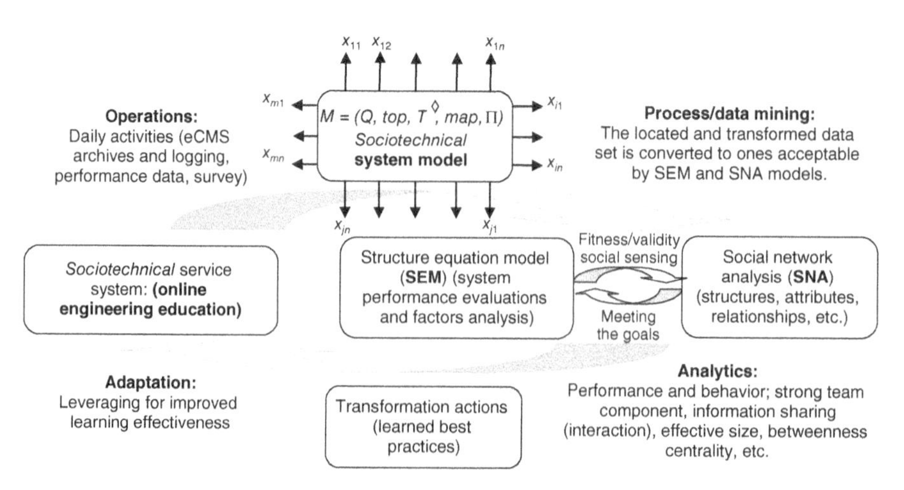

FIGURE 8.2 *A detailed approach to leverage collaborative learning.*

temic approach (Qiu, 2009; Qiu, 2010), aimed at leveraging collaborative learning for improved learning effectiveness in an online engineering education setting.

In general, factors that considerably impact online education include instructor's attitude toward and control of the implemented e-learning technology, instructor's teaching style, student motivation and technical competency, social interaction and collaboration, and course coverage and structure (Beldarrain, 2006; Selim, 2007; Department of Education, 2008; Dymalski, 2011; Parry, 2012). When compared to resident instruction (RI)-based education, online education is disruptively confronted with unprecedented challenges, such as how to nurture and nourish students' self-discipline, encourage social interaction and collaborative learning, and develop a system that can monitor course progression and suggest swift actions to enhance the positive learning atmosphere in an online setting.

In this chapter, we focus on exploring a framework to help monitor and facilitate students' online learning experience so that a pleasant and positive environment is maintained for effective online learning. Without the loss of generality, we mainly show an approach to conduct longitudinal empirical studies of collaborative learning. Initially, the focus is on understanding students' self-disciplines and their interactions and collaborations throughout a course. We collect data related to student profiles, course materials, learning activities, class interaction, performance, as well as their responses to questionnaires. We then apply an integrative approach using principal component analysis (PCA), SEM, and SNA to pinpoint any weakness in the formed online learning networks and identify appropriate actions that the instructor might take to enrich the students' learning experiences within an online learning setting. All of this can be applied to further improve learning outcomes over time.

Self-discipline plays a key role in online education. Frequently, online students make a learning plan at the very beginning of a course, and then follow that plan to realize the course learning objectives. The literature (Dillenbourg, 1999; Qiu, 2010; Djenic et al., 2011; Thoms, 2011; Wyatt, 2011) clearly demonstrates that collaborative learning substantively compensates for the missing conventional social interactions and on-campus engagements. For an online course, collaborative learning, which essentially relies on a variety of interactive and collaborative activities throughout the semester, is the best predictor of positive learning outcomes (Qiu, 2010).

Conceptually, Figure 8.3 illustrates how the collaborative learning network dynamics for a given week, in a given class, exists in a virtual setting. As indicated in Figure 8.3a, the learners' profiles, online activities (e.g., class interaction, discussion activities, communications, etc.), performance, and their responses to designated questionnaires are quantitatively aggregated, algorithmically mined, and analytically processed, which can then be modeled using a SNA tool. Furthermore, the class interaction social network model can be augmented by adding learning activities related to assignments (e.g., Q1: video clip and online question, Q2: online quiz, and Q3: hands-on), which provides the data needed to understand the evolving nature of social networks varying with different topics/assignments from week to week. By developing a sequence of models throughout a semester, model variations might be indicative of an impeding problem, and appropriate actions to help can be suggested to retune teaching/learning practices in a virtual learning environment.

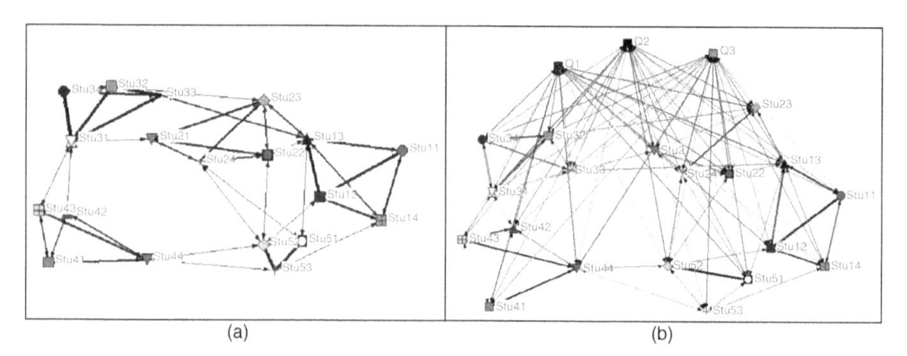

FIGURE 8.3 *Models of learners' social interaction in an online collaborative learning network. (a) PCA/SNA: class interaction social network; (link thickness: tie strength (interaction); darkness and shape: personal profile; size of nodes: class activities). (b) PCA/SNA: augmented class interaction (adding Q1, Q2, and Q3 assignments).*

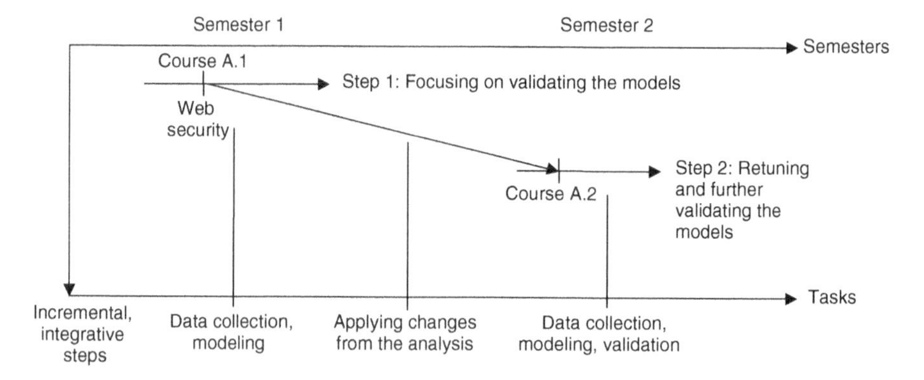

FIGURE 8.4 *Exploratory tasks and milestones.*

The above-mentioned exploratory scenario based on an *integrative* and *interactive* approach to achieve a *better understanding* of the online learning settings can be extended semester by semester (Figure 8.4). As an illustrative case study for this book, we focus on identifying a couple of mechanisms for fostering self-motivation and learning the best practices for leveraging collaborative learning to improve online education. Therefore, only relevant rather than comprehensive data will be presented hereafter.

We focus on showing how the framework presented in Figure 8.2 can be well applied in practice. In Figure 8.4, we provide the stages for this conducted exploration and what the relevant tasks must be completed during each of these stages.

- *Exploring an Online Course Twice*. One engineering graduate course is studied. The selected course is then taught again during semester II (Figure 8.4).

Data for courses are fully collected through the eCMS, including class activities, interaction, student profiles, and learners' performance. For each class, data are logged week by week throughout a semester. In addition, a precourse survey is used to get a better understanding of the general background of each class; peer and self-reviews are conducted at the end of each course to know how students feel about the adopted collaborative learning practice.

- *Performing Analytics With Scientific Rigors*. A suite of mathematical models in the form of integrated SEM, PCA, and SNA are applied. First (in Section 8.3), SEM is used to conduct confirmatory factor analysis for the selected online course during semester I (Qiu, 2010). PCA and SNA are also applied to analyze the teaching/learning dynamics and performance to pinpoint the strength and weakness of studied online course. Appropriate actions are identified for the studied course in semester II. Secondly (in Section 8.4), we make some changes from what we learned from semester I. Instead of exploring system-level performance and factor analysis, we will focus on data and network analysis to decipher how students interact within teams and in class, aimed at finding ways to change team structures and help individuals be better engaged within teams and in class to enhance their learning experience. In addition, an enriched data (or instruments) set is used for the second time in order to show if the list of identified changes actually makes a difference if applied.

In this demonstrative example, we show how to explore mechanisms that can be used to cultivate and/or nourish the needed self-discipline and determine learn collaborative learning best practices so as to compensate for the missing on-campus interaction in a residential course. Specifically, we show how to find and identify progress and potential problem areas for students in an online class in a real-time, integrative, and evolving manner. Ultimately, this developed framework can be essentially utilized to adjust and enhance teaching/learning practices and promote students' active and positive engagements in online education so that effective learning outcomes can be systematically achieved.

8.3 COLLABORATIVE LEARNING ANALYTICS: PART I

As discussed earlier, collaborative learning relies on an interactive group-based learning setting to promote the positive outcomes for both the group and group members. Collaborative learning leverages the diversity and richness of a group and the different strength of individuals. Compared to subject-based assignments, problem-based assignments can considerably motivate and encourage varieties of interaction and knowledge sharing in such a learning environment. When collaborative learning and problem-based assignments are properly combined, the most predictive of positive outcomes in the graduate professional education, particularly in an online setting, should result (Astin, 1997; NSSE, 2003; Hansen, 2006; Antonis et al., 2011). This chapter presents a model, theory, and framework for quantitatively analyzing the systemic dynamics of an online class, aimed at developing *rules of thumbs* for online classes to retune collaborative protocols for improved learning outcomes.

Once again, it is fully understood that a one-size-fit-all thinking should be avoided as different paradigms might be more appropriate for certain circumstances. Therefore, by simply addressing a fundamental understanding of collaborative learning, the following essential three exploration steps, data collection, SEM for learning effectiveness evaluation (Qiu, 2010), and SNA for analyzing the interaction dynamics within teams and in class to pinpoint the strength and weakness of teams' behavior in an online learning setting, will be mainly studied.

8.3.1 Data Collection

An online class was used to validate the applicability of the proposed approach. As mentioned earlier, an eCMS was used to manage the online courses under study. The web system logged all data on how each student accessed learning materials and how they interacted with instructors and each other. The logged students' access data, archived teaching/learning activities, and student profiles and performance are first collected, and then appropriately transformed; all are thus made ready for conducting analytics.

The learning outcomes of teams and individuals largely depend on how team members perform collectively, how they collaborate with each other, and how a team as a whole acts when challenges are confronted from time to time. In general, the following aggregated core measurements collectively reflect how a team is doing at the point of measure: norms (indicated by respects, preferred learning styles, personalities, trust, etc.), communication and mutuality (indicated by learning setting, communication tools, betweenness, information sharing methods and tools, etc.), team capability (indicated by self-learning capability, skills/knowledge on the subject, team competency, etc.), teaching/learning methods, and others (Astin, 1997; Thompson et al., 2009; Qiu, 2010).

As known, adult education is collaborative or participatory in nature (Dillenbourg, 1999; Marks et al., 2005). Thus, collaborative learning should be fully leveraged for graduate professional online education. According to the "known" knowledge of effective adult online learning (Imel, 1991; Smith and Smarkusky, 2005; Marks et al., 2005; Thompson et al., 2009), when leveraging online students' engagement to improve learning effectiveness in graduate professional engineering degree education is the focus, the following learning manifest variables and performance indicators should be essentially included:

- Instructor–student pedagogical engagement
  - Q1. The instructor provided sufficient questions through email, online board, social media, etc.
  - Q2. The instructor answered sufficient questions through email, online board, etc.
  - Q3. The instructor always responded to student inquiries in a timely manner.
  - Q4. The adequacy of the instructor's knowledge of the subject matter.
  - Q5. The appropriateness of the instructor's encouragement of student discussion.

- Student's background, readiness, and commitment
  - Q6. Student's aptitude for collaborative learning before he/she took the course.
  - Q7. Student's background for the course subject before he/she took the course.
  - Q8. Student's readiness before he/she took the course.
- Student–student pedagogical engagement
  - Q9. Student's aptitude for collaborative learning before he/she took the course.
  - Q10. Student's background for the course subject before he/she took the course.
  - Q11. Student's readiness before he/she took the course.
  - Q12. The student always participated in team activities when working on assignments within his/her team.
- Learning materials and supports
  - Q13. The usefulness of team problem-based assignments.
  - Q14. The effectiveness of the integration of instructional materials (textbooks, lecture notes, papers, online forums, etc.).
  - Q15. The adopted teaching that helps to maintain a climate conducive to learning.
  - Q16. The adopted learning methods that help to maintain a climate conducive to learning.
- Perceived learning
  - Q17. The amount of information provided in this online course is adequate.
  - Q18. This online course experience meets the student's learning expectations.
  - Q19. The appreciation of online collaborative learning activities after taking this course.
  - Q20. *Performance*. Individual's contribution and quality of work (peer evaluation).
  - Q21. *Performance*. Learned and individual expectation met (self-evaluation).
  - Q22. *Performance*. Final score (instructor's evaluation).

In this studied course, three to four students were randomly assigned to teams and in total five teams were created. After the teams were formed, the eCMS automatically logged data on how each team and individuals participated in class activities and interacted with each other and the instructor during the study period (i.e., one semester). One precourse survey collected students profile information, which is mainly concerned with individual's backgrounds, work experience, and expectations. One postcourse survey at the end of the semester was used to collect information on what and how teams and individuals performed using peer reviews and self-evaluations (Smith and Smarkusky, 2005), which also asked whether their expectations were met.

TABLE 8.1 Data Sources Information

eCMS	Precourse Survey	Postcourse Survey	Peer and Self-Reviews	Instructor Evaluation
Q1, Q2, Q6, Q7, Q9, Q10, Q11, Q12	Q6, Q8, Q11, Q12, Q20, Q21	Q2, Q3, Q4, Q5, Q6, Q8, Q13, Q14, Q15, Q16, Q17, Q18, Q19	Q6, Q8, Q11, Q12, Q20, Q21	Q22

The instructor evaluated the quality of the interactions (e.g., postings, responses, etc.). Table 8.1 summarizes all the data that were essentially collected in this study. If the same category data were collected from more than one source, the final data of that category for each student were the average calculated from all sources.

8.3.2 Evaluating Learning Effectiveness

As discussed in the preceding chapters, SEM has been widely used to study social and/or economic behavior of organizations. Using the indicators (i.e., measurements, or manifest variables) from both social and technical perspectives of a *service system*, SEM can be effectively applied in this service-oriented interdisciplinary field (Chin et al., 2003; Marks et al., 2005; Qiu, 2009; Qiu, 2010). As compared to many covariance-based modeling approaches, the partial least squares approach to structural equation modeling (PLS SEM) is a soft modeling approach with relaxation of measurement distribution assumptions (Figure 8.5). In addition, PLS SEM requires only a small size of measurement samples and tolerates measurement errors.

Let us briefly review the models first. For a reflective measurement model, measurement variables are a linear function of their latent variable ξ plus a residual ε, and π is the loading set, that is,

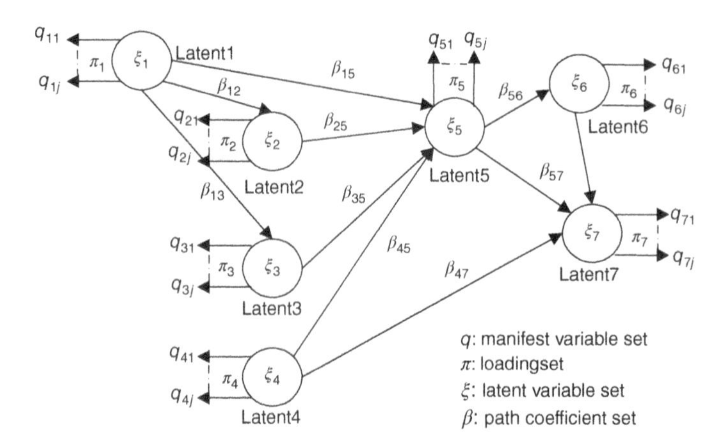

FIGURE 8.5 *A PLS SEM for a service system.*

$$q_h = \pi_{h0} + \pi_h \xi + \varepsilon_h$$
$$E(q|\xi) = \pi_{h0} + \pi_h \xi \tag{8.1}$$

The previous equation implies that the residual ε has a zero mean and is uncorrelated with the latent variable ξ. For a formative measurement model, the latent variable ξ is a linear function of its measurement variable plus a residual δ, that is,

$$\xi = \sum_j \varpi_j q_j + \delta$$
$$E(\xi|q) = \sum_j \varpi_j q_j \tag{8.2}$$

For the structural model, the path coefficients between latent variables ξ is given by

$$\xi_j = \sum_i \beta_{ij} \xi_i + \zeta_j \tag{8.3}$$

where ζ is the vector of residual variance.

With the exception of survey data that will be in Likert scale, logged data in the eCMS are in numbers, regular texts, or standard XML formats. These non-Likert data must be converted into Likert-scale data accepted by SEM (Qiu et al., 2011). An SEM diagram is generated using data of one semester (Figure 8.6). As indicated in Figure 8.6, the coefficient of students' profile is 0.429, standing at the highest influence path. This clearly indicates that the learning outcomes of students mostly depend on students' background, commitment, and diligence. However, the analytical result surely confirms that the student–student (with a path coefficient of 0.276) and student–instructor (with a path coefficient of 0.414) interactions play a key role in improving student's learning experience in graduate professional studies (Marks et al., 2005).

8.3.3 Identifying Best Practices

Varieties of collaborative behavior, such as time to communicate, the frequency of interactions within a team, how team members communicate, and the frequency of an individual's class discussion participation, are highly correlated to the SEM structural and relation attributes at a given time (Selim, 2007). Although it is well proven in empirical studies that group or component structural properties and information flow characteristics substantially impact on the effectiveness of team interaction and collaboration (Imel, 1991; Smith and Smarkusky, 2005; Marks et al., 2005; Thompson et al., 2009; Qiu, 2010), a quantitative understanding of how these factors affect the learning service system dynamics is not seen in the literature. In other words, the outcomes of an online professional class by encouraging positive pedagogical engagement within the class should be explored in a quantitative manner. In this section,

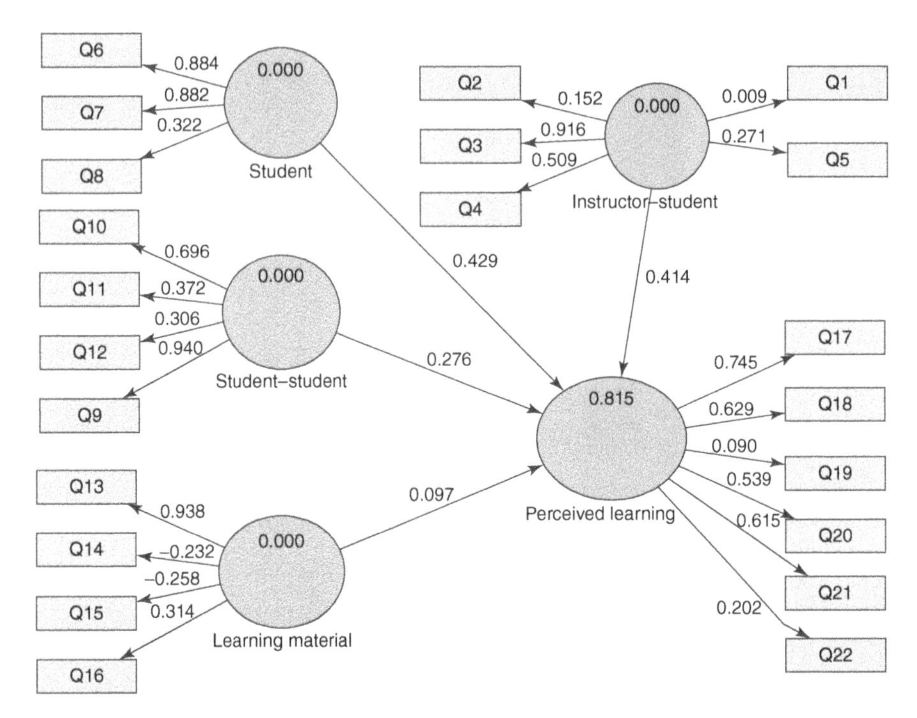

FIGURE 8.6 *An SEM study of online collaborative and problem-based learning.*

we show a systemic approach to identify best practices in leveraging online students' engagement to improve learning effectiveness in graduate professional engineering degree education.

Figure 8.6 shows that many factors substantially affect the final learning outcomes of online students. To understand how pedagogical engagement affects the outcome of online learning in a practical manner, SEM for problem-based and collaborative learning can be further retuned by simplifying its structure by removing these factors that are not directly related to pedagogical engagement. As indicated in Figure 8.7, this focused SEM aims at finding the casual relationships between engagement-related learning factors and the class performance. Table 8.2 shows how a quantitative improvement of each of these engagement-related learning factors would impact the learning performance. The impact score is essentially the potential unit improvement after one unit change in the corresponding learning factor (Chin et al., 2003; Qiu, 2009; Qiu, 2010). In instructor–student interaction category, Q3 (instructor–student interaction after class) substantially and positively influences the students' learning outcomes. In student–student interaction category, Q9 (student–student asking each other questions) and Q10 (student–student interaction, in general) outperform other factors.

Derived from network theory, SNA has been widely used in the study of sociology, anthropology, communication, economics, information science, organizational behavior, and psychology (SNA, 2011). Note that SNA has been recently moved

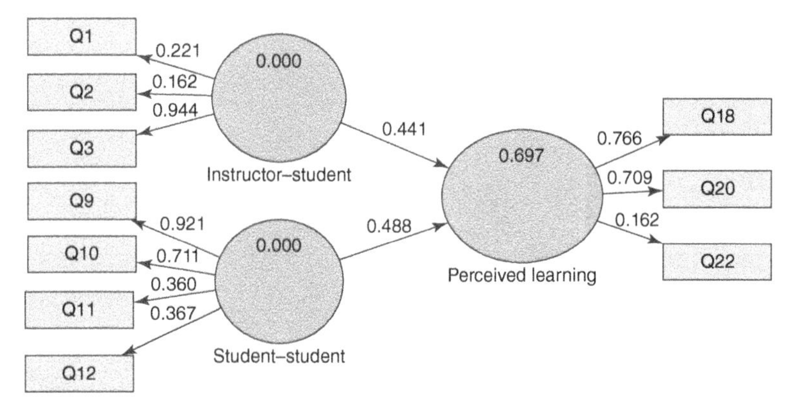

FIGURE 8.7 *An SEM mainly concerned with pedagogical engagement.*

TABLE 8.2 Class Learning Performance Main Factors and their Effects

Learning Factors	>Mean	Loading	Coefficient	Impact Score
Q1 (instructor asking questions)	4.44	0.221	0.441	0.221 × 0.441/1.33 = 0.07
Q2 (instructor–student interaction in class)	4.50	0.162		0.162 × 0.441/1.33 = 0.05
Q3 (instructor–student interaction after class)	3.00	0.944		0.944 × 0.441/1.33 = 0.31
Q9 (student–student asking each other questions)	2.50	0.921	0.488	0.921 × 0.488/2.36 = 0.19
Q10 (student–student interaction, in general)	2.61	0.711		0.711 × 0.488/2.36 = 0.15
Q11 (student's cooperation)	4.50	0.360		0.36 × 0.488/2.36 = 0.07
Q12 (student's participation)	4.56	0.367		0.367 × 0.488/2.36 = 0.08

from being a suggestive metaphor to an analytic approach, focusing on the understanding and management of sociotechnical systems in a quantitative manner. For example, without assumption that groups are the building blocks of society, the approach becomes open to studying less-bounded sociotechnical systems, leading to the research scope extended from bounded communities to virtual connections among websites. How the dynamic structures of physical and virtual ties affect the individuals and their relationships have recently attracted substantial attention worldwide. Conventional analyses assume that socialization into norms determines the behavior; on the contrary, network analysis explores how and why the structure and composition of ties might affect norms (SNA, 2011). Thus, collaboration graphs describing virtual learning settings are used to measure and improve the effectiveness of collaborative relationships between the participants in the formed learning network in this case study.

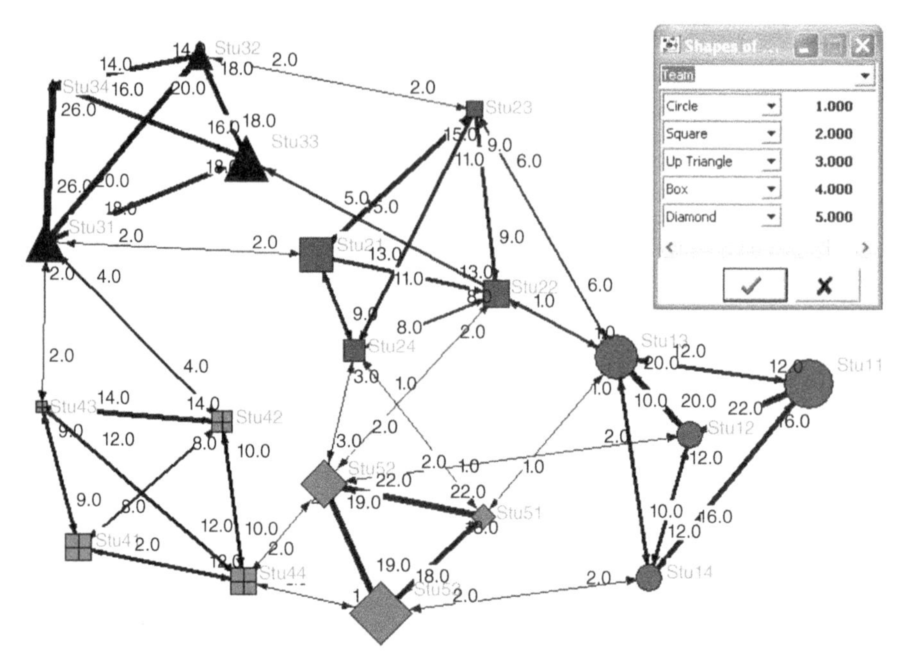

FIGURE 8.8 *An SNA model of collaborative learning network.*

UCINET (Borgatti et al., 2002) is a popular SNA tool. When data on interaction dynamics within teams and in class are loaded into UCINET, a social learning network model is generated (Figure 8.8). In Figure 8.8, link thickness shows the tie strength (i.e., interaction between individuals), darkness and shape indicate team information, and the size of nodes then illustrates individual class participation activities such as posting questions and answers. Detailed explanations of these activities are provided in Table 8.3. Team activities in class (ϕ_1) were derived from Q9 (i.e., student–student asking each other questions); team activities in group (ϕ_2) were derived from Q10 (i.e., student–student interaction when they worked on problem-based projects); cooperation within group (ϕ_3) were derived from Q11

TABLE 8.3 Collaborative Learning Effectiveness

	Weighted Average Grade	Team Activities in Class (ϕ_1)	Team Activities in Group (ϕ_2)	Cooperation Within the Group (ϕ_3)	Participation Within the Group (ϕ_4)	Collaborative Learning Effectiveness ε
Team 1	9	7	6	9	10	0.78
Team 2	8	4	3	9	9	0.58
Team 3	9	6	8	9	9	0.78
Team 4	8	7	2	9	9	0.64
Team 5	10	7	9	10	10	0.88

(i.e., cooperation when working on problem-based projects within the team); and finally participation within group (ϕ_4) were derived from Q12 (i.e., participation when working on problem-based projects within the team).

To compare how individuals in a team collaborate, team-based collaborative learning effectiveness must be defined. On the basis of the previous analyses, collaborative learning effectiveness can be defined as follows

$$E = \sum \Omega\Phi / 10\Sigma\Omega, \text{ that is, } \varepsilon_i = \sum_i \omega_i \phi_i / 10 \sum_i \varpi_i \tag{8.4}$$

where Φ is the set of indicators defined as collaborative team activities. For a given factor ϕ, 10 is the maximum. For a given team, $\phi_j = \sum_i 2q_i/n$, i is the ith team member for the jth factor and n is the number of team members. In Equation 8.4, Ω is the corresponding weights that are defined in Figure 8.5 and computed using SmartPLS (as shown in Figure 8.6), and $0 \leq E \leq 1$. If a given team's ε is close to 1, the team is doing well in terms of collaboration when the team works on the team-based class assignments. Table 8.3 gives how teams collaborated based on collected data during this case study. It clearly indicates that Team 5 is more collaborative than other teams. The collaborative effectiveness is highly correlated to the learning outcomes as Team 5's weighted average grade in 10 (Maximum).

When the outcomes in Tables 8.2 and 8.3 are considered together, it becomes clear that learning effectiveness is positively and highly correlated with students' engagement in an online learning setting. In other words, although online education lacks varieties of in-person interaction affluent in traditional residential programs, best collaborative learning practices in a virtual setting can be appropriately leveraged, resulting in achieving positive and effective learning engagement for improved graduate engineering degree education.

Through this system-based quantitative exploration and with the support of the qualitative feedbacks from the students, the following best collaborative teaching practices are identified:

- Preparing more team-based assignments and activities, so students can be more actively involved in class and team discussions.
- Allowing students to use different communication means. Different teams might prefer to use some different collaborative tools with which they are familiar. When they can choose their own ways to share and interact, they tend to be more active.
- Encouraging a team to have coleaders for each team assignment. By doing this way, the team can stay active without disruption in case that an emergency circumstance with one of team leaders occurs.
- Initiating and maintaining constant communications with students by asking questions, sharing new findings on the Internet, and answering their questions in a timely manner.

All the above-identified best practices essentially focus on encouraging students to get more involved in the activities described in Table 8.3. The identified rules of thumb could be embedded in the eCMS as a teaching/learning guidance. For instance, at the beginning of each week, the system provides a relevant alert message that reminds students/instructors of retuning their collaborative practices in a proactive manner. As a result, we can improve or retain effective learning engagement with the class from week to week.

8.3.4 Brief Remarks on Part I of Collaborative Learning Analytics

By taking advantage of the rich data in the eCMS, a web-based course management system for online engineering degree education, this section aimed at providing a systemic approach to help instructors and administrators evaluate and guide problem-based collaborative teaching/learning practices. More specifically, by relying on the collected system data and course surveys, an integrated SEM and SNA approach was proposed to describe a variety of learning circumstances, which helps to identify certain rules of thumb that could be used to guide students/instructors to retune their collaborative practices in a proactive manner for improving or retaining effective learning engagement.

As compared to many traditional empirical studies that focus on hypotheses testing, this study relied on the "known" in the pedagogies of engagement in college education, and then proposed a systemic approach to explore effective mechanisms of leveraging students' engagement in pursuit of effective online learning practices. More specifically, this exploration confirmed the general norms in the pedagogies of engagement in education and innovatively presented a system-based framework that could be practically and fully integrated in online education settings.

However, this exploration is limited because of the limited availability of online classes during the first study period. More experiments to further validate the presented model should be conducted. The limits of this exploration and corresponding further works are summarized as follows:

- The data sample size was small. Other new findings might be revealed using empirical studies, which can enrich the presented effectiveness index model. Indeed, this is what we discuss in the next section.
- The simplified model illustrated in Figure 8.6 requires further investigation. For example, probabilistic-based analysis methodology such as SEM-based and semisupervised Bayesian networks can be included to improve the accuracy of analyses. This will be further discussed in Chapter 9.
- The identified best practices were applied manually in the following courses offered online. This should be ultimately applied in real time. In other words, the framework proposed in Figure 8.1 should be fully implemented in a closed-loop manner. As a matter of fact, this is exactly the purpose of Service Science research in the long run, which was discussed in Chapter 7 and is further articulated in Chapter 9.

In summary, instead of simply finding the facts and proving the hypotheses using empirical studies, this exploration took a system-based approach, aimed at resulting in better qualitative and quantitative guidelines that can be applied and integrated in the eCMS or the like to help students/instructors retune collaborative practices in a proactive manner for improving or retaining effective learning engagement. Thus, online learning could overall have the same effect as that of classroom-based education in light of the appreciation of pedagogies of engagement, although the traditional on-campus and face-to-face interactions largely disappear over the Internet.

8.4 COLLABORATIVE LEARNING ANALYTICS: PART II

We conducted our second exploration as planned. During the second semester, we had more students. More importantly, we made certain changes based on the "rules of thumb" we developed in our previous analysis. Hence, it will be intriguing to see if any improvement has been made in light of improving collaborative learning effectiveness.

8.4.1 Individual's Profile, Learning Activities, and the Learning Outcomes

We first check how the class was doing overall in terms of their participation in the discussion forums. It is well known that learning outcomes are positively correlated to learning objectives. Table 8.4 shows the number of individual activities throughout the semester. It is clear that individuals' learning outcomes are highly correlated with their participations on the discussion forums. In other words, students' interactions in a virtual learning setting play a critical role in helping them achieve their learning objectives.

We can look into insight of students' online interactions by analyzing the correlations between different kinds of participations, including submitting an original post, responding to others' posts and receiving responses from classmates. Three corresponding correlations are depicted in Figure 8.9. Evidently, if one is active in submitting posts, one replies others and also receives more responses.

Team assignments were mainly used to encourage students to get engaged pedagogically. It is well recognized that team projects can significantly promote and facilitate collaborative learning. Thus, we secondly check how individuals performed within their team settings. People might think that one has more team experience in the past could contribute more to the team work. The same speculation is also often applied to team attitude. We found that neither was true (Table 8.5), although we did find that individual's team experience is highly correlated to one's team attitude in this second exploration. Dots are sporadic in Figure 8.10, which indicates that individual's team experience and attitude are not well correlated to one's contribution to the quality of team work.

Thirdly, we study how individuals performed in their team settings. We were particularly interested in how individual's team involvements, in general, are correlated to their contributions to the team work. Table 8.6 shows three measurements we used to evaluate individuals' involvements within their teams.

TABLE 8.4 The Overall Online Interactive Discussions in Terms of Individual's Participation

User ID	Final Grade	Total Posts	Posts	Replies	Peer replies
S201	89	16	9	7	5
S202	92	16	7	9	17
S203	84	21	12	9	10
S204	97	74	19	55	47
S205	89	14	7	7	5
S206	94	28	10	18	27
S207	98	47	21	26	43
S208	91	16	8	8	3
S209	94	16	8	8	13
S210	92	17	10	7	5
S211	94	46	10	36	20
S212	86	15	9	6	6
S213	90	21	11	10	11
S214	91	32	14	18	22
S215	82	18	9	9	7
S216	96	51	19	32	35
S217	87	27	13	14	9
S218	90	17	8	9	10
S219	98	31	14	17	12
S220	93	16	7	9	17
S221	76	15	9	6	6
S222	92	18	9	9	19
S223	93	19	8	11	12
S224	87	13	7	6	4
S225	96	23	8	15	14
S226	78	18	9	9	10
S227	90	57	18	39	34
S228	96	33	15	18	28
S229	98	52	20	32	24
S230	99	63	31	32	58
S231	93	16	7	9	15
S232	100	83	11	72	33
S233	92	36	10	26	23
S234	89	15	7	8	5
S235	98	39	10	29	22
Correl (X_i, Final Grade)		0.597	0.438	0.57785	0.6266

It comes with no surprise. We found that individual's participations in a team setting are also highly correlated to one's contribution to the quality of team work. Figure 8.11 shows how individual's team involvements correspond to one's contribution to the team work, which clearly shows that both individuals' dependability of finishing assignments within teams and their participations are correlated to their contributions to the quality of team work. Cooperation was measured in a subjective way. We did not find that this perceived measurement of individual's team attitude was well correlated to one's contribution to the quality of team work.

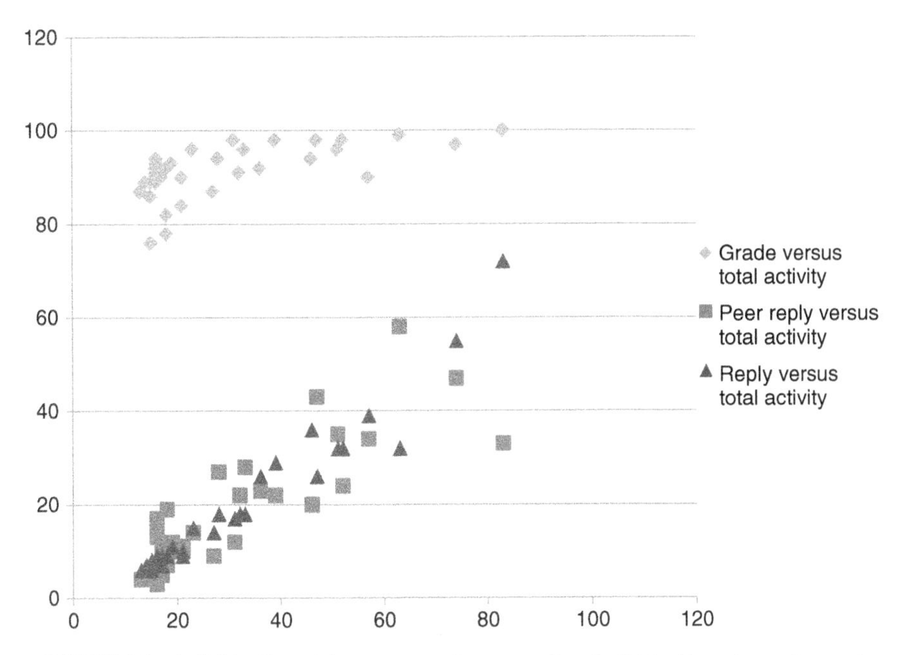

FIGURE 8.9 *Individual's activity versus performance (i.e., individual learning outcomes).*

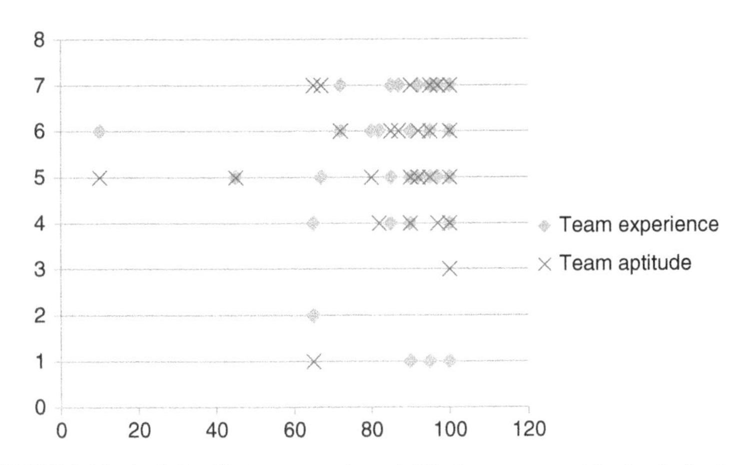

FIGURE 8.10 *Individual's team experience/attitude versus contribution to the team.*

Finally, we look into how individual's backgrounds impact one's online learning performance in the class and team settings. The course under exploration was Web Security. Therefore, we mainly explore how individual's knowledge of web servers and web programming and general experience in programming are related to one's class participation and final learning achievements. For this particular course, we found that individual's knowledge of web servers had a strong correlation to one's

TABLE 8.5 Individual's Team Contribution Versus Team Experience and Attitude

User ID	Team Experience	Team Attitude	Contribution to the Team
S201	5	4	97
S202	5	5	100
S203	6	7	90
S204	7	6	92
S205	7	6	72
S206	6	6	95
S207	7	6	87
S208	6	5	95
S209	1	5	95
S210	5	5	92
S211	6	6	72
S212	1	5	100
S213	4	4	90
S214	6	5	80
S215	5	5	45
S216	6	4	82
S217	5	4	90
S218	7	6	85
S219	7	5	95
S220	7	7	97
S221	5	7	67
S222	6	5	95
S223	5	4	100
S224	5	6	85
S225	4	5	100
S226	2	1	65
S227	4	7	65
S228	4	3	100
S229	6	5	10
S230	5	7	95
S231	6	6	100
S232	5	5	100
S233	1	5	90
S234	4	6	85
S235	7	7	100
Correl (X_i, Team Quality	−0.054	0.028	
Correl (Experience, Aptitude)		0.422	

class participation and final learning achievements (Table 8.7). However, individual's experience in neither web programming nor general programming mattered. Amazingly, these findings are true in the whole class or within their team settings (Figures 8.12 and 8.13).

In summary, we applied the "rules of thumb" identified in Semester I to Semester II's teaching/learning management and operations. Students seemed to be

TABLE 8.6 Individual's Team Contribution Varying With Participations Within the Team

User ID	Cooperation	Dependability	Participation	Contribution to the Team
S201	95	90	90	97
S202	100	100	97	100
S203	90	85	90	90
S204	90	95	87	92
S205	85	80	70	72
S206	95	90	90	95
S207	90	90	95	87
S208	100	97	92	95
S209	100	100	95	95
S210	95	95	95	92
S211	100	100	85	72
S212	95	97	95	100
S213	90	82	87	90
S214	92	75	90	80
S215	65	45	45	45
S216	95	95	90	82
S217	85	75	70	90
S218	90	90	90	85
S219	95	95	95	95
S220	100	100	100	97
S221	95	77	62	67
S222	100	100	95	95
S223	95	95	90	100
S224	85	85	85	85
S225	100	100	100	100
S226	70	60	65	65
S227	95	65	65	65
S228	80	95	95	100
S229	100	100	95	10
S230	100	95	85	95
S231	100	97	100	100
S232	100	100	100	100
S233	95	95	90	90
S234	100	95	90	85
S235	95	97	95	100

more active than those who took the same course in Semester I. More importantly, we found the following compelling facts in the second exploration:

- Students' learning outcomes are highly correlated with their active participations in class activities. Simply put, students' active interactions in a virtual setting play a critical role in helping them improve their learning outcomes.
- Individual's team experience and attitude are not well correlated to one's contribution to the quality of team work. However, one's active involvement positively impacts one's contribution to team work.

- Both individuals' dependability of finishing assignments and their active participations within teams are truly correlated to their contributions to the quality of team work.
- Individuals' certain education background and work experience could influence one's involvement in both the class and team settings.

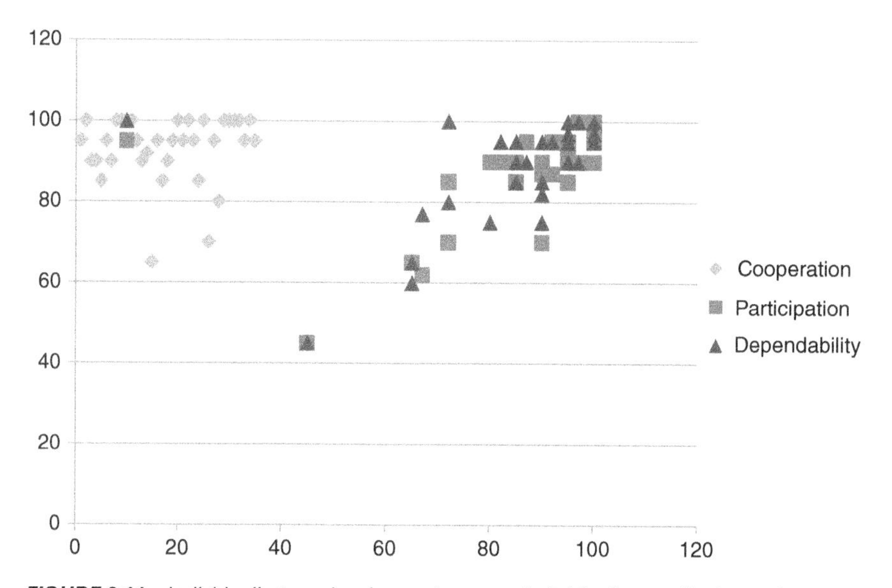

FIGURE 8.11 *Individual's team involvements versus individual's contribution to the team.*

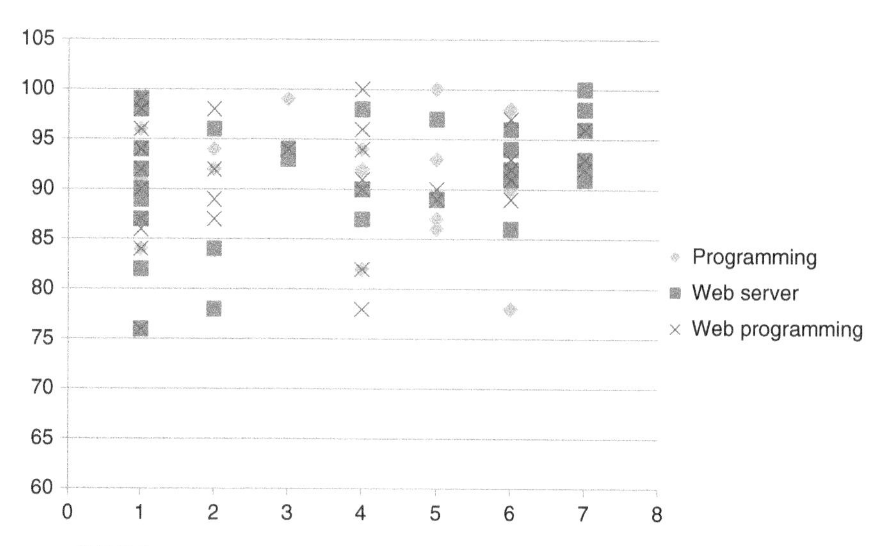

FIGURE 8.12 *Background versus individual overall performance for the class.*

TABLE 8.7 Background Versus Performance and Participation in Team Activities

User ID	Final Grade	Total Posts	Programming	Web Server	Web Programming	Participation
S201	89	16	5	5	6	90
S202	92	16	7	7	6	97
S203	84	21	1	2	1	90
S204	97	74	5	5	6	87
S205	89	14	5	1	5	70
S206	94	28	4	3	3	90
S207	98	47	1	4	1	95
S208	91	16	7	6	6	92
S209	94	16	2	1	1	95
S210	92	17	4	1	2	95
S211	94	46	2	6	4	85
S212	86	15	5	6	1	95
S213	90	21	1	1	1	87
S214	91	32	1	7	4	90
S215	82	18	4	1	4	45
S216	96	51	1	2	1	90
S217	87	27	5	4	2	70
S218	90	17	1	1	5	90
S219	98	31	4	1	1	95
S220	93	16	5	3	6	100
S221	76	15	1	1	1	62
S222	92	18	1	1	7	95
S223	93	19	7	3	7	90
S224	87	13	1	1	1	85
S225	96	23	6	7	7	100
S226	78	18	6	2	4	65
S227	90	57	6	4	4	65
S228	96	33	6	6	4	95
S229	98	52	6	1	1	95
S230	99	63	3	1	1	85
S231	93	16	7	7	6	100
S232	100	83	5	7	4	100
S233	92	36	2	6	1	90
S234	89	15	1	1	2	90
S235	98	39	1	7	2	95
Correl (X, Final Grade)		0.6	0.090512863	0.297	0.061591291	0.68140137
Correl (X, Team Participation)			0.019446771	0.361	0.093930016	

8.4.2 Pedagogical Engagements and Learning Outcomes in the Network Perspective

Surely, instructors must take all the available means to get pedagogically engaged with students. Usually, sending out messages to guide students through each lesson and assignment is much appreciated by students. Students can feel the presence of instructors. Synchronous communications through live chats, virtual meetings, etc., help to improve the faculty–student pedagogical engagement. In collaborative learn-

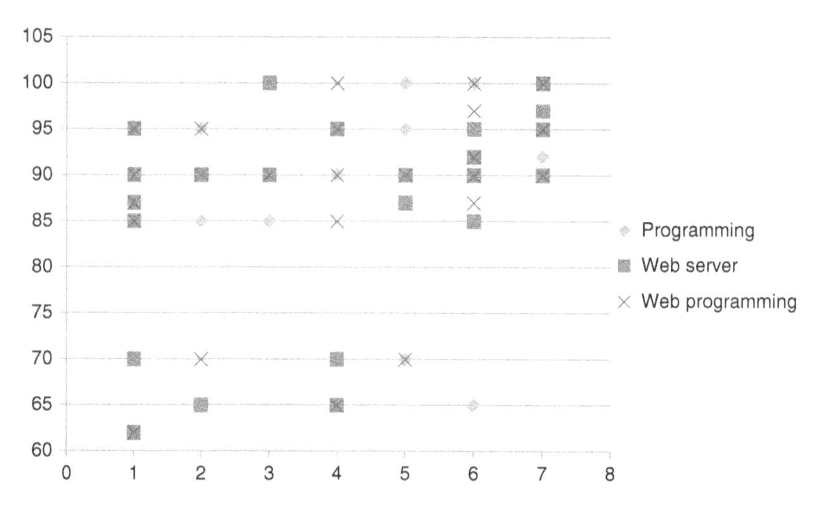

FIGURE 8.13 *Background versus individual participations within teams.*

ing, the frequent discussions among students play a critical role in improving learning outcomes. Therefore, we must explore students' interaction analytics, aimed at getting more insights of the interactive learning dynamics in both the class and team settings.

As an example, Table 8.8 shows online interactive activities throughout the course during the exploration in Semester II. When the interactive activities are depicted using NetDraw (Borgatti et al., 2002), we can find appealing results (Figure 8.14). Figure 8.14 illustrates interactive learning network based on individual's class activities, background, and performance. In Figure 8.14, we use darkness levels to represent the web server knowledge levels, the thickness to indicate the tie strength between students, and different shapes to tell their team assignments, and node size to provide individual's final grade information (i.e., individual's performance evaluated by the instructors).

A brief summary of identified insights of the learned network can be stated as follows: individuals who are active tend to receive better grades; individuals who have better general understandings of web servers tend to be more active; although individuals who have better knowledge of web servers tend to interact with each other, some students who have no good experience using web servers like to interact with those who have better knowledge of web servers would lead to improved learning outcomes.

From the previous observation, it might make sense to balance all the teams with an equivalent number of students who have good backgrounds in the field, which could ensure that the interactive dynamics within the teams and in the class are well balanced and retained. Let us check how the teams under study were formed in terms of their backgrounds. Table 8.9 shows the team distribution of individuals who has prior knowledge of web servers above the average in the class. Apparently, based on team formation information (Table 8.9), we observed that the team formation

TABLE 8.8 Online Interactive Activities Throughout the Course

	S201	S202	S203	S204	S205	S206	S207	S208	S209	S210	S211	S212	S213	S214	S215	S216	S217
S201	9	0	1	0	0	0	1	0	0	0	3	0	0	0	0	0	1
S202	0	7	0	2	0	0	2	2	0	0	0	1	0	0	0	1	2
S203	1	0	12	1	0	0	1	0	0	1	0	0	0	1	1	0	1
S204	0	2	1	19	0	2	7	0	0	0	5	1	1	2	1	10	0
S205	0	0	0	0	7	0	0	0	0	0	0	1	1	0	2	0	0
S206	0	0	0	2	0	10	5	0	0	0	2	1	0	2	0	4	0
S207	1	2	1	7	0	5	21	0	0	0	5	0	0	1	0	3	0
S208	0	2	0	0	0	0	0	8	0	0	0	0	0	0	1	0	0
S209	0	0	0	0	0	0	0	0	8	0	0	1	1	1	0	1	0
S210	0	0	1	0	0	0	0	0	0	10	0	0	1	0	0	1	0
S211	3	0	0	5	0	2	5	0	0	0	10	0	2	0	0	3	0
S212	0	1	0	1	1	1	0	0	1	0	0	9	0	0	0	1	0
S213	0	0	0	1	1	0	0	0	1	1	2	0	11	2	0	0	0
S214	0	0	1	2	0	2	1	0	1	0	0	0	2	14	1	3	1
S215	0	0	1	1	2	0	0	1	0	0	0	0	0	1	9	2	0
S216	0	1	0	10	0	4	3	0	1	1	3	1	0	3	2	19	0
S217	1	2	1	0	0	0	0	0	0	0	0	0	0	1	0	0	13
S218	1	0	0	0	0	0	0	1	5	0	0	1	0	0	1	1	0
S219	0	0	2	0	0	1	3	1	1	1	0	0	0	1	0	0	1
S220	0	0	0	4	0	1	2	0	0	0	0	0	0	1	0	1	3
S221	0	0	0	0	0	0	0	0	0	0	2	0	0	1	0	1	0
S222	1	0	1	2	1	0	2	2	2	0	2	0	0	1	0	0	0
S223	0	1	1	5	0	0	0	0	0	0	0	0	1	0	0	3	1
S224	0	1	2	0	0	0	0	0	0	0	0	0	0	0	0	1	2
S225	0	0	0	0	2	3	2	0	0	0	3	0	0	0	0	0	1
S226	1	0	0	0	0	1	2	1	1	1	1	0	1	1	0	0	0
S227	0	0	2	8	0	2	6	0	1	0	6	0	0	3	2	6	0
S228	0	1	0	6	0	5	3	1	0	0	0	0	2	1	0	2	3
S229	1	0	3	6	1	0	1	1	2	2	1	0	2	3	0	10	2
S230	0	0	0	9	0	7	11	0	1	0	6	2	1	5	2	4	0
S231	1	1	0	0	1	0	2	0	0	0	0	0	0	1	0	0	0
S232	1	10	0	19	1	3	3	0	2	3	6	0	3	6	3	6	1
S233	0	1	0	0	1	0	1	0	0	0	0	0	0	1	0	0	1
S234	0	0	1	1	0	0	2	1	0	0	2	1	0	0	0	0	1
S235	0	0	0	7	1	1	2	0	0	1	3	2	3	1	0	0	1

(continued)

TABLE 8.8 ***(Continued)***

	S218	S219	S220	S221	S222	S223	S224	S225	S226	S227	S228	S229	S230	S231	S232	S233	S234	S235
S201	1	0	0	0	1	0	0	0	1	0	0	1	0	1	1	0	0	0
S202	0	0	0	0	0	1	1	0	0	0	1	0	0	1	10	1	0	0
S203	0	2	0	0	1	1	2	0	0	2	0	3	0	0	0	0	1	0
S204	0	0	4	0	2	5	0	0	0	8	6	6	9	0	19	0	1	7
S205	0	0	0	0	1	0	0	2	0	0	0	1	0	1	1	1	0	1
S206	0	1	1	0	0	0	0	3	1	2	5	0	7	0	3	0	0	1
S207	0	3	2	0	2	0	0	2	2	6	3	1	11	2	3	1	2	2
S208	1	1	0	0	2	0	0	0	1	0	1	1	0	0	0	0	1	0
S209	5	1	0	0	2	0	0	0	1	1	0	2	1	0	2	0	0	0
S210	0	1	0	0	0	0	0	0	1	0	0	2	0	0	3	0	0	1
S211	0	0	0	2	2	0	0	3	1	6	0	1	6	0	6	0	2	3
S212	1	0	0	0	0	0	0	0	0	0	0	0	2	0	0	0	1	2
S213	0	0	0	0	0	1	0	0	1	0	2	2	1	0	3	0	0	3
S214	0	1	1	1	1	0	0	0	1	3	1	3	5	1	6	1	0	1
S215	1	0	0	0	0	0	0	0	0	2	0	0	2	0	3	0	0	0
S216	1	0	1	1	0	3	1	0	0	6	2	10	4	0	6	0	0	0
S217	0	1	3	0	0	1	2	1	0	0	3	2	0	0	1	1	1	1
S218	8	0	1	1	0	0	0	0	1	1	1	0	1	1	0	1	0	0
S219	0	14	0	0	0	0	0	1	1	0	1	1	2	0	0	1	0	4
S220	1	0	7	0	0	2	0	0	0	0	0	2	1	1	6	0	1	0
S221	1	0	0	9	0	0	0	0	0	0	0	1	1	1	2	1	1	0
S222	0	0	0	0	9	0	1	1	0	3	1	2	2	1	1	0	1	0
S223	0	0	2	0	0	8	0	0	0	0	3	0	1	0	2	1	0	1
S224	0	0	0	0	1	0	7	0	0	0	0	0	0	0	1	0	0	1
S225	0	1	0	0	1	0	0	8	4	0	1	0	3	0	4	0	0	2
S226	1	1	0	0	0	0	0	4	9	1	0	0	0	0	1	0	0	0
S227	1	0	0	0	3	0	0	0	1	18	4	3	6	0	5	1	1	9
S228	1	1	0	0	1	3	0	1	0	4	15	0	2	1	6	1	0	1
S229	0	1	2	1	2	0	0	0	0	3	0	20	3	0	4	2	1	1
S230	1	2	1	1	2	1	0	3	0	6	2	3	31	1	8	5	0	5
S231	1	0	1	1	1	0	0	0	0	0	1	0	1	7	12	0	0	0
S232	0	0	6	2	1	2	1	4	1	5	6	4	8	12	11	4	1	1
S233	1	1	0	1	0	1	0	0	0	1	1	2	5	0	4	10	0	0
S234	0	0	1	1	1	0	0	0	0	1	0	1	0	0	1	0	7	0
S235	0	4	0	0	0	1	1	2	0	9	1	1	5	0	1	0	0	10

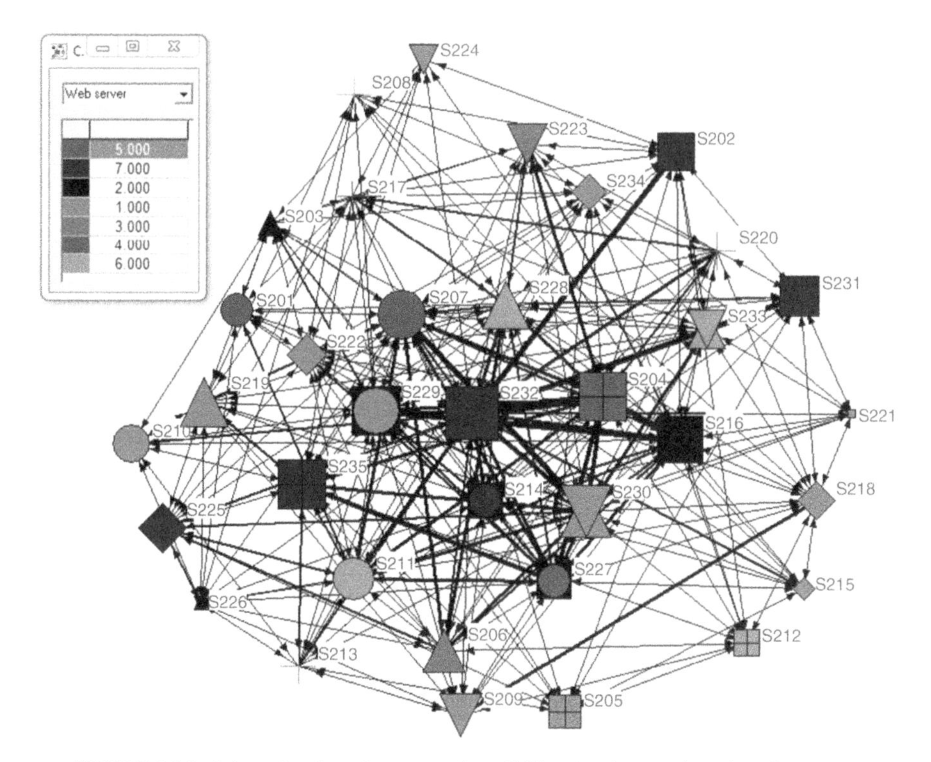

FIGURE 8.14 *Interactive learning network: activities, background, and performance.*

TABLE 8.9 Background and Team Information in the Studied Course During Semester II

Level of Web Server Knowledge	Team 1	Team 2	Team 3	Team 4	Team 5	Team 6	Team 7	Team 8	Team 9
7		S232, S202, S231		S235			S214		
6	S211		S228	S212				S208	S233
5	S201			S204					

might have some unbalanced issues between teams in terms of the average of prior knowledge of web servers before they took the course. Therefore, to make the whole class achieve improved learning outcomes, we might have to have better ways to assign or form teams.

In general, what could we do differently to make collaborative learning more effective? The above-discussed team formation issue can truly be one of the actionable items during the course offering process. Surely, we could ask ourselves a variety of similar or different questions under different circumstances. However, as each class is different, we must look into the specifics by focusing on actionable items to make effective and positive changes. Hence, it will be much more effective if we can take appropriate actions real time and from beginning to end than simply a few adjustments at the beginning of an offered course.

8.4.3 Guiding Individual's Participations in Real Time for Improved Learning Outcomes

From the discussions in Sections 8.4.1 and 8.4.2, we know that both data and networks analytics help us to understand the dynamics of learning networks. The formed learning networks vary with a variety of factors. More importantly, the networks evolve over time as these factors keep changing. Service Science should help service organizations design, execute, monitor, and improve their service systems. Scientific approaches thus are essential for providing guidance and control of their formed service networks and ensure that service networks evolve in such a way through the service lifecycle that optimal and sustainable service is always delivered to customers.

Let us revisit the exploratory example in Semester II. At the end of the previous section, we had one interesting observation. We articulate that we might have better ways to assign teams, which could lead to achieve better learning outcomes than what students actually accomplished in the course. As indicated in Table 8.9, we now know that both Teams 5 and 6 had no students who had good levels of understanding of web servers. We could change the way we assigned teams. For instance, we could make sure that each team would have at least one student who had at least level 5 of web server knowledge. By doing so, we know that the interactions within each team could be more dynamic and intensive as we know that one who has a better general understanding of web servers tends to be more active and interact with others. As a result, we might make collaborative learning in the course during Semester II more effective.

On the basis of the concept and principles of Service Science discussed in this book, we know that it will be much effective if we can take appropriate actions real time throughout the process of service transformation. For an online course like the one we just discussed, for example, we can divide the course—Web Security—into seven learning modules (Figure 8.15). To make sure that we will get sufficient information of students who are enrolled in the course, we let them individually fill in a preclass survey (which mainly collects individual education background and work experience related to the Web technologies), do online assignments and homework, and take part in online discussions within the class and teams. As indicated in Figure 8.15, based on data and network analytics, as shown in Sections 8.4.1 and 8.4.2, we will then assign teams by fully taking into consideration their backgrounds and behaviors in class before teams are officially assigned (Figures 8.16 and 8.17). Real-time guidance to individuals, teams, and the class will be provided as soon as a new module completes. The iteration of monitoring, analytical, and guiding process continues until the end of the course under exploration. Ultimately, the approach to promote and facilitate students' pedagogical engagement leads to improved collaborative learning in overall.

8.4.4 Brief Remarks on Part II of Collaborative Learning Analytics

According to Hall (2013), one of the attractive assets of virtual courses is that they are quite more flexible in terms of working around learners' schedules. In a

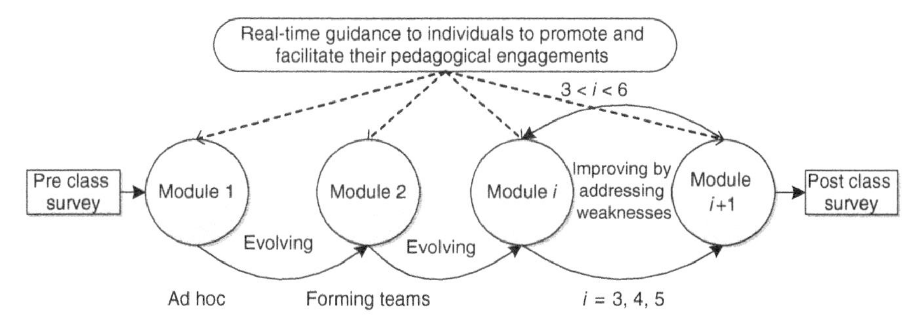

FIGURE 8.15 Interactive learning network: activities, background, and performance.

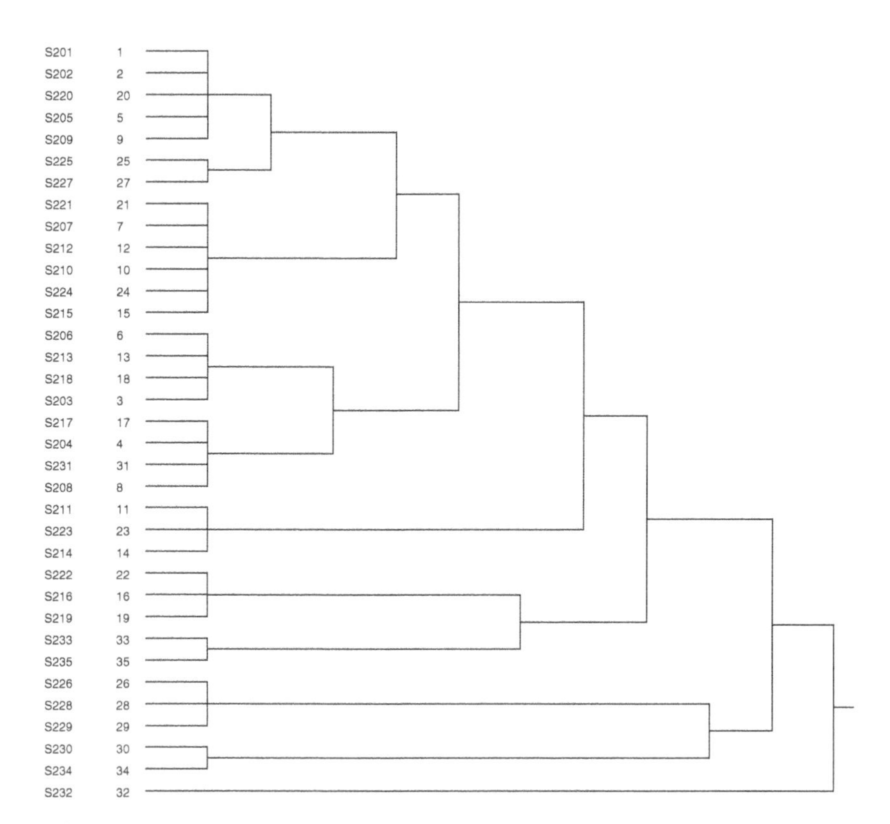

FIGURE 8.16 Class cluster diagram at the end of Module 2: activities and background.

residential class on campus, students and instructors must meet at the same time in assigned classrooms or laboratories. While in an online class over the Internet, students are simply required to complete their assignments before they are due. Students can learn lectures and work on assignments at any time, avoiding conflicts with their work or home responsibilities. This is particularly true for adult learners. However, lacking physical interactions in a timely manner, students who are not self-disciplined

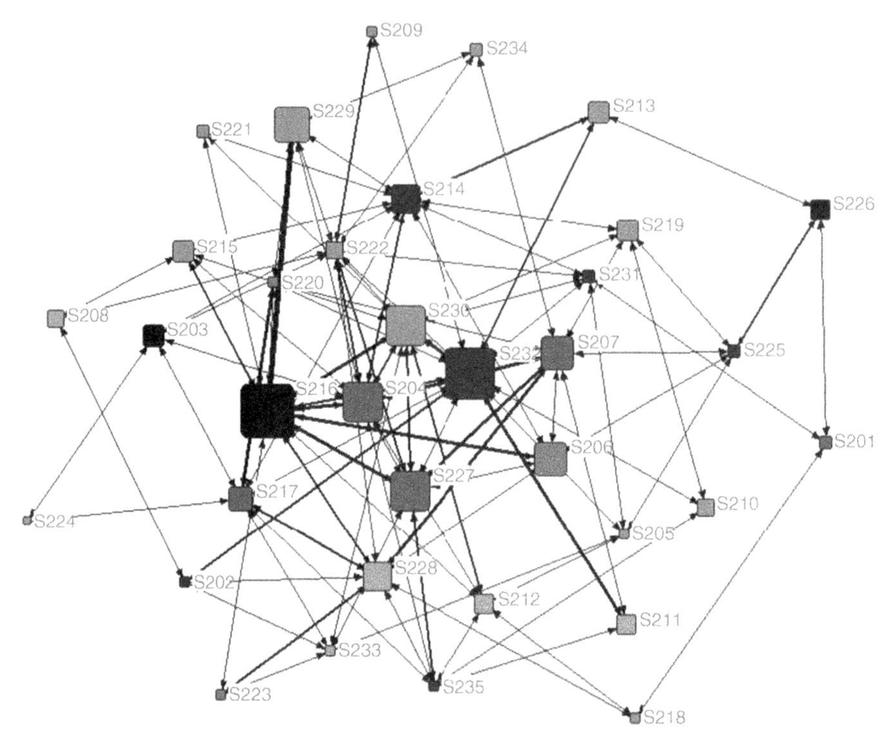

FIGURE 8.17 *Interactive learning network at the end of Module 2: activities, background, and quality of work (the meanings of symbols and labels are similar to those used in Figure 8.14).*

might be left behind. Therefore, if collaborative learning is the focus in an offered virtual course, we must take account of numerous possible ways to change how the course is delivered, design what pedagogical engagements should be enabled, and adjust the learning materials and assignments if needed. Because of different mix of education backgrounds and work experience of a given class, instructors might also need different mechanisms to help and guide students throughout their learning processes, individually or collectively.

Surely, there are many more challenges in either an online or a brick-and-mortar school. In this section, we used an example to show how we can address certain issues in collaborative learning. Indeed, this section surely provides much more insights into the students' pedagogical engagements in an online Web Security class than those in Part I of collaborative learning analytics we conducted in Section 8.3. We could clearly see how individuals behave and how their behaviors in both the class and team settings evolve. In particular, we concluded that an instructor could provide real-time guidance to individuals, teams, and the class if the monitoring, analytical, and guiding process is equipped with transformative mechanisms for the class that is well constructed and deployed appropriately. As a result, schools can design, develop, and deliver competitive online education services as promised and desired.

8.5 THE SIGNIFICANCE OF THIS ILLUSTRATED CASE STUDY

Indeed, pedagogies of engagement in residential education have been studied for decades as discussed earlier. Many researchers have focused on three of the main principles, namely, that good practice encourages student–faculty interaction, cooperation among students, and active learning. The interaction among students and between faculty and students are the two learning practice factors best predictive of a positive change in students' academic development, personal development, and satisfaction (Astin, 1997; NSSE, 2003; Smith et al., 2005). However, online education misses typical while strict class/assignment schedules and in-person interactions that are taken for granted on campus, resulting in unprecedented challenges, such as how to cultivate and nourish students' self-motivation and which approach to be applied to encouraging social interaction and collaborative learning.

There has been little scientific research work to address the challenges in a scientific and comprehensive manner. Therefore, the significance of this case study will be the groundbreaking development of a framework to provide an integrative and quantitative process that helps professors and students adjust and enhance the teaching/learning practices and facilitate the students' development in an online educational environment (i.e., online degree programs, training modules, and MOOC courses) by generating pleasant and positive interactions.

Indeed, the adopted framework could provide a new, innovative, and scientific avenue for addressing a variety of other online education challenges (Burnsed, 2011). For example, by relying on the developed approaches, the quality of online education from any service provider can be fully studied. The computing and network technology has transformed the way we teach and learn. Blended learning—a combination of online and in-class instruction—has been found to be a more effective teaching method in many leading universities (Ripley, 2012). Thus, the framework and related models/tools that have being developed have the promise of helping to enrich residential education programs, to train instructors to develop and teach online courses, and introduce a new pedagogy for online or blended learning (Burnsed, 2011).

8.6 CONCLUSIONS

Like it or not, nearly all colleges and universities in the United States are now offering some online or distance learning courses. One benefit for residential students will be that students could leverage the best components of both brick-and-mortar and online curricula (Hall, 2013). Therefore, one potential and promising use of MOOC is to help brick-and-mortar schools improve their ongoing residential education by incorporating the useful insights from online courses. Moreover, with the help of big data technologies, schools can surely get a better understanding of the ever-changing education market (Ahlquist and Saagar, 2013) and how generation Y prefers to be pedagogically engaged in this new millennium.

Using a specific online course delivery example, this chapter showed how a service system can be analyzed beyond the SEM-based causality and performance model.

A network perspective truly becomes critical for us to understand the dynamics of a people-centric service network that essentially consists of extremely dynamic, complex, and interwoven service encounters.

Broadly speaking, for a service system, small or large, simple or complex, it has different business goals and objectives. At a given time and location, the resources available to engineer and manage services that are offered and delivered by the service system can be limited, technologically, financially, socially, or even politically. A viable approach to address the challenges that are confronted by the service system as a whole could vary with circumstances. In other words, because the mix of 8 Ps varies, so does its corresponding solution to ensure the success of service provision.

Simply put, we could rely on the fast advances in technologies to collect the necessary measurements of a service system that could help describe the true dynamic behaviors of the service system (Figure 8.18). By capturing the insights of the service system in real time, the presented approach can surely provide effective data, network, and business analytics in support of ongoing service engineering and management throughout the service lifecycle across all internal business domains in the service system and external collaborators across the service delivery networks. Ultimately, all stakeholders (e.g., customers, employees, and collaborators) can make informed decisions and take prompt and optimal actions in cocreating and keeping the offered and delivered services attractive, competitive, and satisfactory.

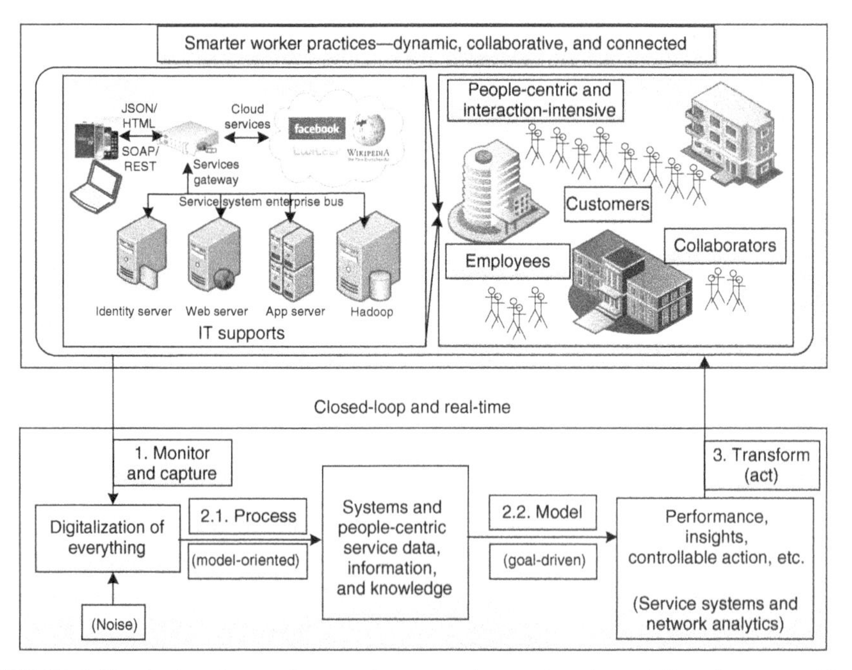

FIGURE 8.18 *Engineering and managing a service system in real time for competitive advantage.*

REFERENCES

Ahlquist, J., & Saagar, K. (2013). Comprehending the complete customer. *Analytics—INFORMS Analytics Magazine*, May/June, 36–50.

Andriessen, J., & Verburg, R. (2004). A Model for the Analysis of Virtual Teams. Chapter XV in *Virtual and Collaborative Teams: Process, Technologies and Practice*, ed. by S. Godar and S. Ferris. Hershey, PA: Idea Group Inc.

Antonis, K., Daradoumis, T., Papadakis, S., & Simos, C. (2011). Evaluation of the effectiveness of a web-based learning design for adult computer science courses. *IEEE Transaction on Education*, 54(3), 374–380.

Astin, A. (1997). *What Matters in College? Four Critical Years Revisited*. San Francisco, CA: Jossey-Bass.

Baker, M. J., & Lund, K. (1997). Promoting reflective interactions in a computer-supported collaborative learning environment. *Journal of Computer Assisted Learning*, 13, 175–193.

Beldarrain, Y. (2006). Distance education trends: integrating new technologies to foster student interaction and collaboration. *Distance Education*, 27(2), 139–153.

Borgatti, S., Everett, M., & Freeman, L. (2002). *UCINET for Windows: Software for Social Network Analysis*. Harvard, MA: Analytic Technologies.

Burnsed, B. (2011). Online education may transform higher ed. *US News and Report*. Retrieved on Oct. 10, 2012 from http://www.usnews.com.

Carley, K., & Lin, Z. (1997). A theoretical study of organizational performance under information distortion. *Management Science*, 43(7), 976–997.

Chea, T. (2012). Elite colleges transform online higher education. *Associated Press*. Retrieved on Oct. 10, 2012 from http://people.uis.edu/rschr1/et/?p=5070.

Cheng, Y. (2005). *New Paradigm for Re-engineering Education, Globalization, Localization and Individualization*. The Netherlands: Springer.

Chin, W., Marcolin, B., & Newsted, P. (2003). A partial least squares latent variable modeling approach for measuring interaction effects: result from a Monte Carlo simulation study and an electronic-mail emotion/adoption study. *Information Systems Research*, 14(2), 189–217.

Coursera. (2013). Coursera. Retrieved on Mar. 5, 2013 from https://www.coursera.org.

Department of Education. (2008). *Evaluating Online Learning: Challenges and Strategies for Success*. Retrieved on Mar. 5 2013 from http://www2.ed.gov/admins/lead/academic/eval online/evalonline.pdf.

Dillenbourg, P. (1999). *Collaborative-Learning: Cognitive and Computational Approaches*. Oxford: Elsevier.

Djenic, S., Krneta, R., & J. Mitic, (2011). Blended learning of programming in the Internet age. *IEEE Transaction on Education*, 54(2), 247–254.

Duerr, M., Zajonc, A., & Dana, D. (2003). Survey of transformative and spiritual dimensions of higher education. *Journal of Transformative Education*, 1(3), 177–211.

Dymalski, S. (2011). How to overcome the top 5 online education challenges. *Career Advice at ClassesandCareers.com*. Retrieved on Mar. 5, 2013 from http://www.classesandcareers.com/.

Ebrahim, N., Ahmed, S., & Taha, Z. (2009). Innovation and R&D activities in virtual team. *European Journal of Scientific Research*, 34(3), 297–307.

edX. (2012). edX. Retrieved on May 2, 2012 from https://www.edx.org/.

Fincher, M. (2010). Adult student retention: a practical approach to retention improvement through learning enhancement. *The Journal of Continuing Higher Education*, 58(1), 12–18.

Garrison, R. (2000). Theoretical challenges for distance education in the 21st century: a shift from structural to transactional issues. *International Review of Research in Open and Distance Learning*, 1(1), 1–17.

Garrison, D., & Kanuka, H. (2004). Blended learning: uncovering its transformative potential in her education. *The Internet and Higher Education*, 7, 95–105.

GHFP. (2006). *A vision for transformative education. The 3rd Vittachi Conference "Rethinking Educational Changes"*, Retrieved on Oct. 10, 2009 from http://www.transformedu.org/Home/tabid/158/language/en-US/Default.aspx.

Gokhale., A. (1995). Collaborative learning enhances critical thinking. *Journal of Technology Education*, 7(1), 22–30.

Goodwin, G., Medioli, A., Sher, W., Vlacic, L., & Welsh, J. (2011). Emulation-based virtual laboratories: a low-cost alternative to physical experiments in control engineering education. *IEEE Transaction on Education*, 54(1), 48–55.

Hall, P. (2013). Brick & mortar vs. virtual school. *eHow*. Retrieved on Mar. 5, 2013 from http://www.ehow.com/info_8690282_brick-mortar-vs-virtual-school.html.

Hansen, R. (2006). Benefits and problems with student teams: suggestions for improving team projects. *Journal of Education for Business*, 82(1), 11–19.

Illeris, K. (2004). Transformative learning in the perspective of a comprehensive learning theory. *Journal of Transformative Education*, 2(2), 79–89.

Imel, S. (1991). Collaborative learning in adult education. *ERIC Clearing on Adult Career and Vocation Education*, Columbus, OH, ERIC Digest No. 113.

Jin, Y., & Levitt, R. (1996). The virtual design team: a computational model of project organizations. *Computational & Mathematical Organization Theory*, 2(3), 171–196.

Larson, R. C. (2009). Editorial Column-Education: Our Most Important Service Sector. *Service Science*, 1(4), i–iii.

Levitt, R., Thomsen, J., Chritiansen, T., Junz, J., Jin, Y., & Nass, C. (1999). Simulating project work processes and organizations: towards a micro-contingency theory of organizational design. *Management Science*, 45(11), 1479–1495.

Lurey, J., & Raisinghani, M. (2001). An empirical study of best practices in virtual teams. *Information and Management*, 38, 523–544.

Marks, R., Sibley, S., & Arbaugh, J. (2005). A structural equation model of predictors for effective online learning. *Journal of Management Education*, 29(4), 531–563.

Mayo, P. (2003). A rational for a transformative approach to education. *Journal of Transformative Education*, 1(1), 38–57.

Moore, J. (2005). Is higher education ready for transformative learning? A question explored in the study of sustainability. *Journal of Transformative Education*, 3(1), 76–91.

NAP. (2009). *Developing Metrics for Assessing Engineering Instruction: What Gets Measured is What Gets Improved*. Washington, DC: The National Academies Press.

NSB. (2007). *Moving Forward to Improve Engineering Education*. National Science Board and Workshops at NSF. Retrieved from May 10, 2011 from http://ww.nsf.gov/pubs/2007/nsb07122/nsb07122.pdf.

NSSE. (2003). *National Survey of Student Engagement: The College Student Report*—2003 Annual Report. Bloomington, IN: Center for Postsecondary Research, Indiana University.

Orange, A., Heineche, W., Berger, E., Krousgrill, C., Mikic, B., & Quinn, D. (2012). An evaluation of HigherEd 2.0 technologies in undergraduate mechanical engineering courses. *Advances in Engineering Education*, 3(1), 1–29.

O'Sullivan, E. (1999). *Transformative Learning: Educational Vision for the 21st Century*. Toronto, Canada: University of Toronto Press.

Parry, M. (2012). 5 ways that edX could change education. *The Chronicle of Higher Education*. Retrieved on Mar. 5, 2013 from http://chronicle.com/article/5-Ways-That-edX-Could-Change/134672/.

Pauleen, D. (2004). *Virtual Teams: Projects, Protocols and Processes*. Idea Group Publishing, Hershey, PA.

Qiu, R. G. (2009). Computational thinking of service systems: dynamics and adaptiveness modeling. *Service Science*, 1(1), 42–55.

Qiu, R. G. (2010). A collaborative model of engineering education for complex global environments. *The 40th ASEE/IEEE Frontiers in Education Conference*, Washington, DC, S3J1-S3J5.

Qiu, R., Wu, Z., Wang, D., & Yu, Y. (2011). Tractable approximation approach to improving hotel service quality. *Journal of Service Science Research*, 3(1), 1–20.

Ripley, A. (2012). College is dead. Long live college. *Time Magazine*, 180(18), 33–41.

Selim, H. (2007). Critical success factors for e-learning acceptance: confirmatory factor models. *Computers and Education*, 49, 396–413.

Shachaf, P., & Hara, N. (2005). Team effectiveness in virtual environments: an ecological approach. *Teaching and Learning with Virtual Teams*, 83–108 (Eds. P. Ferris and S. Godar). Idea Group Publishing, Hershey, PA.

Smith, H., & Smarkusky, D. (2005). Competency matrices for peer assessment of individuals in team projects. *Proceedings of the 6th Conference on Information Technology Education*, 115–162.

Smith, K., Sheppard, S., Johnson, D., & Johnson, R. (2005). Pedagogies of engagement: classroom-based practices. *Journal of Engineering Education*, 94(1), 87–102.

SNA. (2011). Social Network Analysis. Retrieved on May 10, 2011 from http://en.wikipedia.org/wiki/Social_network#Social_network_analysis.

Thompson, A., Perry, J., & Miller, T. (2009). Conceptualizing and measuring collaboration. *Journal of Public Administration Research and Theory*, 19(1), 23–56.

Thoms, B. (2011). A dynamic social feedback system to support learning and social interaction in higher education. *IEEE Transactions on Learning Technologies*, 4(4), 340–352.

Thomsen, J., Levitt, R. E., & Nass, C. I. (2005). The virtual team alliance (VTA): Extending Galbraith's information-processing model to account for goal incongruency. *Computational & Mathematical Organization Theory*, 10(4), 349–372.

Tryggvason, G., & Apelian, D. (2006). Re-engineering engineering education for the challenges of the 21st century. *JOM Journal of the Minerals, Metals and Materials Society*, 37, 14–17.

Vickery, C., Clark, T., & Carlson, J. (1999). Virtual positions: an examination of structure and performance in *ad hoc* workgroups. *Information Systems Journal*, 9, 291–312.

Wong, S., & Burton, R. (2000). Virtual teams: what are the characteristics, and impact on team performance? *Computational & Mathematical Organization Theory*, 6, 339–360.

Wyatt, L. (2011). Nontraditional student engagement: increasing adult student success and retention. *The Journal of Continuing Higher Education*, 59(1), 10–20.

9

The Science of Service Systems and Networks

Manufacturing dominated the global economy during the last couple of centuries. Both academics and practitioners thus paid significant attention to the design, development, production, and innovation of physical products. With their contributions to the development of manufacturing science and technology, the manufacturing industry has considerably improved its production productivity and the quality of made products. In the second half of the twentieth century, in particular, the world witnessed a long period of prosperity in all aspects of well-being that were mainly driven by the spread of industrialization and substantially increased manufacturing productivities around the world.

Today, the quality of life has taken into account not only the material standard of living but also other intangible values of living that are recognized to be mainly service-oriented. As discussed in Chapter 2, the global economy has shifted its focus from manufacturing to services to meet the changing needs of human beings. Indeed, entering the information era has accelerated the shift, which created unfilled gaps in the service science and technology. Indeed, service organizations have been on the hunt for appropriate methodologies and tools that can help them engineer and manage their service offering and delivery throughout the service lifecycle at the scale they would like to reach, efficiently, cost-effectively, and globally (Spohrer and Riechen, 2006; Qiu, 2012).

As discussed earlier, the effectiveness (E) of a service as a solution to meet the changing needs of customers is equal to the product of the quality (Q) of the technical attributes of the solution and the acceptance (A) of that solution by the customers, that is, $E = Q \times A$. However, the acceptance of customers changes rapidly, varying with

Service Science: The Foundations of Service Engineering and Management, First Edition. Robin G. Qiu.

time, places, cultures, and service contexts. Because people's acceptance is largely subjective, manufacturing mindsets with a focus on physical attributes indeed become ineffective when applied in the field of service engineering and management. Hence, to address the discussed change acceleration phenomena with scientific rigor, we must develop service science based on people-centric and service mindsets.

Promisingly, the introduction of putting employee and customers first in 1990s made the first breakthrough in developing people-centric and service mindsets. Since then, service organizations have begun to develop, operate, and manage businesses and measure their successes by focusing on both customers' satisfaction and employees' job satisfaction, resulting in an operational philosophy shift in business operations and management. This book essentially presented such a new perspective of service study. We took a holistic view of the service lifecycle to explore the dynamics of service systems and the structure and behavior of people-centered service networks.

By defining service as a cocreation transformation process enabled and executed by a service system, we discussed how the performance of the service system could be quantitatively analyzed using a holistic approach. By leveraging the advances in computing and network technologies, social science, management science, and other relevant fields, we demonstrated that service networks in light of service encounters could be comprehensively explored in a closed-loop and real-time manner. The presented science of service should help service organizations understand and capture market trends, design and engineer service products and delivery networks, operate service operations, and control and manage the service lifecycles for competitive advantage.

In this final chapter, we first summarize this book by providing some final thoughts on the development of service science in a comprehensive manner. We strongly advocate that the service research and practice community must appreciate and continue to develop a variety of methodologies and tools that can be well derived and evolved from the well-known theories and principles in systems theory, operations research, marketing science, organizational behavior and theory, network theory, social computing, and analytics. In Section 9.2, we then conclude this chapter by articulating that innovative approaches to the development of service science are truly on demand. The science of service will be well developed by the scholars and practitioners worldwide in an evolutionary and collective manner.

9.1 THE SCIENCE OF SERVICE SYSTEMS AND NETWORKS

Holistically, a service organization is a service system, essentially consisting of service providers, customers, products, and processes. As compared to a producing-goods system, a service system must be people-centered. Therefore, a service system surely is sociotechnical. On the basis of the earlier discussion, we understand that it is the transformation process that ties all other system constituents together and cocreates the respective values for both service providers and customers. Whether the values can be fully met relies on the efficient, effective, and smart business operations, which must be engineered, executed, and managed with scientific rigor across the service system.

Now it is crystal clear that service is people-centric, truly cultural and bilateral. The type and nature of a service dictates how a service is performed, which accordingly determine how a series of service encounters could occur throughout its service lifecycle. The type, order, frequency, timing, time, efficiency, and effectiveness of the series of service encounters throughout the service lifecycles determine the quality of services perceived by customers who purchase and consume the services. On the basis of the discussions in the preceding chapters, we understand that people-centered, interactive, and behavioral activities in a service system essentially engender a service interaction cocreation network or simply service network. Indeed, as the velocity of globalization accelerates, the changes and influences are more ambient, quick, and substantial, impacting us as providers or customers in dynamic and complex ways that have not seen before. The understanding of service networks becomes essential for service providers to be able to design, offer, and manage services for competitive advantage.

Because of the sociotechnical nature of a service system, we use a systems approach to evaluate the performance of the service system, aimed at capturing both utilitarian functions and sociopsychological needs that characterize service systems. However, the true people's behavioral and attitudinal dynamics of a sociotechnical system requires performing real-time social network analytics. As a result, the insights of service interactions in the formed service networks can be truly explored and understood, which assist stakeholders to make respective while cooperative informed decisions at the point of need to improve their service cocreation processes across the service lifecycles in an optimal manner.

Bearing the earlier discussion in mind, we consider a service as a transformation process rather than simply an offered service product. Truly, both provider-side and customer-side people are always involved in an interactive manner, directly or indirectly and physically or virtually, throughout the transformation process. Hence, we view a service as a value cocreation process. For a service, goods are frequently the conduits of service provision; the physical attributes and technical characteristics that specify the goods are indispensable to the service. The quality (Q) of the technical attributes in the service, indeed, mainly defines the quality of the goods. To a service customer, Q is frequently perceived in service provision as the quality of designated service functionalities that are defined in a service specification. As described in the equation of $E = Q \times A$, the value of E also directly depends on the value of A, which is largely related to sociopsychological perceptions of the customer throughout the service lifecycle. A is subjective in nature, varying with people, time, places, cultures, and service contexts (Figure 9.1).

Service is highly heterogeneous. Each service is unique as a unique customer and a service provider agent essentially cocreate the service values that meet the respective needs of the customer and the service provider. The variability of service and the need for measuring sociopsychological perceptions had made extremely challenging the exploration of the service lifecycle, which spans market discovery, engineering,

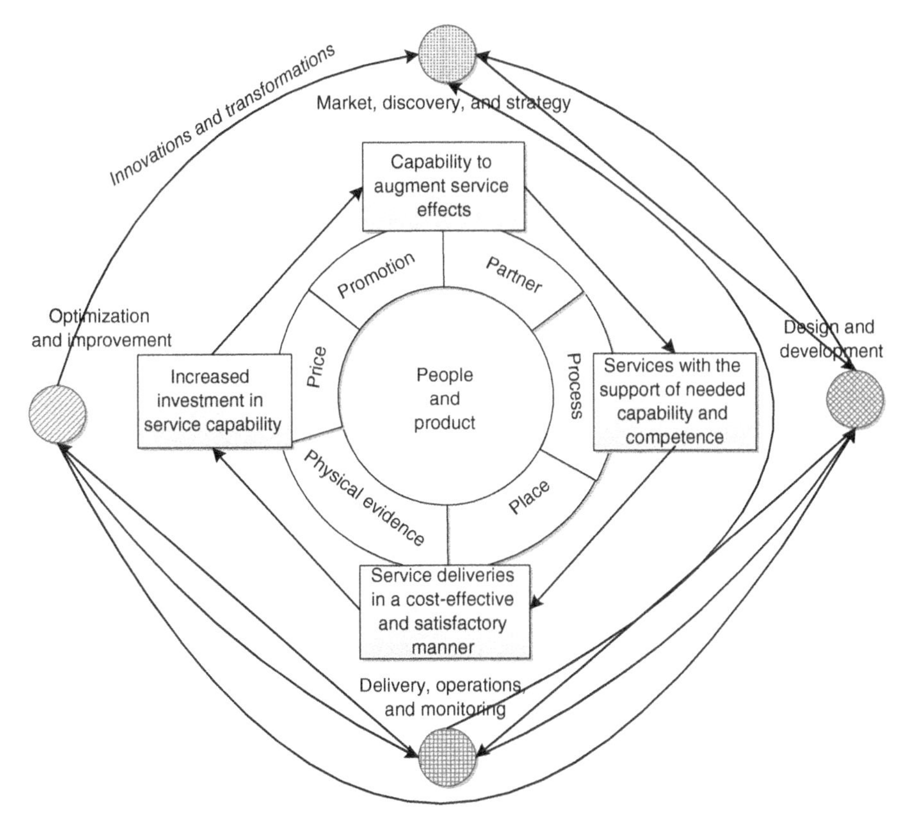

FIGURE 9.1 *Engineering and managing competitive services: holistic and lifecycle perspective.*

delivery, and sustaining, in an integrated and holistic manner. Figure 9.1 highlights a holistic and lifecycle viewpoint of how we should engineer and manage competitive services in the twenty-first century.

It is well understood that the science of service is essential for a service organization to achieve the ultimate goal of engineering and managing competitive services in its service marketplace. As discussed earlier in this book, the prior lack of means to monitor and capture people's dynamics throughout the service lifecycle has prohibited us from gaining insights into the service engineering and management in a service organization for years. However, we believe that the convergence of the following advances in science and technology has made possible the design and development of the needed methods and tools that can facilitate service organization to monitor and capture people's dynamics throughout the service lifecycle:

- Digitalization
- Networks and telecommunications

- Collaborative methods and tools
- The fast advances in social network media
- Big data technologies and analytics methods and tools

Figure 9.2 shows how in a systems and operations perspective a service organization can be successively and real-time transformed for competitive advantage by fully leveraging the convergence of the above-mentioned advances in science and technology.

"People-centric sensing will help drive this trend by enabling a different way to sense, learn, visualize, and share information about ourselves, friends, communities, the way we live, and the world we live in" (Campbell et al., 2008). From the discussions in the preceding chapters, we understand that voluminous, real-time, and heterogeneous data on the service cocreation dynamics of both service providers and customers can be comprehensively captured and analyzed if service systems are well planned, designed, and operated as illustrated in Figure 9.2. In other words, when the enabling technologies are appropriately implemented, we can surely create and execute smarter working and consuming practices so that we can make service cocreation processes not only beneficial but also enjoyable. As a result, services are competitive and satisfactory.

Because of the enablement of people sensing and computational thinking with the support of the above-mentioned advances in science and technology, enormous

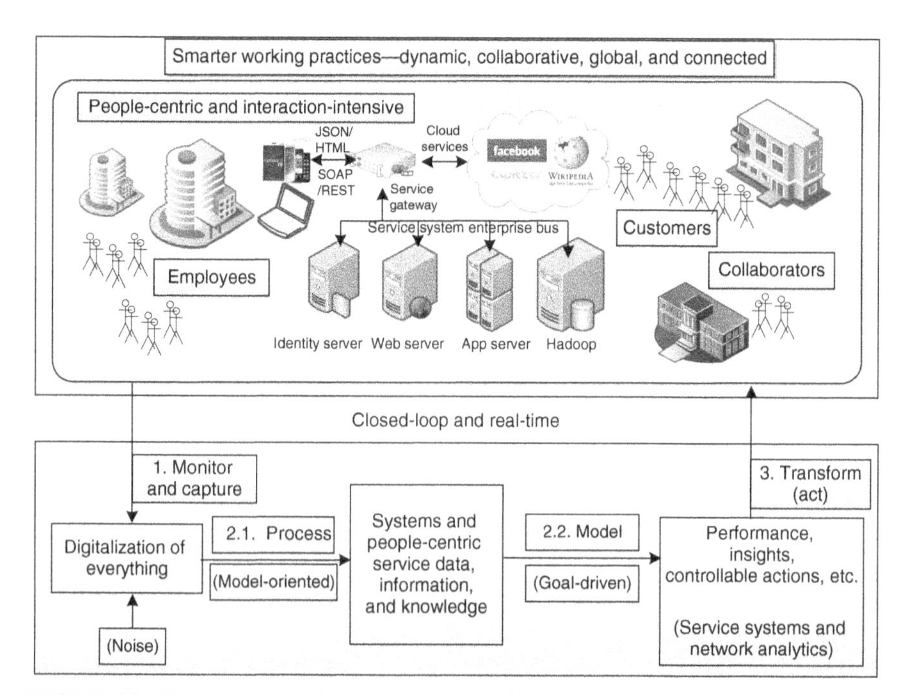

FIGURE 9.2 *Engineering and managing competitive services: systems and operations perspective.*

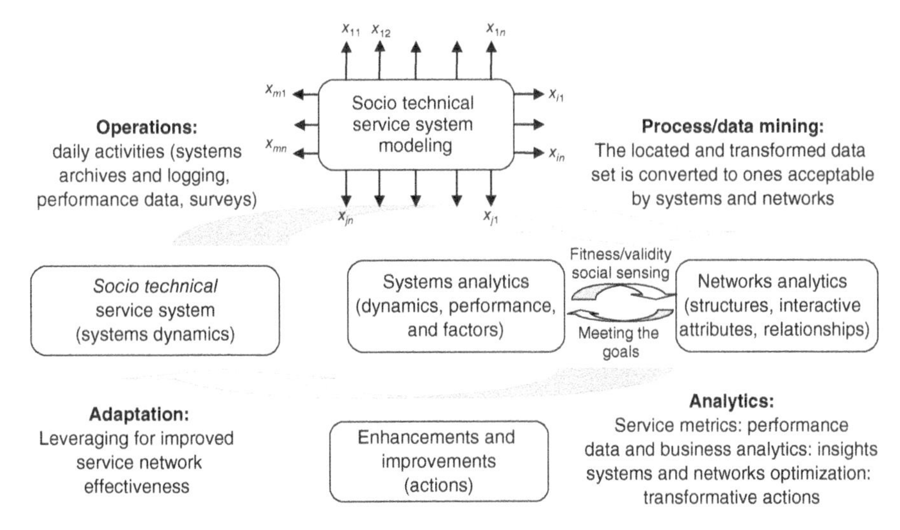

FIGURE 9.3 *Engineering and managing competitive services: scientific perspective.*

opportunities truly lie ahead of us. However, if the science of service is not well developed, we cannot ensure that service systems will perform in such a way that the respective values for both service providers and customers can be optimally met, at present as well as in the long run. By leveraging both systems methods and networks analytics, in this book we essentially present one promising approach to develop the needed methods and tools, making a contribution to the body of knowledge in service science (Figure 9.3).

As shown in Figure 9.3, we advocate that a combined systems and network approach can help service organizations engineer and manage their competitive services. The presented approach in this book fundamentally focuses on identifying actionable areas for service improvements across all service system constituents in a holistic, comprehensive while cost-effective and efficient manner. The presented framework is integrative, quantitative, and closed-loop in nature. As a result, a service system with cocreation processes can be modeled, explored, monitored, and controlled with scientific rigor.

Innovatively, systems and network approaches are integrated in this book. When combined and applied to the field of service engineering and management, they are complementary. A systems approach to gain the fundamental understanding of how a service system as a whole behaves must be first investigated. Specifically, we apply structural equation models (SEM) to describe the systems' performance and/or conduct necessary hypothesis testing and/or confirmatory factor analyses. Secondly, social and collaborative network approaches to explore the interactions and insights of people-centered service networks can be employed. For instance, we apply social network analysis (SNA) models to explore how service networks across the service lifecycle are formed and behaved and understand how the service networks might evolve over time. Consequently, we as service providers can always make optimal decisions at the point of need, strategically and tactically, so that we can innovate, market, engineer, and execute services in a competitive and satisfactory manner.

9.1.1 Enhancing the Approaches to Explore Service Systems and Networks

As explained earlier, we present one approach to develop the needed methods and tools through leveraging the strengths of both system methods and network analytics. Indeed, this could be a perfect starting point for us to analyze the systems' behavior, the network structures, and dynamics of a sociotechnical service system. In the preface, we articulate that this book focuses on the development of a real-time and closed-loop framework to help service organizations engineer and manage their service systems. That is to say, developing an approach to model service systems while allowing performing continual improvements is surely unique, differentiating this book from others. However, we truly understand that the presented approach can be further enhanced and developed. More importantly, we are sure that there must be many other approaches to develop the science of service.

Regardless of the variability of services and the complexity and heterogeneity of service systems, the discovery, design, engineering, and delivery of services must be fully supported by the science of service if service organizations wish to stay competitive from time to time. In general, the foci of decision-making change with the mix of 8 Ps that substantially varies with the progression of service offering and delivery. Therefore, depending on circumstances, we have to customize and further enhance known approaches to explore service systems and networks. As a matter of fact, we must frequently develop new approaches to engineer and manage services in order to meet the needs of service providers and customers over time.

We can take the simplified model illustrated in Figure 8.6 as an example. When significant variations of online classes exist, an SEM based on the prior knowledge might be substantially deviated from the reality. To ensure that we can validate the SEM, we have to find an appropriate way to enhance the modeling. For instance, we could apply the probabilistic-based analysis methodology such as SEM-based and semisupervised Bayesian networks to the exploration of class-dependent collaborations, which might help to improve the accuracy of analyses if significant variations of online classes do exist.

Generally speaking, an identified best practice can be effectively adopted as a general guideline by a service organization in its daily service engineering and managerial operations. However, certain ongoing changes must be applied in the process of service offering and execution for optimal outcomes as each service is unique. Hence, the service industry is looking for practical and scientific service engineering and management approaches that can be applied in a gradual and evolutionary manner. Ultimately, the framework proposed in Figure 9.3 should be fully implemented in a real-time and closed-loop manner, which is graphically illustrated in Figure 9.2. Indeed, there is a long way to go in the service academia and industry before the science of service gets well developed. A full exploration of the science of service is surely necessary in both the academia and industry. A brief discussion in this regard is provided in Section 9.2.

9.1.2 A Pragmatic Approach to Explore Service Systems

On second thought, if people are not the focus of a study in a service system, an alternative approach to explore service systems might be more appropriate than one illustrated in Figure 9.3. This is particularly true when a practically applicable transition in service operations and management is crucial for the time being for a service organization to survive in a fiercely challenging and competitive marketplace. In other words, by applying well-known methods and tools to explore and address the ongoing changes in the marketplaces, service organizations can make swift and appropriate changes and actions to transform operations and management in an evolutionary manner so that they can continue to engineer and execute quality and satisfactory services to meet the needs of their customers.

For example, the performance of a service system is frequently related to business units' operational efficiencies from a managerial perspective. If a study of business units' operational efficiencies is indeed critical for a service organization at a given business period, well-known methods and tools can be practically adopted. For instance, we can take advantage of the following modeling technologies, analytical hierarchy process (AHP), data envelopment analysis (DEA), principal component analysis (PCA), and partial least squares (PLS), to collectively study the dynamics of service systems with a focus on exploring service operations and management on the service provider's behalf. Figure 9.4 illustrates how AHP, DEA, PCA, and PLS can be seamlessly and integratively applied in support of this alternative investigation. Note that this alternative approach highlights a viewpoint of pragmatism as rich data on operational functions and decisions in a service organization are most likely available at present.

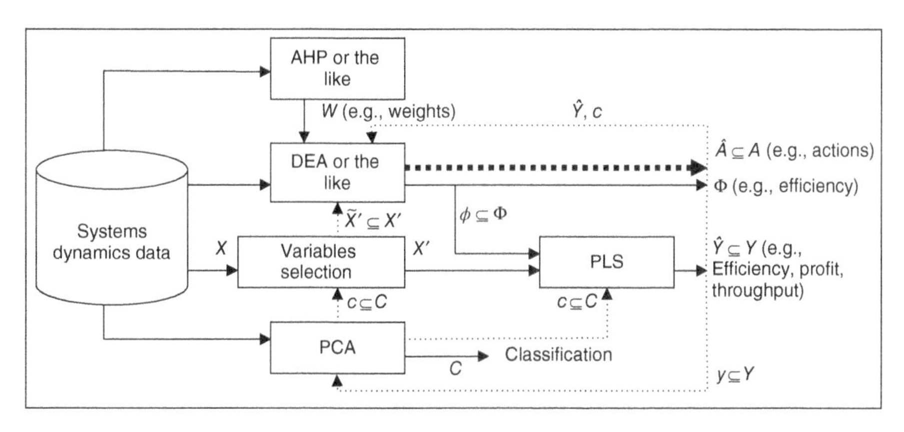

FIGURE 9.4 *An integrated approach for improving daily service business operations.*

As shown in Figure 9.4, AHP is typically used for comparing a list of objectives or alternatives where the problem elements are structured in an organizational hierarchy. AHP depends on experts' knowledge to provide rankings that eventually lead to weights being assigned for the relative importance of different categories defined based on the problem elements and targeted organizational hierarchy. These weights are then input to a DEA procedure to determine corporate and business units' efficiencies. PCA and PLS are strictly data-driven modeling technologies. By leveraging these previously mentioned data-driven technologies and expert-based operations exploratory models, decision makers can gain insights into service systems and hence operate them in a competitive manner.

In practice, PLS methods can facilitate the identification of operational weakness, which considerably relies on the output of the DEA analysis as well as other available systems dynamics data (Figure 9.4). However, these data could conceivably be exceedingly large, and much of it is probably of little value in generating an analytical model. It is crucial to use variables that truly affect the analytical output of the model. Data that does not influence the exploration only serves to degrade the performance of the model and needs to be eliminated.

In the integrated model shown in Figure 9.4, genetic algorithms can be used to select meaningful variables. From the potential candidate variables, a process of random selection of variables can be used to generate a set of models forming an initial population. The choice of variables is defined by a binary word with a one in the bit corresponding to a variable used in the model, and a zero in the position of the variable not used. Next, this population is evaluated to obtain an estimate of the standard error of prediction for each member. Those models with low values of the standard error are better than those with higher ones. The models are rank-ordered from lowest to highest standard error of prediction. A probability can be then assigned to each model that is inversely proportional to the standard error of prediction. Two models are chosen at random with the probability of selection equal to the assigned probability of the model to be used in a breeding process to produce the next generation. One or two random integers are chosen from one to the number of variables. These integers define the crossover points that are used with the binary words to define the next generation of models. The crossover points define where the binary words defining the two models chosen are broken. The broken pieces are then rejoined to define the set of variables to be used in the next generation of models. In addition, some of the bits in the binary are randomly flipped, representing a mutation. In this manner, a new generation is produced. Sometimes, a small fraction of the best models from the previous generation is carried forward so that if an exceptional model is generated, that model is not lost by the breeding process. This procedure is repeated until a stopping criterion, such as a certain standard error of prediction or the number of generation, is reached.

As a result, these selected variables are used to generate a PLS model. The model can be further used in a manner described earlier to help service systems improve the performance. However, this approach does not take into account that there is a resource availability issue (e.g., cost) involved in any improvement scheme. Eventually, an approach that seeks to optimize this approach such as determination of the lowest cost approach to improve profits by a fixed percentage. The chosen variables may not directly provide the lowest cost to make improvements. Using PCA, we can learn the structure of the variables and with this information, remove certain variables and replace them with other variables that might provide a solution with a lower cost. Alternatively, the genetic algorithm might be used to find replacement variables that are less costly. From a set of candidate variables, the genetic algorithm would search to find those variables that are the best predictors. This search can be guided by PCA by choosing those variables related to more expensive ones.

In summary, this alternative approach could help service organizations evaluate, compare, and optimize service business operations when they are facing severe competition in the presence of massive uncertainty and risk in their operating environments. The ultimate goal of this approach is to help service organizations transform their practices for competitive advantage with the support of the following well-developed analytical scenarios in sequence:

1. AHP depends on existing algorithms and/or new inputs from experts to provide the knowledge for weights assignment for the relative importance of different input/output variables in the organizational and operational hierarchy.
2. The AHP output provides weights used by DEA to generate the organization's operational efficiency. Apparently, the quantified outputs combined with those identified weak areas in service business operations better help the service organization understand where they stand in competition and what they could address in improving their performance in terms of operational efficiency.
3. Genetic algorithms can be employed in preprocessing the inputs. Through PCA, the structure of the variables is learned. With a better understanding of the circumstances, on one hand, certain variables can be removed; on the other hand, the identified variable correlations can be utilized in facilitating the prediction generations of quantities of the primary interest in the next step.
4. PLS can then be used to generate predictions of quantities of primary interest under the circumstances. The primary interest, for example, can be profit, throughput, or more sophisticated definable systems outcomes.
5. Comparisons of generated predictions can be conducted through sensitivity analysis by selecting highly influential input variables. When facing massive uncertainty, this integrated model can be utilized in quantifying the consequences when different transformations in operational practices could occur under different circumstances, assisting management in making informed decisions (e.g., a series of optimal changes or transformation actions) to improve systems performance while minimizing potential risks.

9.2 THE SCIENCE OF SERVICE IN THE TWENTY-FIRST CENTURY

Generally speaking, best practices in service engineering and management in the service industry can be effectively adopted as operational and managerial guidelines by service organizations to support and manage their daily operations and business activities. Because each service is unique, it is necessary for both the service providing-side people and consuming-side customers to cocreate the respective values of services in a practical, viable, and competitive manner. To make this happen in a satisfactory manner in both the short term and the long run, the framework illustrated in Figure 9.3 must be well incorporated into the service lifecycle shown in Figure 9.1. As a result, a service organization, with the support of effective service engineering and management that is enabled in a real-time and closed-loop manner shown in Figure 9.2, can offer and deliver competitive services throughout the service lifecycle.

The science of service is still in its early infancy stage although it emerged in the early 2000s (Qiu, 2012). Without question, a well-defined and more developed service science would better facilitate service organizations in conducting service engineering and management across service value-added networks. In reality, capable and competitive service systems must be highly adaptable and sustainable to their service environment (when, where and who to deliver, and whom to be served). Therefore, the developed science of service must span all service offering and delivery areas from engineering and/or managing service marketing, conceiving, design, quality assurance, regulatory compliance, operations, to innovation throughout the lifecycle of service.

Regardless of methods and tools that can be utilized at each stage of the service lifecycle, meeting the needs of people at the point of need is what actually matters in operating competitive service systems. People involved in service are unique and truly different from each other, for example, individuals as customers who have different needs, individuals on the service provider side who are assigned with certain roles and responsibilities, managers who are in charge of designated business domains, executives who are overseeing service organizations, and collaborators who are contributing to the service offering and delivery networks. To ensure that each service of a service system can be well executed, people involved in service must collaborate with each other well throughout the service lifecycle, which can be accomplished only if four interdependent and essential flows (Qiu, 2013) in support of the service system and formed service networks are engineered and managed with scientific rigor (Figure 9.5).

Let us start with the customer dynamics flow. We know that meeting both the utilitarian and psychological needs of customers by focusing on a chain of interactive service encounters is the key to explore the customer dynamics flow. Hence, the customer dynamics flow must be explored with the support of behavioral science, consumer behavior and dynamics, and cognitive science. Understanding the customer dynamics flow becomes essential for service organizations to capture market trends and get ready for and capable of offering and delivering excellent customer experience.

The organizational behavior flow plays a key role in forming functional service networks. The organizational behavior flow focuses on organizational capability development and competence alignment in support of meeting the customers' utilitarian and psychological needs. Organizational behavior flows must ensure that

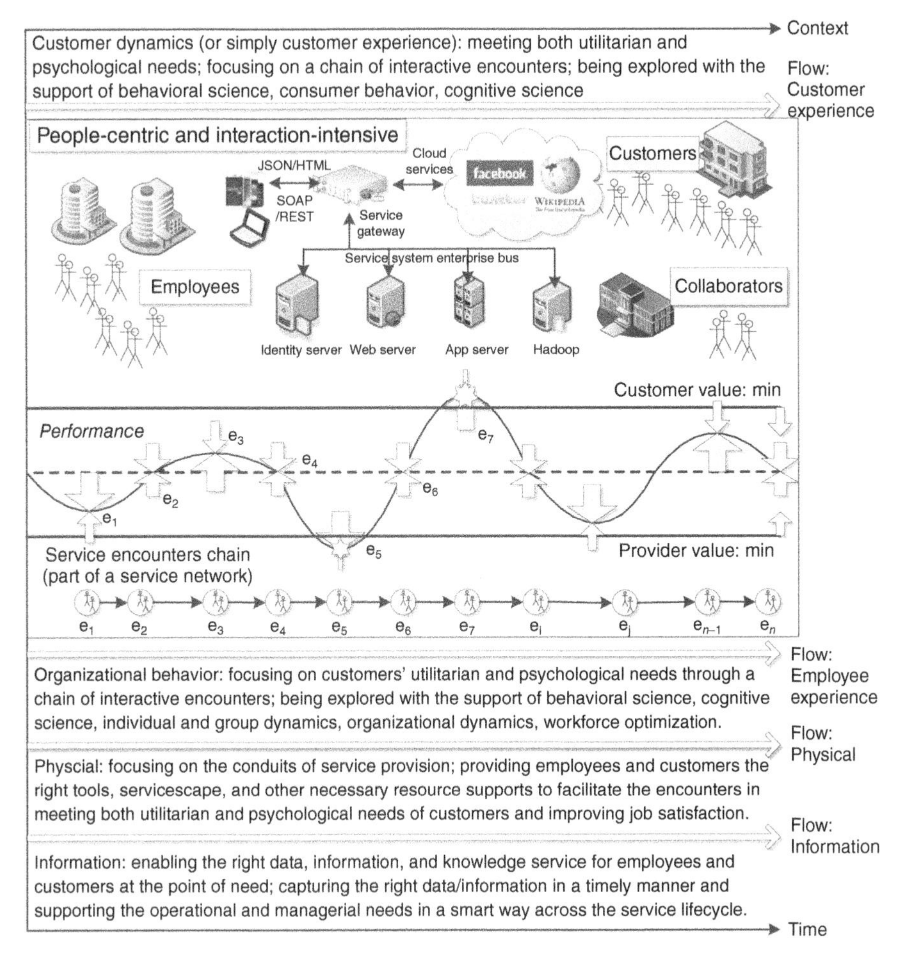

FIGURE 9.5 *Four interdependent and essential flows in support of service systems and networks.*

service organizations can offer and deliver a chain of interactive and positive service encounters while realizing a competitive level of employees' job satisfaction. Therefore, the organizational behavior flow must be explored with the support of behavioral science, cognitive science, individual and group dynamics, organizational dynamics, operations management, and workforce optimization, making sure that service organizations can continuously improve their job satisfaction and organizational behavior.

The physical flow focuses on the conduits of service provision. An efficient and effective physical flow can provide employees and customers with the right tools, servicescape, and other necessary resource supports to facilitate service encounters in meeting both utilitarian and psychological needs of customers while improving job satisfaction. In today's information era, the effectiveness and efficiency of a physical flow considerably rely on the effectiveness and efficiency of a corresponding information flow. An information flow must capture right data/information in a timely manner and then support the operational and managerial needs of employees and customers in an intelligent way across the service lifecycle. An optimal information flow shall promptly enable the right data, information, and knowledge service for employees and customers at the point of need.

This book takes an innovative and unique approach to contributing to the development of service science. Under a given circumstance, one step at a time, we explore a unique research area in a given service context. Collectively, the service research community must explore the defined four flows across service systems in a comprehensive and holistic way. Theoretically, the dynamics of service systems in terms of both systems performance and service networks behavior must be fully explored, understood, and controlled so that the respective values for service providers and customers can be optimally cocreated. In practice, it simply becomes how different methods and tools can be made available at the point of need in real time so that individuals, managers, executives, and collaborators can interactively, effectively, and collectively perform their responsibilities and duties in the processes of transformation in meeting their respective needs.

Once again, we now fully recognize people as the focus during the service production and consumption process in service provision. We learn that different people have their personal traits in the physiological and psychological perspectives, different cognitive abilities, and unique sociological constraints. It has been exceedingly challenging for the service research and practice community to investigate methods and tools that can be well applied for modeling and exploring people's behaviors in service because people-sensing mechanisms in service were hardly enabled not long ago. The recent and fast advances in sensor-based networks, pervasive and mobile computing, online social media, and big data methodologies and tools indeed have changed this (Figure 9.6). Therefore, we are sure that it is time for the service research and practice community to develop service theories and principles that can be applied in effectively managing and controlling systemic behavior, leveraging sociotechnical effects, and stimulating innovations throughout the service lifecycle (marketing, design and engineering, operations, delivery, benchmarking, and optimization for improvement).

Without question, advanced descriptive, predictive, and prescriptive service science studies surely rely on the continual development of systems theory, operations research, management science, marketing science, advanced computing and communication technology, network theory, social computing, and analytics. As a matter of fact, the science of service as a metascience of service must build on predecessors' excellent work from many of the above-mentioned disciplines (Larson, 2011; Qiu, 2012). However, a variety of innovative approaches for the development of the science of service are truly on demand in today's global service-led economy. This book takes an innovative and unique approach to contributing to the development

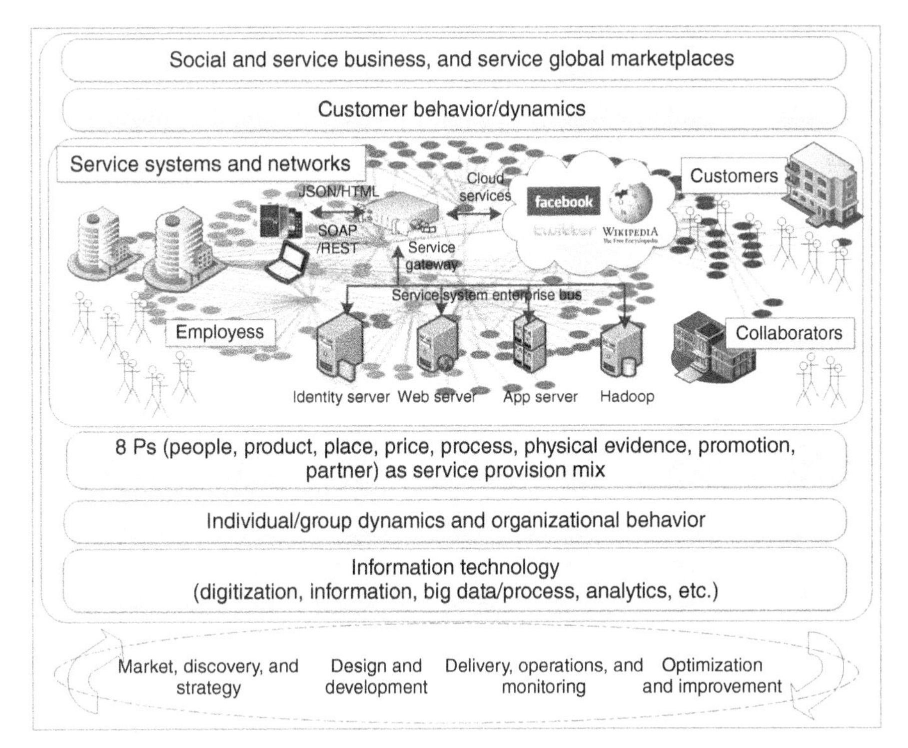

FIGURE 9.6 *A holistic and integrated approach to contributing to the development of service science.*

of service science (Figure 9.6). Specifically, we take a holistic view of the service lifecycle and explore the real-time dynamics of service systems and networks.

In conclusion, the service industry is in need of descriptive, predictive, and prescriptive research of service in a holistic, integral, and quantitative manner. There is a marvelous Chinese saying, "cast away a brick and attract a jade stone." Hopefully, this book serves such a purpose. We are confident that the science of service will be well developed by the scholars and practitioners worldwide in an evolutionary and collective manner. Ultimately, the developed body of knowledge and tools in this emerging interdisciplinary field can be effectively applied by service organization to address their service challenges in the twenty-first century's service-led economy.

REFERENCES

Campbell, A., Lane, N, Miluzzo, E., Peterson, R., Lu, H., Zheng, X., Musolesi, M., Fodor, K., Eisenman, S., & Ahn, G. (2008). The rise of people-centric sensing. *IEEE Internet Computing*, July–August, 12–21.

Larson, R. (2011) Foreword in *Service Systems Implementation*, ed. by H. Demirkan, J. Spohrer, and V. Krishna. Springer.

Qiu, R. G. (2012). Editorial column—launching service science. *Service Science*, 4(1), 1–3.

Qiu, R. G. (2013). We must rethink service encounters. *Service Science*, 5(1), 1–3.

Spohrer, J., & Riechen, D. (2006). Services science. *Communications of the ACM*, 49(7), 30–34.

Index

Service Science: The Foundations of Service Engineering and Management, First Edition. Robin G. Qiu.